Statistics for Chemical and Process Engineers

Yuri A. W. Shardt

Statistics for Chemical and Process Engineers

A Modern Approach

Second Edition

 Springer

Yuri A. W. Shardt
Department of Automation Engineering
Technical University of Ilmenau
Ilmenau, Germany

ISBN 978-3-030-83192-9   ISBN 978-3-030-83190-5   (eBook)
https://doi.org/10.1007/978-3-030-83190-5

This Springer imprint is published by the registered company Springer Nature Switzerland AG
The registered company address is: Gewerbestrasse 11, 6330 Cham, Switzerland

Preface

The need for the development and understanding of large, complex data sets exists in a wide range of different fields, including economics, chemistry, chemical engineering, and control engineering. In all these fields, the common thread is using these data sets for the development of models to forecast or predict future behaviour. Furthermore, the availability of fast computers has meant that many of the techniques can now be used and tested even on one's own computer. Although there exist a wealth of textbooks available on statistics, they are often lacking in two key respects: application to the chemical and process industry and their emphasis on computationally relevant methods. Many textbooks still contain detailed explanations of how to manually solve a problem. Therefore, the goal of this textbook is to provide a thorough mathematical and statistical background to regression analysis through the use of examples drawn from the chemical and process industries. The majority of the textbook presents the required information using matrices without linking to any particular software. In fact, the goal here is to allow the reader to implement the methods on any appropriate computational device irrespective of their specific availability. Thus, detailed examples, that is, base cases, and solution steps are provided to ease this task. Nevertheless, the textbook contains two chapters devoted to using MATLAB® and Excel®, as these are the most commonly used tools both in academics and in industry. Finally, the textbook contains at the end of each chapter a series of questions divided into three parts: conceptual questions to test the reader's understanding of the material; simple exercise problems that can be solved using pen, paper, and a simple, handheld calculator to provide straightforward examples to test the mechanics and understanding of the material; and computational questions that require modern computational software that challenge and advance the reader's understanding of the material.

This textbook assumes that the reader has completed a basic first-year university course, including univariate calculus and linear algebra. Multivariate calculus, set theory, and numerical methods are useful for understanding some of the concepts, but knowledge is not required. Basic chemical engineering, including mass and energy balances, may be required to solve some of the examples.

The textbook is written so that the chapters flow from the basic to the most advanced material with minimal assumptions about the background of the reader. Nevertheless, multiple different course can be organised based on the material presented here depending on the time and focus of the course. Assuming a single semester course of 39 h, the following would be some options:

(1) **Introductory Course to Statistics and Data Analysis**: The foundations of statistics and regression are introduced and examined. The main focus would be on Chap. 1: Introduction to Statistics and Data Visualisation, Chap. 2: Theoretical Foundation for Statistical Analysis, and parts of Chap. 3: Regression, including all of linear regression. This course would prepare the student to take the Fundamentals of Engineering Exam in the United States of America, a prerequisite for becoming an engineer there.

(2) **Deterministic Modelling and Design of Experiments**: In-depth analysis and interpretation of deterministic models, including design of experiments, is introduced. The main focus would be on Chap. 3: Regression and Chap. 4: Design of Experiments. Parts of Chap. 2: Theoretical Foundation for Statistical Analysis may be included if there is a need to refresh the student's knowledge of background information.

(3) **Stochastic Modelling of Dynamic Processes**: In-depth analysis and interpretation of stochastic models, including both time series and prediction error methods, is examined. The main focus would be on Chap. 5: Modelling Stochastic Processes with Time Series Analysis and Chap. 6: Modelling Dynamic Processes. As necessary, information from Chap. 2: Theoretical Foundation for Statistical Analysis and Chap. 3: Regression could be used. The depth in which these concepts would be considered would depend on the orientation of the course: either a theoretical emphasis can be made, by focusing on the theory and proofs, or an application emphasis can be made, by focusing on the practical use of the different results.

As appropriate, material from Chap. 7: Using MATLAB® for Statistical Analysis and Chap. 8: Using Excel® to do Statistical Analysis could be introduced to show and explain how the students can implement the proposed methods. It should be emphasised that this material should not overwhelm the students nor should it become the main emphasis and hence avoid thoughtful and insightful analysis of the resulting data.

The author would like to thank all those who read and commented on previous versions of this textbook, especially the members of the process control group at the University of Alberta, the students who attended the author's course on process data analysis in the Spring/Summer 2012 semester, the members of the Institute of Control Engineering and Complex Systems (Institute für Automatisierungstechnik und komplexe Systeme) at the University of Duisburg-Essen, the members of the Department of Automation Engineering (Fachgebiet Automatisierungstechnik) at the Technical University of Ilmenau (Technische Universität Ilmenau), and the students who attended the course "System Identification" at the Technical University of Ilmenau. The author would specifically wish to thank Profs. Steven X. Ding and

Biao Huang for their support, Oliver Jackson from Springer for his assistance and support, and the Alexander von Humboldt Foundation for monetary support. As well, the author would like to thank Ying Deng, Mike Eichhorn, Benedikt Oppeneiger, and M.P. for their help in improving the English version.

Downloading the data: The data sets, MATLAB® files, and Excel® templates can be downloaded from https://link.springer.com/book/9783030831899.

Ilmenau, Germany   Yuri A. W. Shardt

The original version of this book has been revised. There are some scientific errors in this publication from Chapter 1 to 7. The correction to this book is available at https://doi.org/10.1007/978-3-030-83190-5_9

Symbols and Abbreviations

This section summarises the key symbols and abbreviations found in the book. Please note that despite attempts to limit symbols to single meaning and use, it is occasionally necessary to have multiple meanings assigned to a given symbol. This cases are clarified by pointing out which meaning is used in which chapter. Symbols are always written in italics or using special fonts, while abbreviations are written in uppercase, not in italics, and using normal Latin letters. Symbols and abbreviations are in normal alphabetical order.

Symbols

$\lfloor \cdot \rfloor$	Round-down function
A	A-polynomial
$\mathcal{A}$	Regression matrix (Chaps. 3, 4); state matrix (Chap. 5)
B	B-polynomial
$\mathcal{B}$	Input matrix
$\mathfrak{B}(1, q)$	Bernoulli distribution
$\mathfrak{B}(n, q)$	Binomial distribution
C	C-polynomial
$\mathbb{C}$	Complex numbers
$\mathcal{C}$	Output matrix
D	D-polynomial; seasonal differencing order (both Chap. 5)
d	Differencing order
$\mathcal{D}$	Throughput matrix
D_i	Cook's Distance
E	Expectation operator
e	(White, Gaussian) noise
e_t	Disturbance signal
F	F-polynomial
f	Frequency

$\mathscr{F}$	Fisher information matrix
$\mathbb{F}$	Space of all possible events
$f(x)$	Probability density function; probability mass function
f_{X1}	Marginal probability density function
$f_{Y|x}$	Conditional probability density function
$\mathfrak{F}$	Fourier transform
$\mathfrak{F}(v_1, v_2), F_{v1, v2}$	F-distribution
(ω)	Power spectrum; spectral density
G	General transfer function
G_a	Actuator model
G_c	Controller model
G_l	Disturbance model
G_p	Process model
G_s	Sensor model
$g_{\hat{\theta}}$	Efficiency
h	Impulse response coefficients
H_0	Null hypothesis
H_1	Alternative hypothesis
I	Identity matrix
$\mathcal{J}$	Jacobian matrix
$\mathcal{J}'$	Grand Jacobian matrix
$\mathcal{J}_t$	Kalman smoother gain
k	Factor (Chap. 4); discrete time $\epsilon\ \mathbb{N}$ (Chaps. 5, 6)
l	Level
$L(\theta|x)$	Likelihood function
$\ell(\theta|x)$	Log-likelihood function
m	Number of data points for regression, in a time series
m_i	Uncentred moment
$\bar{m}_i$	Centred moment
n	Number of samples, parameters
$\mathbb{N}$	Set of natural numbers
$\mathfrak{N}(\mu, \sigma_2)$	Gaussian (normal) distribution
n_C	Number of centre point replicates
n_R	Number of replicates
P	Probability measure function; order of the seasonal autoregressive polynomial (Chap. 5)
p	Order of the autoregressive polynomial (Chap. 5)
(λ)	Poisson distribution
$P(Y|X)$	Conditional probability
p_l	Left probability
p_r	Right probability
q	Order of the moving-average polynomial
Q	Order of the seasonal moving-average polynomial
r	Residual
$\mathbb{R}$	Set of real numbers

R^2	Pearson's coefficient of regression
$r_{critical}$	Critical value
r_t	Reference signal
S	Sensitivity
s	Seasonal order
$\mathbb{S}$	Sample space
SSE	Sum of squares due to the error
SSR	Sum of squares due to regression
SSR_i	Sum of squares due to regression for the ith parameter
$t(\nu), t_\nu$	Student's t-distribution
TSS	Total sum of squares
u	Input
u_t	Input signal
w	Weight
X	Data point; random variable
x	Regressor; state (Chaps. 5, 6)
y	Observation
$\hat{y}_\infty$	Infinite-horizon predictor / infinite-step-ahead predictor
y_t	Output signal
$\hat{y}_{t+\tau\|t}$	τ-step-ahead predictor
z	Forward shift operator/z-operator
Z	Standard normal distribution; Z-score
$\mathbb{Z}$	Set of integers
z^{-1}	Backward shift operator/z^{-1}-operator
α	False positive rate; confidence level
β	False negative rate (Chap. 2); parameter (Chaps. 3, 4)
β_0	Mean response
γ	Autocorrelation; skew (Chap. 2)
Γ	Autocovariance matrix
γ_{ij}	Experimental coefficients (Chap. 4)
$\gamma_{XY}(\tau)$	Cross-covariance
δ	Bias
Δ	Difference
ε	Error
$\varepsilon_{t+\tau\|t}$	Prediction error
θ	Regressive coefficients; time delay (Chap. 6)
κ	Basis function
μ	Mean
ν	Degrees of freedom
ρ	Autocovariance
$\rho_{X\|Z}(\tau)$	Partial autocorrelation
$\rho_{X1, X2}$	Correlation
$\rho_{XY\|Z}$	Partial correlation
Σ	Covariance matrix
σ	Standard deviation

σ^2	Variance
σ_{MAD}	Median absolute deviation
τ_s	Sampling time
ϕ	Moving-average coefficients
$\chi^2(\nu)$, χ^2_ν	χ^2-distribution
ψ	Sensitivity function
Ω	Probability space
$\bar{\circ}$ (U+0305)	Mean value
$\hat{\circ}$ (U+0302)	Estimated value
$\vec{\circ}$ (U+20D7)	Vector
$\tilde{\circ}$ (U+0303)	Normalised value

Abbreviations

AIC	Akaike's Information Criterion
ANOVA	Analysis of variance
AR	Autoregressive model
ARMA	Autoregressive, moving-average, exogenous model
ARX	Autoregressive exogenous model
BIC	Bayesian or Schwarz Information Criterion
BJ	Box–Jenkins Model
CCD	Central composite design
CDF	Cummulative distribution function
CI	Confidence interval
CTL	Central limit theorem
I	Integrating model
IR	Impulse response model
MA	Moving-average model
ME	Mean error
MSE	Mean-squared error
MVE	Minimum variance estimator
NLARX	Nonlinear autoregressive exogenous model
OE	Output-error model
OLS	Ordinary least squares
PACF	Partial autocorrelation function
pdf	Probability density function
RBS	Random binary signal
SARIMA	Seasonal, autoregressive, integrated, moving-average model
SNR	Signal-to-noise ratio
tf	Transfer function
WLS	Weighted least squares

Contents

List of Figures

List of Tables

Chapter 1
Introduction to Statistics and Data Visualisation

Εἰκὸς γὰρ γίνεσθαι πολλὰ καὶ παρὰ τὸ εἰκός.
It is likely that unlikely things should happen.
Aristotle, Poetics, 1456a, 24

Although it is a common perception that statistics seeks to quantify and categorise uncertainty and unlikely events, it is actually a much broader and more general field. In fact, statistics is the science of collecting, analysing, interpreting, and displaying data in an objective manner. Built on a strong foundation in probability, the application of statistics has expanded to consider such topics as curve fitting, game theory, and forecasting. Its results are applied in many different fields, including biology, market research, polling, economics, cryptography, chemistry, and process engineering.

Basic statistical methods have been traced back to the earliest times in such forms as the collection of data regarding a farmer's livestock, the amount, quality, and type of grain in the city granaries, or the phases of the moon by early astronomers. With these simple data sets, graphs could be created, summary values computed, and patterns could be detected and used. Greek philosophers, such as Aristotle (384−322 B.C.), pontificated on the meaning of probability and its different realisations. Meanwhile, ancient astronomers, such as Ptolemy (*c.* A.D. 90−168) and Al-Biruni (973−1048), were developing methods to deal with the randomness and inherent errors in their astronomical measurements. By the start of the late Middle Ages around 1300, rudimentary probability was being developed and applied to break codes. With the start of the seventeenth century and spurned by a general interest in games of chance, the foundations of statistics probability were developed by Abraham de Moivre (1667−1754), Blaise Pascal (1623−1662), and Jacob Bernoulli (1655−1705). These scientists sought to resolve and determine optimal strategies for such games of chance. The nascent nation-states also took a strong interest in the collection and interpretation of economic and demographic information. In fact, the word *statistics*, first used by the German philosopher Gottfried Achenwall (1719−1772) in 1749, is derived from the Neolatinate term *statisticum collegium*, meaning *council of the state*, referring to the fact that even then the primary use of the collected information was to provide insight (*council*) about the nation-state (Varberg 1963). In the early nineteenth century, among others, works by Johann Carl Friedrich Gauss (1777–1855), Pierre-Simon Laplace (1749–1827), and Thomas Bayes (1701–1761) led to the development of new theoretical and practical ideas. Theoretically, the grounding of statistics in probability theory, especially the development of the Gaussian distribution, allowed for many practical applications, including curve fitting and linear regression. Subsequent work, by such researchers as Andrei Kolmogorov (1903–1987) and Andrei Markov (1856–1922), solidified the theoretical underpinning and developed new ways of understanding randomness and methods for quantifying its behaviour. From these foundations, Karl Pearson (1857–1936) and Ronald Fisher (1890–1962) developed hypothesis testing, the χ^2-distribution, principal component analysis, design of experiments, analysis of variance, and the method of maximum likelihood, which continue to be used today. Subsequently, these ideas were used by George Box (1919–2013), Gwilym Jenkins (1932–1982), and Lenart Ljung (1946–) to develop stochastic modelling and advanced probabilistic models with applications in economics, biology, and process control. With the advent of computers, many of the previously developed methods can now be realised efficiently and quickly to analyse enormous amounts of data. Furthermore, the increasing availability of computers has led to the use of new methods, such as Monte Carlo simulations and bootstrapping.

Even though statistics still remains solidly applied to the study of economics and demographics, it has broadened its scope to cover almost every human endeavour. Some of the earliest modern applications were to the design and analysis of agricultural experiments to show which fertilisers and watering methods were better despite uncontrollable environmental differences, for example, the amount of sunlight received or local soil conditions. Later these methods were extended to analyse various genetic experiments. Currently, with the use of powerful computers, it is possible to process and unearth unexpected statistical relationships in a data set given many thousands of variables. For example, advertisers can now accurately predict changes in consumer behaviour based on their purchases over a period of time.

Another area where statistics is used greatly is the chemical process industry, which seeks to understand and interpret large amounts of industrial data obtained from a given (often, chemical) process in order to achieve a safer, more environmentally friendly, and more profitable plant. The process industry uses a wide range of statistics, ranging from simple descriptive methods through to linear regression and on to complex topics such as system identification and data mining. In order to appreciate the more advanced methods, there is a need to thoroughly understand the fundamentals of statistics. Therefore, this chapter will start the exploration with some fundamental results in statistical analysis of data sets coupled with a thorough analysis of the different methods for visualising or displaying data. Subsequent chapters will provide a more theoretical approach and cover more complex methods that will always come back to use the methods presented here. Finally, as a side note, it should be noted that the focus of this book is on presenting methods that can be used with modern computers. For these reasons, heavy emphasis will be made on matrices and generalised approaches to solving the problems. However, except for the last two chapters dedicated to MATLAB® and Excel®, little to no emphasis will be placed on any specific software as a computational tool; instead, the theoretical and implementation aspects will be examined in depth.

1.1 Basic Descriptive Statistics

The most basic step in statistical analysis of a data set is to describe it descriptively, that is, to compute properties associated with the data set and to display the data set in an informative manner. A data set consists of a finite number of **samples** or data points. In this book, a data set will be denoted using either set notation, that is, $\{x_1, x_2,\ldots, x_n\}$ or vector notation, that is, as $\vec{x} = \langle x_1, x_2, \ldots, x_n \rangle$. Set notation is useful for describing and listing the elements of a data set, while vector notation is useful for mathematical manipulation. The size of the data set is equal to n. The most common descriptive statistics include measures of **central tendency** and **dispersion**.

1.1.1 Measures of Central Tendency

Measures of central tendency provide some information about the central or typical value in the data set. The basic measures of central tendency include the **mean**, **mode**, and **median**. Since the most common such measure is the **mean**, which is often colloquially called the average, all of these measures are often referred to as **averages**. A summary of the basic properties of these measures is provided in Table 1.1.

The **mean** is a measure of the central value of the set of numbers. It is often denoted as an overbar ($\bar{o}$) over a variable, for example, the mean of $\vec{x}$ would be written as $\bar{x}$. The most common **mean** is simply the sum of all the values divided by the total number of data points, n, that is,

$$\bar{x} = \frac{\sum_{i=1}^{n} x_i}{n}. \tag{1.1}$$

Table 1.1 Summary of the main properties of the measures of central tendency

Measure	Formula	Advantages	Disadvantages
Mean	$\bar{x} = \frac{\sum_{i=1}^{n} x_i}{n}$	Easy to compute and interpret	Can easily be influenced by extreme values
Mode	Most common entries in the data set	Easy to interpret	May do not accurately represent the data set
Median	Middle entry of the ordered data set	Robust and easy to interpret	Not necessarily easy to compute

Alternatively, a weighted mean can be computed, where for each value a weight w is assigned, that is,

$$\bar{x} = \frac{\sum_{i=1}^{n} w_i x_i}{\sum_{i=1}^{n} w_i}. \tag{1.2}$$

The weighted mean can be used when the accuracy of some of the values is suspected to be less than that of others. Although the mean is a commonly used measure of central tendency and hence widely reported when describing data, it is not necessarily a robust measure, that is, the mean can be heavily skewed by one or two numbers that are significantly different from the others. For example, if we have the data set of three numbers, $\{2, 3, 4\}$, whose mean is $\bar{x} = 3$, and replace the 4 by 10, the mean becomes $\bar{x} = 5$, which is larger than two of the other numbers.

The **mode** represents the most common entry in a given data set. Multiple entries can be tied for the mode, in which case, the data set is said to be **multimodal**.[1] For the following set of numbers, $\{2, 4, 5, 5, 5, 6, 7, 10, 10, 10, 11\}$, there are two modes: 5 and 10, as both occur exactly three times. Although, in general, the mode is less sensitive to minor changes in the data set, it is still relatively easy to skew the results by adding too many identical values to create a new modal value. Furthermore, the most common entry need not be in any way descriptive of the overall properties of the data set. This can especially be the case if one of the extreme values occurs slightly more often than the other numbers and hence becomes the modal value.

The **median** represents the middle value of an ordered data set. If the number of data points is odd, then the median will represent the middle value. On the other hand, if the number of data points is even, then the median will be the mean value of the two middle values. Although it can happen that the median value is equal to a value in the data set, this is not necessarily always true. For the set given as $\{2, 4, 5, 10, 14, 14, 16, 17\}$, the median value would be 12 $(= \frac{1}{2}(10 + 14))$. The main advantage of the median value is that it represents the middle value of a given set and is robust to single extreme values.

1.1.2 Measures of Dispersion

Measures of dispersion seek to provide some information about how the values in a given data set are distributed, that is, are all the values clustered about one number or are they spread out across a large range of numbers. The basic measures of dispersion include **range, standard deviation** or **variance, skew,** and **median absolute deviation (MAD)**. A summary of the basic properties of these measures is provided in Table 1.2.

---

[1] If the specific number of tied entries is known, then the data set can be referred to by that number, for example, **bimodal** for a data set with two modes or **trimodal** for three modes.

Table 1.2 Summary of the main properties of the measures of dispersion

Measure	Formula	Advantages	Disadvantages	Comment
Range	max − min or [min, max]	Easy to compute	Can be easily influenced by extreme values	
Standard deviation	$\hat{\sigma} = \sqrt{\frac{\sum_{i=1}^{n}(x_i - \bar{x})^2}{n-1}}$	Commonly used and easy to interpret	Can be easily influenced by extreme values	Squaring it gives the variance
Median absolute difference	$\hat{\sigma}_{MAD} = \text{median}(|x_i - \bar{x}_{median}|)$	Robust estimate		Can be converted to an estimate of the standard deviation
Skew	$\hat{\gamma} = \dfrac{\frac{1}{n}\sum_{i=1}^{n}(x_i - \bar{x})^3}{\left(\frac{1}{n}\sum_{i=1}^{n}(x_i - \bar{x})^2\right)^{1.5}}$	Measures the spread of the extreme values		Rarely used in practice

The **range** of a data set is simply defined as the difference between the largest and smallest values within the data set. It is also possible to report the range as the two numbers representing the extreme data set values. It provides a simple, but not very meaningful, interpretation of the spread of the values. The larger the range, the more spread out the values are. Clearly, the range is affected adversely by large extreme values, since they would be directly used in its computation.

The **standard deviation**, σ, and **variance**, σ^2, are two related measures of the spread of the data set. The variance is always equal to the second power of the standard deviation. The larger the standard deviation, the more spread out the data set is. The variance can be computed as

$$\hat{\sigma}^2 = \frac{\sum_{i=1}^{n}(x_i - \bar{x})^2}{n-1}. \tag{1.3}$$

The standard deviation can then be computed by taking the square root of the value obtained using Eq. (1.3). In statistics, the circumflex ($\hat{\ }$) over a value shows that it is estimated or computed from a data set, rather than some theoretical value, for example, in Eq. (1.3), $\hat{\sigma}^2$ is the estimated value of the true variance, σ^2, given the data set. Even if the variance for the data set were the same, taking different data points will lead to some variation in the computed value. It can be noted that the variance is sensitive to extreme values. Occasionally, the variance can be denoted as the function *var*, for example, var(x) is the variance of x.

A method to avoid the sensitivity of the standard deviation to extreme values is to compute the **median absolute deviation (MAD)**, denoted by σ_{MAD}, which replaces the mean by the robust median. It can be computed as follows:

$$\hat{\sigma}_{MAD} = \text{median}(|x_i - \overline{x}_{median}|), \qquad (1.4)$$

where *median* is the function that determines the median value given a data set and $\overline{x}_{median}$ is the median value for the data set. It is possible to convert $\hat{\sigma}_{MAD}$ to a robust estimate of the standard deviation. However, it requires knowing the underlying distribution in order to compute the conversion factor. For a normal distribution, the robust estimate of the standard deviation can be written as

$$\hat{\sigma} = 1.4826\hat{\sigma}_{MAD}. \qquad (1.5)$$

The **skew**, denoted by γ, measures the amount of asymmetry in the distribution. Skewness is determined by examining the relationship in the clustering of extreme values, that is, the tails. If more of the data set is clustered towards the smaller extreme values, then it is said that the system has **positive** or **right skewness**. On the other hand, if the data set is clustered towards the larger extreme values, then it is said that the system has **negative** or **left skewness**. The skew of a data set can be computed as

$$\hat{\gamma} = \frac{\frac{1}{n}\sum_{i=1}^{n}(x_i - \overline{x})^3}{\left(\frac{1}{n}\sum_{i=1}^{n}(x_i - \overline{x})^2\right)^{1.5}}. \qquad (1.6)$$

Graphically, the skewness can be seen from a histogram, which plots the frequency of a value against the value. Examples of left and right skewness are shown in Fig. 1.1.

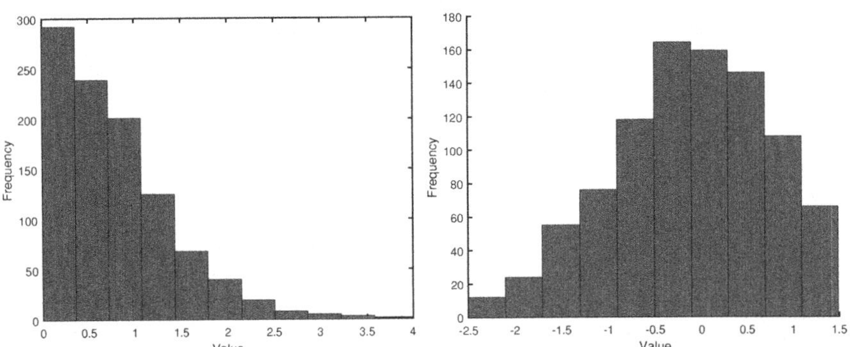

Fig. 1.1 (Left) Right-skewed and (right) Left-skewed data set

1.1.3 Other Statistical Measures

In addition to the measures of central tendency and dispersion, there exist other ways of quantifying a particular data set. This section will briefly review the two most common such methods: **quantiles** and **outliers**.

1.1.3.1 Quantiles

A **quantile** is a way of dividing the data set into segments based on the ordered rank of the data set. Common quantiles are the median (two segments with the split at 50%), **quartiles** (four segments at 25, 50, and 75%), **quintiles** (five segments at 20, 40, 60, and 80%), and **percentiles** (100 segments). In order to obtain a meaningful division, there should be at least as many different data points as there are segments.

Partitioning a data set into quantiles can be accomplished using the following steps:

(1) **Order** the data set from smallest to largest.
(2) **Obtain** an estimate of the boundaries for each of the segments using the following formula (Hyndman and Fan 1996):

$$h = (n-1)p + 1$$
$$Q_p = x_{\lfloor h \rfloor} + (h - \lfloor h \rfloor)\left(x_{\lfloor h \rfloor + 1} - x_{\lfloor h \rfloor}\right), \tag{1.7}$$

where n is the number of data points, $p = k/q$, k and q are defined as the kth q-tile, x_i is the ith data point of the ordered data set, and $\lfloor \cdot \rfloor$ is the floor function, that is, round down any number to its nearest integer. When $p = 1$, then $Q_p = x_n$.

Different formulae for computing the sample quantile can be obtained by changing the equation for h. Two other common formulations are:

(1) **Exclusive Formulation**: $h = (n+1)p + 1$, with both p and Q_p computed the same way as before.
(2) **Linear Interpolation Formulation**: $h = np + \frac{1}{2}$, with both p and Q_p computed the same way as before.

The differences in the estimated values are in most cases quite minimal. A comparison of the above methods is given in Table 1.6 in the context of an example (see Sect. 1.3.3, p. 28).

It can be noted that in all versions of Excel® the method defined by Eq. (1.7) is available (as either the function quartile or quartile.inc). Newer versions of Excel® (2010 or newer) also support option 1 (as quartile.exc). All versions of MATLAB® implement option 2.

1.1.3.2   Outliers

Outliers are data points that seem to be quite different from surrounding values and expected behaviour. Outliers can be caused by many different factors, including data entry or data collection errors or caused by randomness inherent in the system. Whenever a point is suspected to be an outlier, it is always useful to check that it has been correctly recorded and collected. Determining whether a point is an outlier is ultimately subjective and relies on intuition. Common rules for determining outliers include (Lin et al. 2007):

(1)   **Visual Tests**: Visual inspection to determine which values are located far from the bulk of the data, for example, in the set $\{1, 2, 1, 2, 3, 0, 2, -10\}$, -10 could be considered to be an outlier. Displaying the data using graphs can be a very useful approach. Graphs, such as the box-and-whisker plot, line charts, and scatter plots, can be useful for determining outliers.

(2)   **3σ Edit Rule**: Data points whose Z-score is large (>3), where the Z-score is given as

$$Z_i = \frac{x_i - \bar{x}}{\sigma}, \tag{1.8}$$

x_i is the data point of interest, Z_i is the corresponding Z-score, $\bar{x}$ is the mean value of the data set, and σ is the standard deviation of the data set. This approach only works if it can be assumed that the data set comes from a normal distribution. It is not very robust.

(3)   **Hampel identifier** (Davies & Gather, 1993): The Hampel identifier assumes that points lie outside the band $x_{median} \pm 3\hat{\sigma}_{rob}$, where $\hat{\sigma}_{rob}$ is defined as

$$\hat{\sigma}_{rob} = 1.4826 \text{median}(|x_i - x_{median}|) \tag{1.9}$$

and *median* is the function that determines the median value of the given data set. This equation represents the median absolute difference and is a robust manner of estimating the spread of the data. The constant is selected such that $\hat{\sigma}_{rob}$ is equal to σ for a normal distribution. In fact, for a normal distribution, the Hampel identifier and the 3σ edit rules will produce the same results.

1.2   Data Visualisation

Data visualisation is the science and art of displaying information in a visual manner that not only displays the relevant information accurately but also in a visually appealing way. There exist many different methods for visualising a given data set, including graphs and tables. Each method has its advantages and disadvantages when it comes to displaying the data. In general, the following principles can be followed to determine which method is best to display the data:

(1) **Density of Information**: How much information is to be presented? Are there only a few points that need to be summarised, or are there multiple points that need to be shown?

(2) **Comparison**: What is the point of showing the values? What types of relationships between the data are to be highlighted?

(3) **Efficiency**: Which method shows the desired relationships the best? How well is the information displayed? Are the desired relationships visible clearly?

(4) **Display Scheme**: What kind of display scheme will be required? Will you need to use different colours? If so, how many? Will you need to use multiple different symbols? If so, which ones? Can they all be distinguished easily in the figure? What if the figure is printed in black and white? What type of scale will be used: normal or logarithmic?

Irrespective of the method selected, it is important that the following information, as appropriate, be included:

(1) **Titles/Captions**: Each figure or group of figures should have a clear title or caption that briefly explains the information in the figure.

(2) **Labels**: Appropriate labels should be included. This should include, as appropriate, the full name of what is being shown, abbreviations, and units. All axes and legend headings should be considered. For axes, an acceptable and very useful approach would be to use the following label: "*Full Name, Abbreviation (Units)*", for example, "Temperature, T ($^\circ$C)". A legend should be provided if multiple types of information are plotted on the same graph.

(3) **Display**: Are the different symbols used clearly distinguishable? Consider the fact that many figures will end up in black-and-white publications. This implies that relying solely on colour to distinguish different aspects of a figure can be difficult. Furthermore, data points should not be connected by lines unless there is a reason for connecting the points. This implies that experimental data in many cases should be entered as single points, while theoretical values should be connected with a single continuous line.

A good discussion of the art of data visualisation, as well as some ideas on how to implement it, can be found in the books by Tufte (1997, 2001).

1.2.1 Bar Charts and Histograms

A **bar chart** is a graph that contains vertical or horizontal bars whose length is proportional to the value. Bar charts compare by their nature discrete information. One axis will contain the category or discrete item, while the other axis will contain the value axis. Typical bar charts are shown in Fig. 1.2. Although 3-D bar charts are

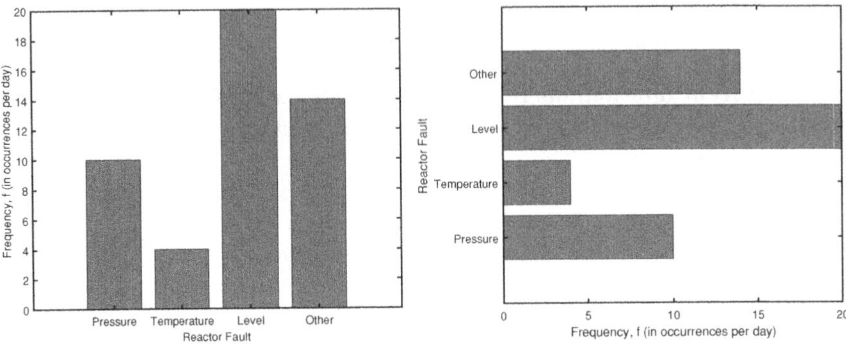

Fig. 1.2   (Left) Vertical bar chart and (right) Horizontal bar chart

possible, they do not provide any advantage for displaying the information accurately or efficiently.

A **histogram**, similar to a bar chart, shows the frequency of a given range of values that occur in the data set. Thus, a histogram records continuous data but presents it in a similar manner. A histogram is constructed by first creating bins or ranges of numbers. Next, the number of times a value from the data set falls within each of the ranges is determined and noted. Once this has been completed, a vertical bar chart is plotted using the bins as the category and the occurrences as the value. It should be noted that the bins are normally assigned so that they are of equal size (except for the two endpoints) and are continuous, that is, two adjacent bins share the same endpoint. A four-bin example could be $x < 3$, $3 \leq x < 5$, $5 \leq x < 7$, and $x \geq 7$. A typical histogram is shown in Fig. 1.3. Not all software provides methods for directly creating a histogram. In some cases, it is necessary to manually bin the data and then create the corresponding bar graph.

Fig. 1.3   Typical histogram

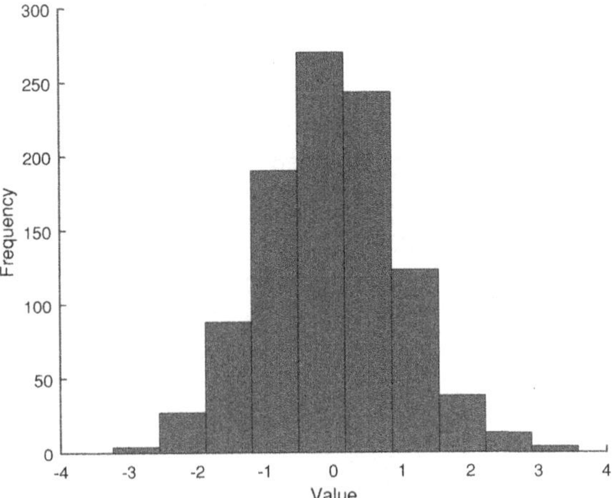

Fig. 1.4 Typical pie chart

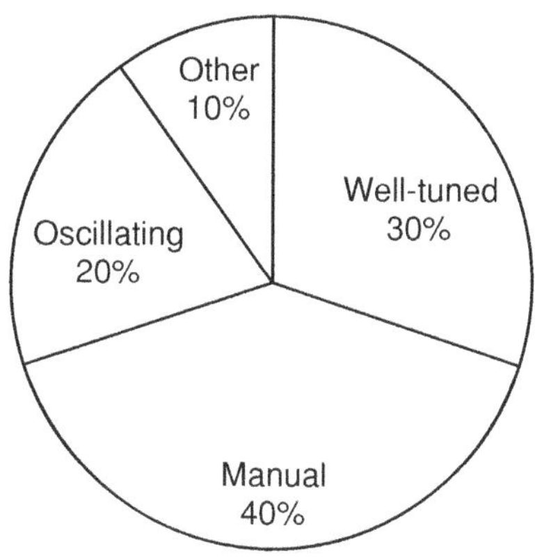

1.2.2 Pie Charts

A **pie chart** is a circle whose arc length has been divided up into different proportions. It is named after how a pie is cut. Pie charts can be used to display the relationships of parts to a whole, for example, components of a budget. However, too many different items in a pie chart can lead to difficulties with representing the items effectively, as the number of available colours and amount of space can be limited. As well, a pie chart tends to require more space than would ideally be needed to display the information. A typical pie chart is shown in Fig. 1.4.

1.2.3 Line Charts

A **line chart** is a graph that contains individual data points connected by a line. Very often, the horizontal, or x-axis, will represent time and the vertical, or y-axis, will represent the value of some variable over time. For this reason, a line chart is often called a **time-series plot**. A line chart is very effective in showing how a variable(s) changes over time. However, too many competing lines can make the figure difficult to read and understand. A typical line chart is shown in Fig. 1.5.

1.2.4 Box-and-Whisker Plots

A **box-and-whisker plot**, or more simply a **boxplot**, is a complex graph that is based on quartiles to conveniently display multiple different properties of the data set. It can conveniently be used to compare different data sets. A box-and-whisker plot consists

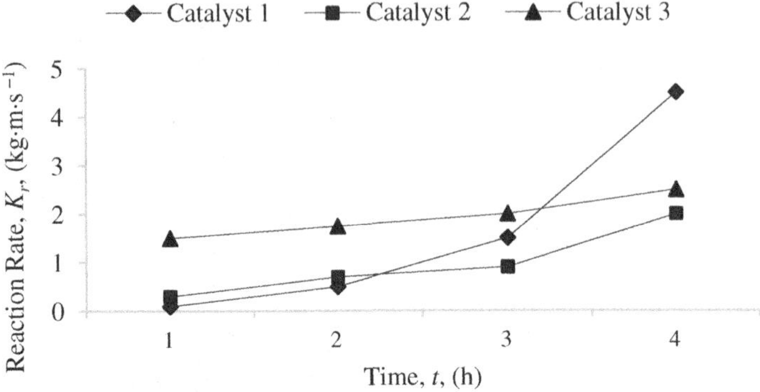

Fig. 1.5 Typical line chart

of two parts: the box and the whiskers. The box is formed by the 25th (Q1) and 75th (Q3) percentile boundaries with the middle line invariably being the median (Q2). The whisker limits are defined using any of the following rules:

(1)   Maximum and minimum of the data set.
(2)   Lowest data point located within 1.5 of the interquartile range from the lower quartile and the largest data point located within 1.5 of the interquartile range above the upper quartile. The interquartile range is defined as the difference between Q3 and Q1. Such a plot is often called a **Tukey boxplot**.
(3)   The 9th and 91st percentiles.
(4)   The 2nd and 98th percentiles.

In all cases, data points lying outside the whisker limits are conventionally denoted by crosses or dots, often in another colour. Such points can be labelled as **outliers**. Of the available definitions, the most commonly encountered box-and-whisker plots use whisker bounds defined by the first two rules. Typical box-and-whisker plots are shown in Fig. 1.6. These box-and-whisker plots were created using the interquartile range for the data points.

1.2.5   Scatter Plots

A **scatter plot** shows the values obtained using some mark. These marks are not connected and hence it looks like all the values are scattered around. A scatter plot is useful when it is desired to show the relationship between two variables, but the values vary quite a bit between each sample. Often, the true or predicted values can be superimposed using a line. The selection of the appropriate mark can be important, especially when there are many data points to show. Ideally, each data

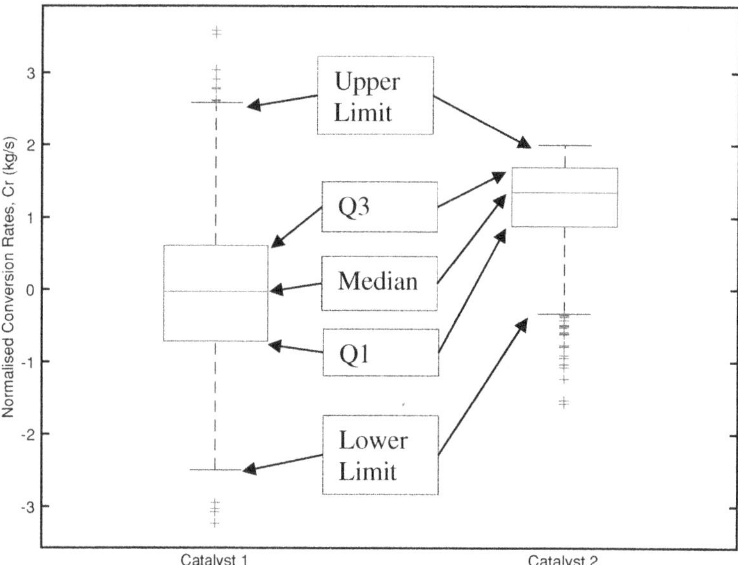

Fig. 1.6 Typical box-and-whisker plots

point should be clearly visible. In some cases, it may be useful to show data from multiple experiments or runs together on a single plot. Again, the various marks need not only be individually distinguishable but also distinguishable from each other. A typical scatter plot is shown in Fig. 1.7.

1.2.6 Probability Plots

A **probability plot** is a graph that compares the data set against some expected statistical distribution by comparing the actual quantiles against the theoretical quantiles. Such probability plots are also often called $Q - Q$ or $P - P$ plots. The most common statistical distribution for comparison is the normal distribution. The exact values plotted on each of the axes depend on the desired graph and software used. In general, the theoretical values are plotted on the x-axis, while the actual values are plotted on the y-axis. Occasionally, the actual values are modified in order to emphasise certain properties. A generalised probability plot can be constructed using the following steps:

(1) For each data point, compute its rank, denoted by R_i.
(2) Compute an approximation of the quantile position using the following formula:

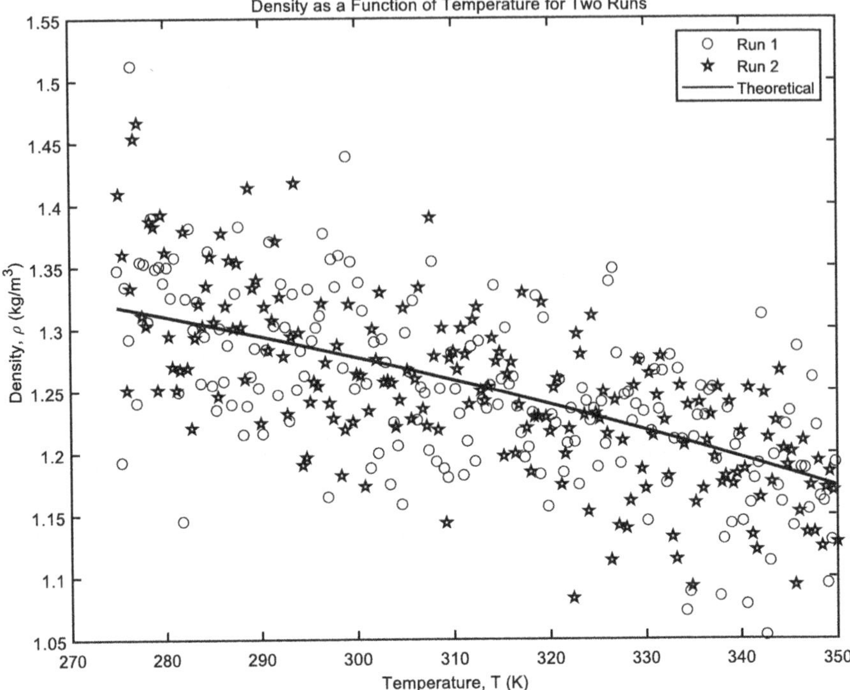

Fig. 1.7 Typical scatter plot

$$U_{Ri} = \begin{cases} 1 - 0.5^{\frac{1}{n}} & i = 1 \\ \frac{i - 0.3175}{n + 0.365} & i = 2, 3, \ldots, n - 1 \\ 0.5^{\frac{1}{n}} & i = n \end{cases}. \tag{1.10}$$

It can be noted that any of the many different formulae can be used here. The simplest formula is given as

$$U_{Ri} = \frac{i - 0.5}{n}. \tag{1.11}$$

The final results will be very similar, irrespective of the actual formula used.

(3) Compute the N-score for each rank, N_{Ri}, using the following formula:

$$N_{Ri} = \text{icdf}(U_{Ri}), \tag{1.12}$$

where *icdf* is the inverse of the cumulative distribution function of the desired distribution (further information about distribution functions can be found in Sect. 2.3).

(4) If desired, transform the actual data. Two common transformations are:

a. **Z-score**: If the theoretical distribution is assumed to be normal, then it can be convenient to transform the data into the corresponding Z-score. This will minimise the need to know the true mean and standard deviation of the data set. The formula for the Z-score is

$$Z_i = \frac{x_i - \bar{x}}{\hat{\sigma}}. \tag{1.13}$$

b. *Quantiles*: Another option is to plot the quantiles corresponding to the data set on the y-axis rather than the actual values. Any of the formulae for computing the quantile can be used. The most common one in this case is Eq. (1.11). This will give a cumulative distribution feel to the data set. Some software, such as MATLAB®, uses this approach to produce its probability plots.

(5) Plot N_{Ri} on the x-axis and x_i on the y-axis to construct the normal probability plot.

The interpretation of this probability plot is based on the following theoretical observations:

(1) The data should lie on a straight line, which, in the ideal case, is $y = x$.
(2) If the straight line given by the data is shifted vertically by a fixed amount, then this represents the difference in the mean between the assumed distribution and the actual, data distribution.
(3) If the straight line given by the data has a different slope ($\neq 1$), then the standard deviation of the data set is different from the assumed distribution's standard deviation.

This is shown graphically in Fig. 1.8, for the case of a normal distribution with different means and variances compared against a normal distribution with a mean of zero and a variance of 1. It can be seen that the straight line's slope and y-intercept match well with the theoretical values. Therefore, based on these observations, it can be useful to include a straight line (line of best fit) to give an estimate of the true mean and standard deviation.

From these theoretical observations, this means that the points in the probability plot should all lie along a straight line. The exact slope and y-intercept are not all that important. Deviations from a straight line are indications that the data may not come from the proposed theoretical distribution. The most common deviations are:

(1) **Outliers**, or extreme values, at the endpoints.
(2) **Tails at the endpoints**, or curvature, that is, one tail is below the straight line and the other is above the straight line. This implies that the true distribution of the data set has a different distribution than the target distribution. Practically, if the left tail is below and the right tail is above the straight line, then the distribution in the tails is larger than in the target distribution. On the other hand, if the left tail is above and the right tail is below the straight line, then the data distribution in the tails is smaller than in the target distribution.

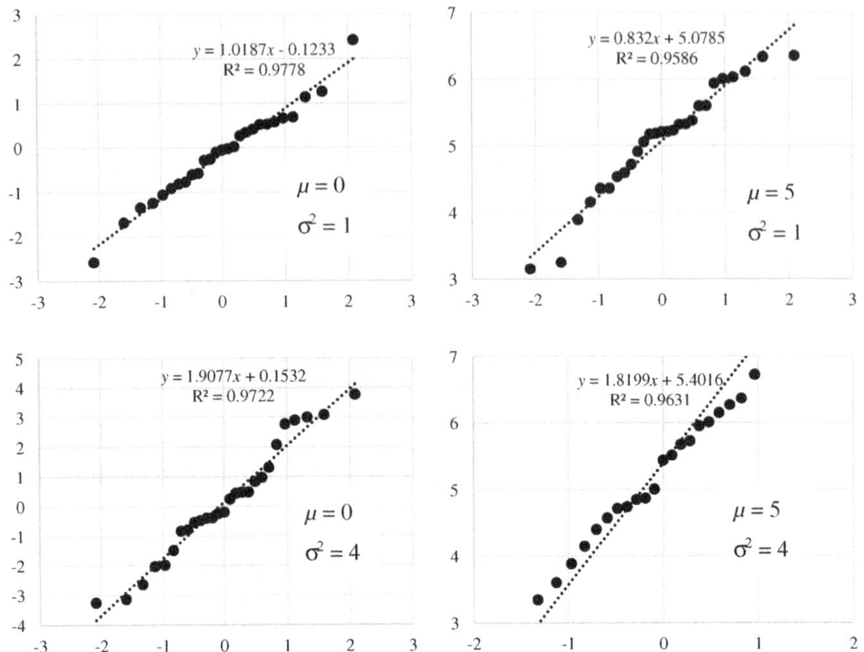

Fig. 1.8 Probability plots and the effect of the location parameters (μ and σ^2)

(3)   **Convex or concave curvature** in the centre, that is, the given data set is not symmetric compared with the target distribution.

(4)   **Plateaus, gaps, or horizontal data**, that is, the data seem to fall only within certain values. This is most likely to be the result of rounding errors introduced during measurement, storage, or data manipulation.

Figure 1.9 shows examples of how these kinds of problems can appear on a probability plot. Figure 1.9a shows a normal probability distribution with a mean of 0 and a variance of 1 with two outliers (circled). Notice how the outliers can cause some of the adjacent points to also be skewed from the ideal location. Figure 1.9b shows the case where the tails of the distribution do not match. In this case, a two-degrees-of-freedom Student's t-distribution was compared against the normal distribution. The t-distribution has larger tails than the normal distribution. This can be clearly seen by the deviations on both sides from the central line. Figure 1.9c shows the case where there is convex curvature in the centre. In this case, the asymmetric F-distribution was compared with the normal distribution. In such a case, drawing the desired straight line can be quite difficult since there can be potentially two or more "best" regions. Figure 1.9d shows the case where there are horizontal plateaus combined with gaps. In this case, the normal distribution with mean of 5 and variance of 4 was rounded down to three decimal places. This clearly shows the gaps and plateaus that rounding can induce in the results. Furthermore, it should be noted that drawing the straight

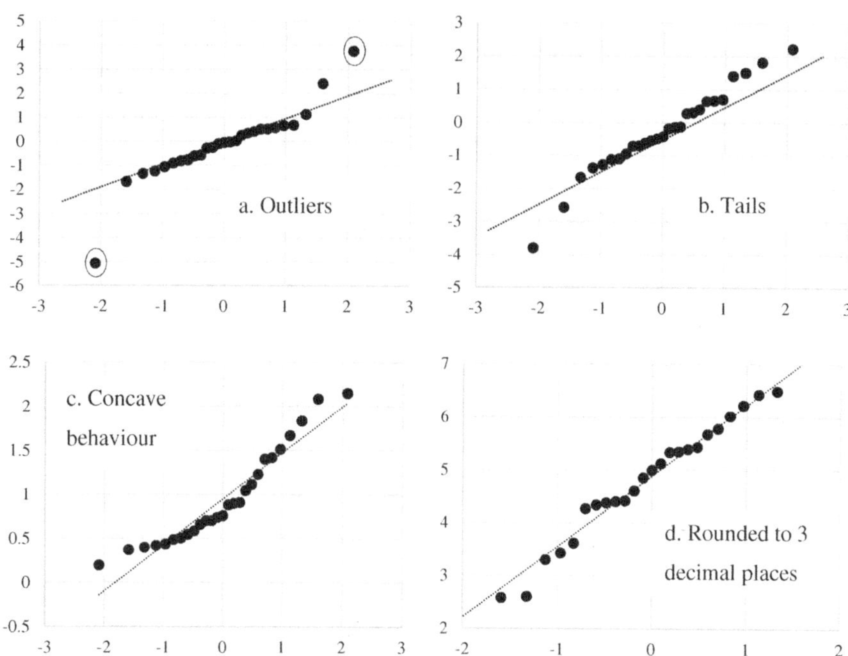

Fig. 1.9 Issues with probability plots

line for comparison can be difficult when the data set does not match the underlying distribution.

Finally, when dealing with small samples (say less than about thirty points), less ideal behaviour in the extreme regions (tails) can be tolerated. The extent and amount of tolerated deviations will depend on where the normal probability plot is being used. Figure 1.10 shows the normal probability plot for nine different realisations of eight data points drawn from the standard normal distribution. It can be seen that all samples show varying amounts of curvature and tails. Detailed comparisons of the effect of data size on normal probability plots can be found in Daniel and Wood (1980).

1.2.7   Tables

A **table** is a textual way of displaying information that consists of rows and columns. A table is useful to present a small amount of data whose exact values are important. It can be used to give information about summary statistics, such as the mean, mode, and standard deviation. Every table should have headers for its columns and rows. This can be formatted similarly to graph axes, by including the name of the variable, its symbol, and units. A well-designed table will contain all the relevant information

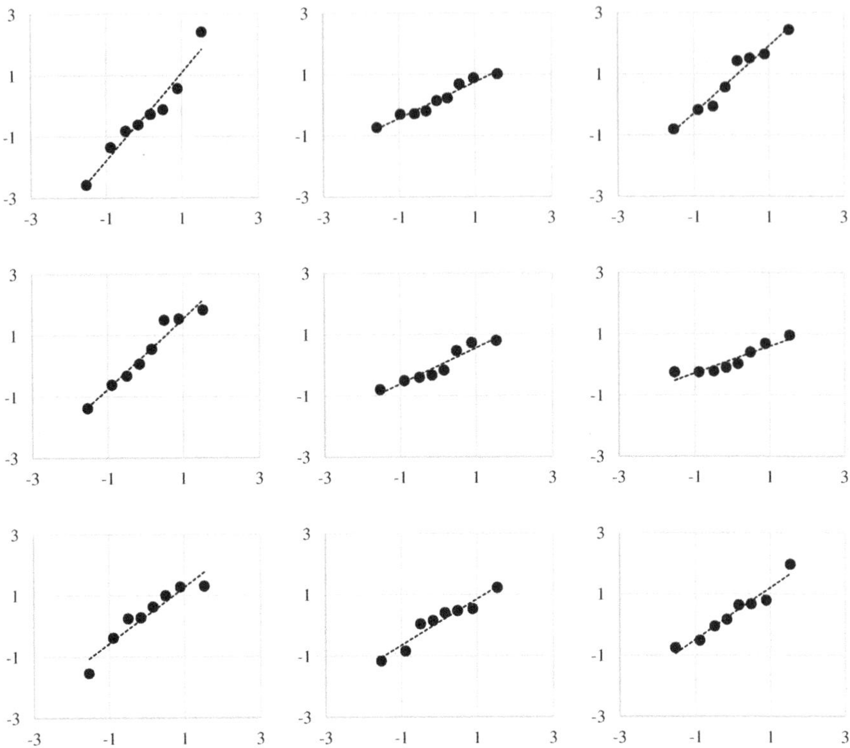

Fig. 1.10 Nine probability plots of eight samples drawn from a standard normal distribution

within it and be self-explanatory. Numbers should be properly formatted and not taken straight from the software used. There is no need to display more than about three or four digits (unless special circumstances warrant) with spacing between groups of three digits (on both sides of the decimal place). Scientific notation should be used as appropriate, for example, the number obtained from a calculator as 1.25896321532e3 could be written as either 1.259×10^3 (using scientific notation) or 1,259. A typical table is shown in Table 1.3.

Table 1.3 Typical table formatting

Treatment	Mean thickness, δ (μm)	Variance σ^2 (μm^2)	Range [lower, upper] (μm)
A	1.25	0.25	[0.25, 5.00]
B	1.50	0.10	[0.50, 2.25]
C	2.25	0.50	[0.50, 10.0]

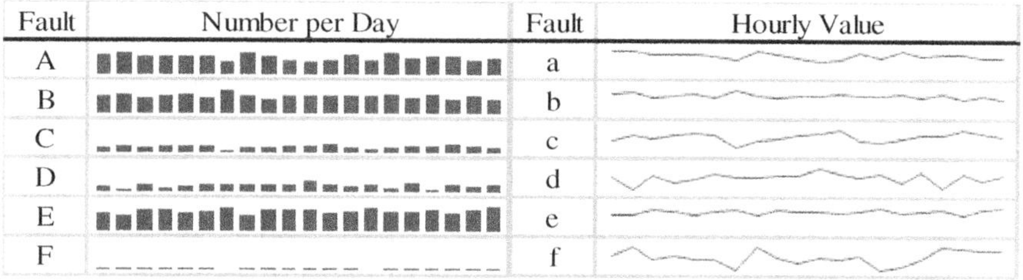

Fault	Number per Day	Fault	Hourly Value
A		a	
B		b	
C		c	
D		d	
E		e	
F		f	

Fig. 1.11 (Left) Spark bar graph showing the number of times a given fault occurs over the course of many days and (right) Sparkline showing the hourly process value for six different variables from a single unit over the course of a day

1.2.8 Sparkplots

Sparkplots or **profile plot** are various ways of summarising information so that only the trends and comparison between different data sets are compactly shown. They often do not have explicit axes or category markings. Sparkplots can be either line graphs (known as sparklines) or bar graphs (known as spark bar graphs). It is common to use sparklines to show the behaviour of multiple process variables in order to understand which variables could be influencing others. Spark bar graphs are often used as histograms to show the distribution of variables and at the same show the individual values. Typical examples are shown in Fig. 1.11.

1.2.9 Other Data Visualisation Methods

The above sections have presented the most common data visualisation methods for a given data set. More complex forms can be created by combining different simple data visualisation methods into a final integrated plot. Alternatively, the data could be transformed (changed in some manner) before being plotted. The different techniques that are available to accomplish this depend strongly on the intended application and will be introduced in the relevant sections in later chapters. Often such plots are created when there is multiple information that needs to be displayed, for example, one is interested in determining which of twenty variables are important for your analysis.

Two typical integrated data visualisation methods are presented in Figs. 1.12 and 1.13. In Fig. 1.12, the linear relationship between 100 different variables is plotted to determine which variables are most related to each other. This plot involves taking the data, transforming it, and then computing the correlation between each pair of the transformed data. A strong linear relationship is denoted by one (or a red colour), while a weak linear correlation is denoted by zero (or a dark blue colour). Obviously, the variables themselves are strongly related to each other and so the diagonal is always equal to one in such plots. More information on creating and

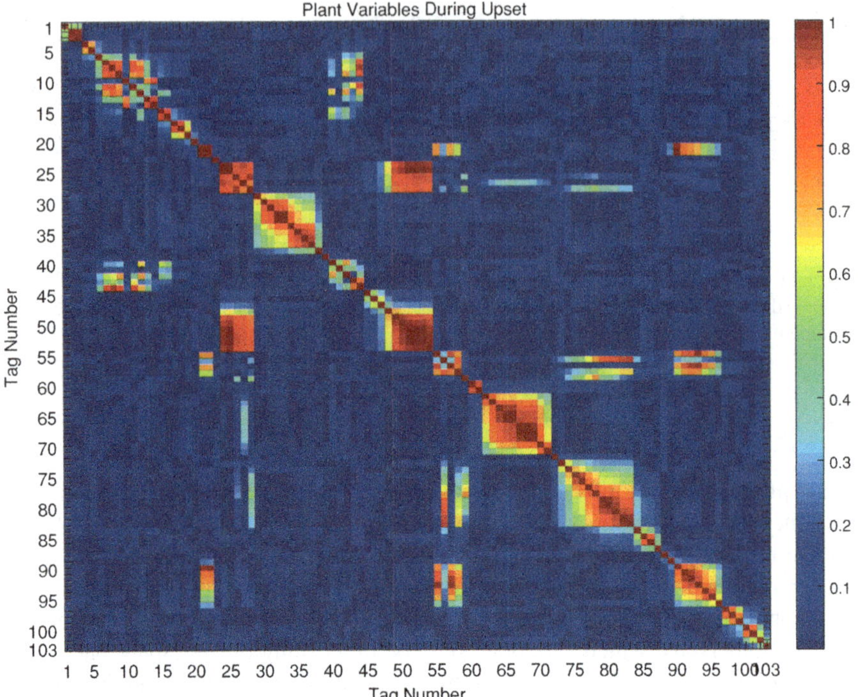

Fig. 1.12 Complex data visualisation example: A cross-correlation plot

plotting such figures can be found in Chap. 5. In Fig. 1.13, two variables are plotted against each other as a scatter plot with histograms to show the distribution of the individual variables. These plots can be useful for seeing and understanding complex interactions between different variables and how best to interpret them later. In this particular example, it can be seen that both variables are skewed to the left, with a rather large right tail.

1.3   Friction Factor Example

In this section, experimental data from a friction factor experiment will be considered. This data set consists of four separate runs performed on different pipe diameters collected on different days (often with a large separation in time).

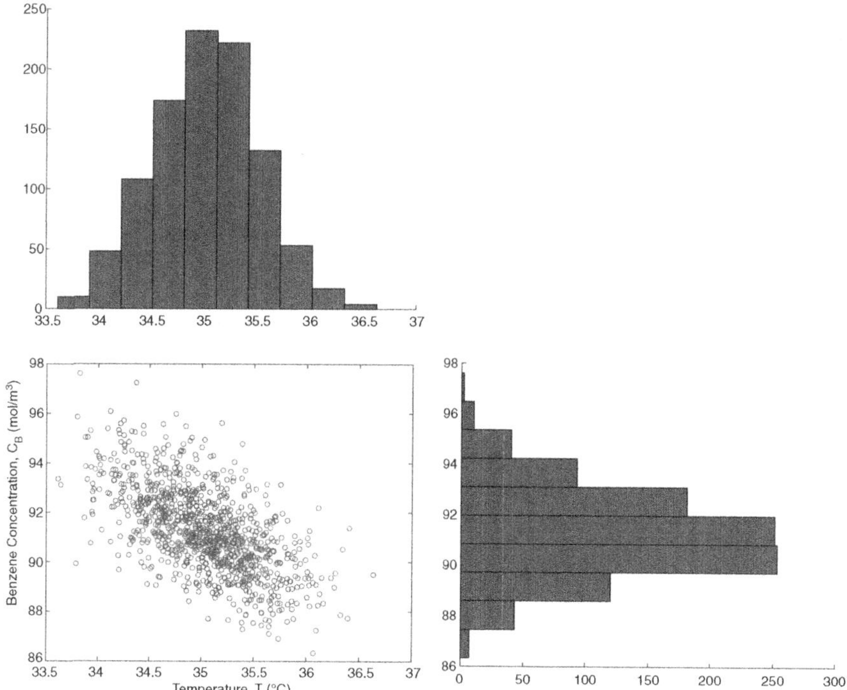

Fig. 1.13 Complex data visualisation example: Combining multiple plot types

1.3.1 Explanation of the Data Set

In the friction factor experiment, the flow of water through a pipe is changed to determine the pressure drop across a length of pipe for pipes with different diameters. In order to compare the results across multiple different diameters, the data are converted into two dimensionless numbers: the Reynolds number (Re), which represents the flow, and is defined as

$$\mathrm{Re} = \frac{\rho v D}{\mu},\tag{1.14}$$

where ρ is the density of the fluid, v is the velocity, D is the pipe diameter, and μ is the dynamic viscosity of the fluid; and the friction factor (f), which represents the pressure drop in the pipe, and is defined as

$$f = \frac{2D\Delta P}{\rho v^2 L},\tag{1.15}$$

where L is the length of the pipe and ΔP is the pressure drop.

The relationship between the friction factor and Reynolds number can be written as (Gerhart et al. 1992)

$$f = K\,\mathrm{Re}^{\beta}, \tag{1.16}$$

where K and β are parameters to be fit. For turbulent flow, where $4{,}000 < \mathrm{Re} < 100{,}000$, the Blasius equation predicts that $K = 0.316$ and $\beta = -0.25$ (Gerhart et al. 1992).

The experiment consisted of data collected on multiple days for different pipe diameters and flow rates using water as the fluid. Sample data are presented in Table 1.4. Runs 1 and 2 were performed on the same day, but with different pipe diameters: 4.9 mm for Run 1 and 6.1 mm for Run 2. Run 3 was performed on another day with a pipe diameter of 7.8 mm. Finally, Run 4 was some historical data obtained six years previously using the same equipment and a pipe diameter of 4.9 mm. The data are presented sequentially in the order in which the experiments were run, that is, for example, in Run 1, the experiment with a Re = 6,478 was run first, followed by the experiment with Re = 11,7785. Replicates were performed at some of the values, for example, in Run 1, there are two cases with a Re $\approx$ 11,800.

Table 1.4 Data from friction factor experiments

Run 1		Run 2		Run 3		Run 4	
Re	f	Re	f	Re	f	Re	f
6,478	0.0355	19,476	0.0268	20,701	0.0251	11,529	0.0308
11,785	0.0303	13,439	0.0293	13,248	0.0286	9,993	0.0318
5,485	0.0369	15,844	0.0281	18,409	0.0266	9,340	0.0329
9,075	0.0321	5,251	0.0369	5,602	0.0351	3,187	0.042
11,815	0.0302	11,980	0.0303	14,251	0.0281	6,248	0.0362
7,246	0.0343	17,732	0.0272	18,978	0.0261	4,838	0.0387
10,403	0.0309	6,366	0.0352	9,787	0.0309	4,427	0.0394
13,364	0.0292	15,115	0.0283	6,638	0.0339	9,567	0.0327
10,811	0.031	7,461	0.0345	10,748	0.0302	7,141	0.0351
7,730	0.0334	10,227	0.0314	16,813	0.027	5,750	0.0371
9,938	0.0316	13,240	0.0296	12,730	0.029	11,187	0.0312
11,581	0.0305	13,987	0.0291	8,794	0.0319	3,925	0.0405
8,432	0.0327	16,606	0.0277	15,041	0.0278		
12,546	0.0297	11,152	0.0307	12,060	0.0292		
9,051	0.0325	5,226	0.0377	6,937	0.0337		
9,470	0.0317			4,895	0.0364		

1.3.2   *Summary Statistics*

The mean, median, standard deviation, range, and median absolute difference will be determined for all four runs. Sample computations will be shown for Run 4 using the Reynolds number values. The results are summarised in Table 1.5.

For Run 4 and the Re values, the mean would be computed using Eq. (1.1) to give

$$\overline{Re} = \frac{\sum_{i=1}^{n} x_i}{n} = \frac{\begin{array}{c} 11{,}529 + 9{,}993 + 9{,}340 + 3{,}187 + 6{,}248 + 4{,}838 \\ +4{,}427 + 9{,}567 + 7{,}141 + 5{,}750 + 11{,}187 + 3{,}925 \end{array}}{12} = 7{,}261.$$

$$(1.17)$$

Similarly, the median would be computed by first ordering the data set from smallest to largest and then finding the average of the two midpoint values (since there is an even number of values present), that is,

$$3{,}187;\ 3{,}925;\ 4{,}427;\ 4{,}838;\ 5{,}750; \qquad \underbrace{6{,}248;\ 7{,}141} \qquad ;\ 9{,}340;\ 9{,}567;\ 9{,}993;\ 11{,}187;\ 11{,}529.$$

$$\boxed{\;\text{median} = \frac{6{,}248 + 7{,}141}{2} = 6{,}694.5\;}$$

$$(1.18)$$

The standard deviation can be computed using a modified form of Eq. (1.3) commonly used for manual computations to give

$$\sigma_{Re} = \sqrt{\frac{\sum_{i=1}^{n} x_i^2 - \frac{1}{n}\left(\sum_{i=1}^{n} x_i\right)^2}{n-1}} = \sqrt{\frac{\left(\sum_{i=1}^{n} x_i^2\right) - n\overline{x}^2}{n-1}}$$

$$= \sqrt{\frac{\left(\begin{array}{c} 11{,}529^2 + 9{,}993^2 + 9{,}340^2 + 3{,}187^2 + 6{,}248^2 + 4{,}838^2 \\ +4{,}427^2 + 9{,}567^2 + 7{,}141^2 + 5{,}750^2 + 11{,}187^2 + 3{,}925^2 \end{array}\right) - 12(7{,}261)^2}{12-1}} \qquad (1.19)$$

$$= 2{,}900$$

Table 1.5 Summary statistics for the friction factor data set

Summary Statistic	Run 1		Run 2		Run 3		Run 4	
	Re	f	Re	f	Re	f	Re	f
Mean	9,700	0.0320	12,200	0.0309	12,200	0.0300	7,260	0.0357
Median	9,700	0.0317	13,200	0.0296	12,400	0.0291	6,700	0.0357
σ	2,300	0.0021	4,500	0.0036	4,900	0.0034	2,900	0.0039
Range	7,880	0.0077	14,300	0.0109	15,800	0.0113	8,340	0.0112
σ_{MAD}	1,900	0.0012	3,000	0.0018	4,000	0.0023	2,700	0.0034

The range can be found by determining the largest and smallest values and subtracting them. Thus, the maximum value is 11,529 and the minimum value is 3,187. Therefore, the range is $11{,}529 - 3{,}187 = 8{,}340$.

The median absolute difference can be computed by first ordering the absolute value of the difference between the data point and the median to give

$$3{,}507.5;\, 2{,}769.5;\, 2{,}267.5;\, 1{,}856.5;\, 944.5;\, 446.5;\, 446.5;\, 2{,}645.5;\, 2{,}872.5;\, 3{,}298.5;\, 4{,}492.5;\, 4{,}834.5. \tag{1.20}$$

The ordered list then becomes

$$446.5;\, 446.5;\, 944.5;\, 1{,}856.5;\, 2{,}267.5;\, \underbrace{2{,}645.5;\, 2{,}769.5}_{\text{median} = 2{,}707.5};\, 2{,}872.5;\, 3{,}298.5;\, 3{,}507.5;\, 4{,}492.5;\, 4{,}834.5. \tag{1.21}$$

The median of the residuals is therefore 2,707.5.

It should be noted that all of the values have been rounded to three decimal places, except for the standard deviation, which has been rounded to two decimal places, in order to improve the presentation. It should be noted that the original mass flow rated and pressure drops used to compute the Reynolds number and friction factor were recorded to only three decimal places.

1.3.3 Data Visualisation

In this particular case, a scatter plot showing all the four runs together and a box-and-whisker plot of each run separately will be plotted. Detailed code for creating these graphs is given in Chap. 7 for MATLAB® and Chap. 8 for Microsoft Excel®. Figure 1.14 shows a scatter plot of the data showing each of the runs separately, while Fig. 1.15 gives the box-and-whisker plots for both the Reynolds number and the friction factor. The theoretical values using the Blasius equation have also been included in Fig. 1.14 to provide some reference point against which to compare the data set.

In order to illustrate the procedure for constructing a box-and-whisker plot by hand and determining the appropriate quartile boundaries, the Reynolds numbers from Run 4 will be used. For a box-and-whisker plot, it is necessary to determine the values located at $Q_{0.25}$, $Q_{0.5}$ (median), and $Q_{0.75}$. Equation (1.7) gives a general formula for computing these values. For $Q_{0.25}$, the first quartile, setting $n = 12$ and $p = 0.25 = \frac{1}{4}$ in the formula gives

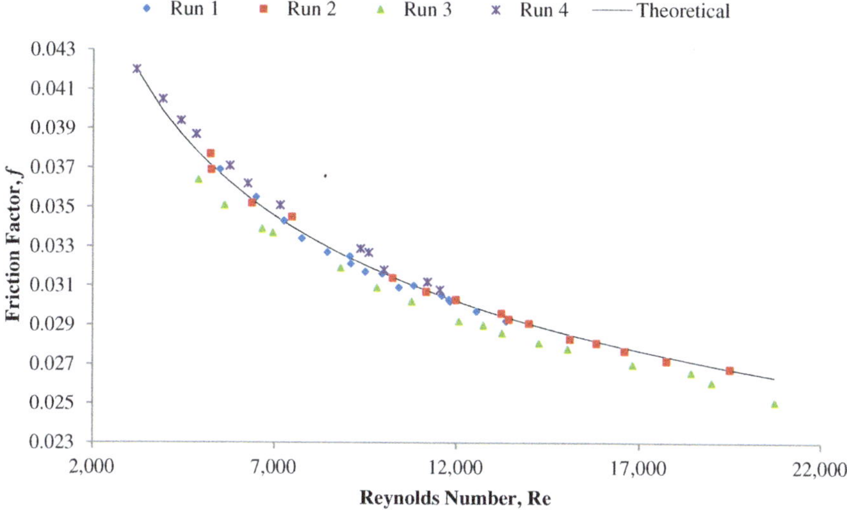

Fig. 1.14 Scatter plot of the friction factor as a function of Reynolds number for all four runs

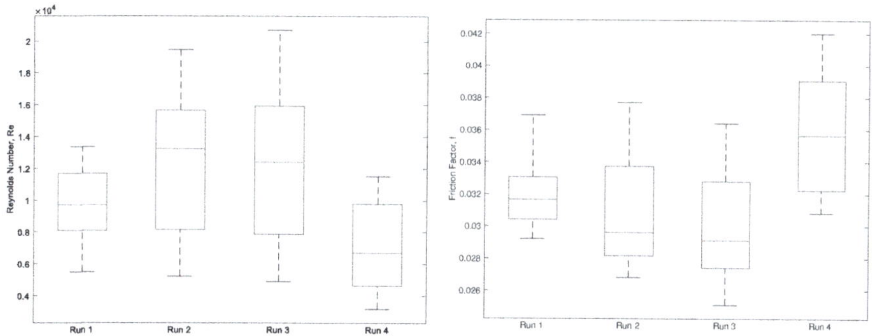

Fig. 1.15 Box-and-whisker plots for the friction factor experiment for the (left) Reynolds number and (right) Friction factor

$$h = (12 - 1)0.25 + 1 = \frac{15}{4} = 3.75$$

$$
\begin{aligned}
Q_{0.25} &= x_{\lfloor 3.75 \rfloor} + (3.75 - \lfloor 3.75 \rfloor)\left(x_{\lfloor 3.75 \rfloor + 1} - x_{\lfloor 3.75 \rfloor}\right) \\
&= x_3 + (3.75 - 3)(x_4 - x_3) \\
&= 4{,}427 + 0.75(4{,}838 - 4{,}427) \\
&= 4{,}735
\end{aligned}
\tag{1.22}
$$

Similarly, $Q_{0.5}$ can be computed as

Table 1.6 Computing quartiles with different software packages

Quartile	Manual	Excel® (quartile.inc)	Excel® (quartile.exc)	MATLAB® 2019
1	4,735	4,735	4,530	4,633
2	6,695	6,695	6,695	6,695
3	9,674	9,674	9,887	9,780

$$h = (12 - 1)0.5 + 1 = \frac{13}{2} = 6.5$$

$$\begin{aligned}
Q_{0.5} &= x_{\lfloor 6.5 \rfloor} + (6.5 - \lfloor 6.5 \rfloor)\left(x_{\lfloor 6.5 \rfloor + 1} - x_{\lfloor 6.5 \rfloor}\right) \\
&= x_6 + (0.5)(x_7 - x_6) \\
&= 6,248 + 0.5(7,141 - 6,248) \\
&= 6,695
\end{aligned} \tag{1.23}$$

It can be noted that, after rounding, this value is identical to that previously computed for the median. This should be always the case.

Similarly, $Q_{0.75}$ can be computed as

$$h = (12 - 1)0.75 + 1 = \frac{37}{4} = 9.25$$

$$\begin{aligned}
Q_{0.75} &= x_{\lfloor 9.25 \rfloor} + (9.25 - \lfloor 9.25 \rfloor)\left(x_{\lfloor 9.25 \rfloor + 1} - x_{\lfloor 9.25 \rfloor}\right) \\
&= x_9 + (0.25)(x_{10} - x_9) \\
&= 9,567 + 0.25(9,993 - 9,567) \\
&= 9,674
\end{aligned} \tag{1.24}$$

For comparison, the values computed above are compared with the values obtained using different software in Table 1.6. It can be seen that each software package can compute the same value differently. In all cases, the median will be computed in the same way since it is a fixed value. As was previously mentioned, this verifies that the quartile.inc function in Excel® is equivalent to the values obtained manually based on Eq. (1.7), while the `quartile.exc` function in Excel® is based on option 1 for Eq. (1.7). Finally, MATLAB® uses option 2 for Eq. (1.7). Nevertheless, all values are relatively close to each other and would not impact too greatly the overall results.

1.3.4 Some Observations on the Data Set

First, consider the results in Table 1.5, which presents the summary statistics for the data set. It can be noted that for Runs 2 and 3, which both have a similar mean Reynolds number, the median is quite different for each. This suggests that the distribution is different. Looking at Fig. 1.15 for these two runs, it can be seen that Run 3 has more extreme values (in both directions) than Run 2, which will balance out both the mean and median values. On the other hand, Fig. 1.15 shows that for Run 2, the size of the Q2–Q3 area is much smaller than for Run 3, suggesting that 25% of the data are compactly located in a small area. On the other hand, for Run 1, the mean and median are more closely aligned, which suggests that the data are more evenly distributed. This is confirmed by looking at Fig. 1.15 for Run 1, where the size of the two boxes is almost equal. Run 4 for the friction factor has a similar even distribution. In all cases, Table 1.5 shows that a larger range implies that the standard deviation will also be larger.

Next, consider the scatter plot shown in Fig. 1.14, where a scatter plot of the data by run and the theoretical values are presented. Note that each run is denoted by a symbol that appears distinct even if there is no colour. From here, it can be observed that Run 3 is consistently below the theoretical value. This suggests that this run could potentially be some sort of outlier. Furthermore, Run 4 seems to have been performed at much lower Reynolds numbers than the rest of the experiments. This difference is even evident from the summary statistics.

1.4 Further Reading

The following are references that provide additional information about the topic:

(1) **History of Statistics**:

    a.   Hald, A. (2003). *A History of Probability and Statistics and Their Application before 1750.* Hoboken, New Jersey, United States of America: John Wiley & Sons.

    b.   Sheynin, O. (2004). *History of the Theory of Probability to the Beginning of the 20th Century.* Berlin, Germany: NG Verlag.

    c.   Varberg, D. E. (1963). The development of modern statistics. *The Mathematics Teacher*, 56(4), 252–257.

(2) **Data Analysis**:

    a.   Barnett, V., & Lewis, T. (1994). *Outliers in Statistical Data* (3rd ed.). Chichester, England, United Kingdom: John Wiley & Sons.

    b.   Daniel, C., & Wood, F. S. (1980). *Fitting Equations to Data* (2nd ed.). New York, New York, United States of America: John Wiley & Sons, Inc.

    c.   Davies, L., & Gather, U. (1993). The Identification of Multiple Outliers. *Journal of the American Statistical Association*, 88(423), 782–792.

    d.   Hawkins, D. M. (1980). *Identification of Outliers.* London, England, United Kingdom: Chapman and Hall.

    e.   Hodge, V. J., & Austin, J. (2004). A Survey of Outlier Detection Methodologies. *Artificial Intelligence Review, 22*, 85–126.

    f.   Hyndman, R. J., & Fan, Y. (1996). Sample Quantiles in Statistical Packages. *The American Statistician*, 50(4), 361–365.

g. Lin, B., Recke, B., Knudsen, J. K., & Jørgensen, S. B. (2007). A systematic approach for soft sensor development. *Computers and Chemical Engineering*, 31, 419–425.

(3) **Data Visualisation**:

a. Tufte, E. R. (1997). *Visual and Statistical Thinking: Displays of Evidence for Making Decisions.* Cheshire, Connecticut, United States of America: Graphics Press LLC.

b. Tufte, E. R. (2001). *The Visual Display of Quantitative Information.* Cheshire, Connecticut, United States of America: Graphics Press LLC.

1.5   Chapter Problems

Problems at the end of the chapter consist of three different types: (a) Basic Concepts (True/False), which seek to test the reader's comprehension of the key concepts in the chapter; (b) Short Exercises, which seek to test the reader's ability to compute the required parameters for a simple data set using simple or no technological aids. This section also includes proofs of theorems; and (c) Computational Exercises, which require not only a solid comprehension of the basic material, but also the use of appropriate software to easily manipulate the given data sets.

1.5.1   Basic Concepts

Determine if the following statements are true or false and state why this is the case.

(1) The mean is a robust measure of central tendency.
(2) A trimodal data set has four modes.
(3) The median measures the middle value of a data set.
(4) The median and the mean will always be the same.
(5) The variance is equal to the standard deviation squared.
(6) The range is a useful measure of the spread of the data.
(7) The median absolute difference is a robust measure of dispersion.
(8) A left-skewed data set has many values in the left tail.
(9) The skewedness of a data set measures how symmetric the data set is.
(10) Sextiles partition a data set into six parts.
(11) Outliers are data points whose values are abnormal.
(12) A graph should have clearly labelled axes and an appropriate legend.
(13) Graphs containing many different symbols distinguished solely by colour are well designed.
(14) Pie charts are the foundation upon which histograms are constructed.
(15) Sparkplots are useful for describing trends and general behaviour of a data set.

(16) Tables are useful for summarising important information, such as mean and variance, of a data set.

(17) Taking a numeric value directly from the software and placing it unformatted into a table is a good idea.

(18) A probability plot is useful for comparing the data set against some theoretical distribution.

(19) Transforming a data set can lead to a more meaningful graph.

(20) Combining different types of graphs together can create a graph with more information.

1.5.2 Short Exercises

These questions should be solved using only a simple, nonprogrammable, nongraphical calculator combined with pen and paper.

(21) For the data set $\{1, 3, 5, 2, 5, 7, 5, 2, 8, 5\}$,

    a.   Compute the mean, mode, and median.
    b.   Compute the variance, median absolute difference, and range.
    c.   Compute the first, second, and third quartiles.
    d.   Plot a box-and-whisker plot.
    e.   Plot a histogram with bins $x < 2$, $2 \leq x < 4$, $4 \leq x < 6$, $6 \leq x < 8$, and $x \geq 8$.

(22) For the data set $\{2.3, 1.2, 3.4, 4.5, 3.4, 1.2, 3.4, 4.0, 1.1\}$,

    a.   Compute the mean, mode, and median.
    b.   Compute the variance, median absolute difference, and range.
    c.   Compute the first, second, third, and fourth quintiles.
    d.   Plot a box-and-whisker plot.
    e.   Plot a histogram with bins $x < 2$, $2 \leq x < 3$, $3 \leq x < 4$, and $x \geq 4$.

1.5.3 Computational Exercises

The following problems should be solved with the help of a computer and appropriate software packages, such as MATLAB® *or* Excel®.

(23) Consider the data in Table 1.7 that shows the different faults (problems) associated with running a reactor over a thirty-day period. A fault can occur multiple times in a given timeframe. Compute appropriate summary statistics and create appropriate graphs to summarise the data. (*Hint: There is no one single correct solution.*)

(24) Consider the data in Table 1.8 that shows the flow rate of steam in kg/h through a pipe. Due to the presence of stiction and other nonlinearities in the control

Table 1.7 Reactor fault types by shift (for Question 23)

Fault Type	Number of Faults by Shift			
	Night (Midnight to 6:00 A.M.)	Morning (6:00 A.M. to Noon)	Afternoon (Noon to 6:00 P.M.)	Evening (6:00 P.M. to Midnight)
High reactor level	5	6	2	6
Abnormal pressure	10	2	2	5
Explosion	2	0	0	0
Low temperature	5	2	10	5
High temperature	5	8	0	10
Others	2	10	5	0

Table 1.8 Steam control data with two different methods (for Question 24)

Time (min)		5	10	15	20	25	30	35	40	45	50	55	60
Base	1 h	8.5	8.7	8.4	8.6	8.2	8.7	8.9	8.5	8.5	8.4	8.3	8.6
	2 h	8.2	8.4	8.3	8.2	8.4	8.5	8.8	8.3	8.6	8.7	8.5	8.3
New	1 h	8.4	8.5	8.4	8.5	8.6	8.3	8.6	8.7	8.2	8.3	8.4	8.5
	2 h	8.5	8.6	8.4	8.3	8.4	8.6	8.7	8.5	8.5	8.5	8.3	8.4

valve, a new control algorithm is being proposed. The engineer in charge of making the change has to evaluate whether the new algorithm is better. A better algorithm is defined as one that reduces the variance of the steam flow rate and can keep the process closer to the desired setpoint of 8.5 kg/h. The original and new control methods are both tested for 2 h and the data are collected every 5 min. Plot the available data and analyse it. Without using any formal statistical tests, suggest whether the proposed control algorithm is better than the original, base case.

(25)   Take any large data set that is of interest to you and analyse it using the methods presented in this chapter. The data set should have at least 1,000 data points and two variables. You can then use this data set in subsequent chapters to perform additional analysis.

Chapter 2
Theoretical Foundation for Statistical Analysis

Having examined briefly the application of statistics to describe and visualise a given data set, it is now necessary to examine and understand the theoretical foundation underpinning most statistical methods. With such a theoretical foundation, it is then possible to apply statistics to solving such problems as regression and design of experiments.

2.1 Statistical Axioms and Definitions

Consider a measurable probability space, Ω, defined by three variables: $\mathbb{S}$, $\mathbb{F}$, and P, denoted as $\Omega = (\mathbb{S}, \mathbb{F}, P)$, which defines a complete σ-algebra for statistical manipulation. Let $\mathbb{S}$ be defined as the **sample space**, which includes all **possible outcomes**. Let $\mathbb{F}$ be defined as the σ-algebra that contains all **possible events** for a given situation. It is most often a power set of $\mathbb{S}$. $\mathbb{F}$ must satisfy the following properties[1]:

(1) It contains the empty set { } or $\emptyset$;
(2) It is closed under complementation, that is, if an event $\mathbb{E} \subseteq \mathbb{S}$ is an element in $\mathbb{F}$, or $\mathbb{E} \in \mathbb{F}$, then the set $\mathbb{S}$ excluding $\mathbb{E}$ is also an element of $\mathbb{F}$, or $\mathbb{S} \backslash \mathbb{E} \in \mathbb{F}$.
(3) It is closed under union, that is, the union of countable many elements of $\mathbb{F}$ is in $\mathbb{F}$.

Finally, let P be the measure function, called the **probability measure function**,[2] that assigns a real number to all the members of $\mathbb{F}$. In most cases, it provides the likelihood that a given event will occur. The measure function must satisfy Kolmogorov's axioms: for any event, $\mathbb{E} \in \mathbb{F}$,

(1) $P(\mathbb{E}) \geq 0$,

---

[1] A review of set theory is provided in Appendix A2, at the end of this chapter.

[2] Also called a *probability measure*, a *probability function*, or even just *probability*.

Y. A. W. Shardt, *Statistics for Chemical and Process Engineers*,
https://doi.org/10.1007/978-3-030-83190-5_2

(2) $P(\mathbb{S}) = 1$,

(3) $P(\mathbb{E}_1 \cup \mathbb{E}_2 \cup \mathbb{E}_3 \cup \cdots \cup \mathbb{E}_n) = \sum_{i=1}^{n} P(\mathbb{E}_i)$, where each event $\mathbb{E}_i$ is pairwise disjoint (that is mutually exclusive).

Example 2.1: *Determining the Probability Space*

Determine the probability space, $\Omega = (\mathbb{S}, \mathbb{F}, P)$, for tossing a fair coin.

Solution: Tossing a coin has two outcomes: either the coin lands *heads* or it lands *tails*. Thus, the sample space $\mathbb{S}$ is defined as $\mathbb{S} = \{\text{heads}, \text{tails}\}$.

The event set, $\mathbb{F}$, is defined as the power set of $\mathbb{S}$. A power set is simply a new set created from all possible combinations (subsets) of the original set, that is, it contains all combinations of the elements of $\mathbb{S}$ drawn singly, doubly, in threes, etc. If there are n elements in the original set, then there will be 2^n elements in the power set. Since $\mathbb{S}$ has two elements, this implies that $\mathbb{F}$ will have $2^2 = 4$ elements. Irrespective of the situation, $\mathbb{F}$ will always contain the empty set, $\{\}$, and $\mathbb{S}$. This is a consequence of the requirements on $\mathbb{F}$. Therefore, in this example, $\mathbb{F}$ would be defined as $\mathbb{F} = \{\{\}, \{\text{heads}\}, \{\text{tails}\}, \{\text{heads}, \text{tails}\}\}$.

For each of the four events in $\mathbb{F}$, the following values would be assigned:

$$P(\{\}) = 0$$
$$P(\{\text{heads}\}) = \tfrac{1}{2}$$
$$P(\{\text{tails}\}) = \tfrac{1}{2}$$
$$P(\{\text{heads}, \text{tails}\}) = 1.$$

It can be noted that when tossing a coin one of two options must occur.[3] Therefore, it is impossible for none of the options to occur, that is, the probability of the empty set is zero! This is the case in all situations. Similarly, the likelihood that either heads or tails occurs is certain, as the coin will land on one of these two options. Therefore, $P(\{\text{heads}, \text{tails}\}) = P(\mathbb{S}) = 1$. This once again will always hold.

Let X be a **random variable** that assigns to each outcome in $\mathbb{S}$ a real number, that is, $X: \mathbb{S} \to \mathbb{R}$ for which $\{s: X(s) \leq x\} \in \mathbb{F}$, $\forall x \in \mathbb{R}$. Let the observed outcome at some given point be denoted by x. It should be noted that by convention random variables are denoted by capital letters, while the observations themselves are denoted by the corresponding lowercase letter. Thus, a random variable allows us to assign to each outcome in $\mathbb{S}$ a numeric value and, hence, compute various statistical properties, such as mean and variance. A random variable can be either discrete or continuous. A discrete random variable can only take certain values within a countable set (for example, flipping a coin), while a continuous random variable can take any values within (some subset of) $\mathbb{R}$. Often for continuous variables, the values assigned to the random variable are equal to the numeric values in $\mathbb{S}$. The process of obtaining

---

[3] At least under normal circumstances!

an observation is called **sampling**. In the simplest case, the random variable can be viewed as assigning to a coin flip a payout based on the outcome, for example, we could assign \$1 for heads and $-\$1$ for tails. The individual observations would be determined based on the flips of the coin, for example, if heads was obtained, then $x_1 = \$1$.

Finally, the probability of obtaining a given observation can still be found using the probability measure function, P. Often, for discrete random variables, we write this probability as $P(X = x)$, where we mean the probability that the random variable X takes the specific value x. For continuous variables, we most commonly deal with intervals. The most common case is to seek the probability that the random variable will be less than or equal to a specific value x, that is, $P(X \leq x)$. For both discrete and continuous variables, other intervals can also be used, for example, $P(x_1 \leq X \leq x_2)$, where we wish to determine the probability that the value of the random variable will lie between two values, x_1 and x_2.

Furthermore, it is now possible to define two functions $f_X(x)$ and $F_X(x)$ whose name and definition depend on whether we are dealing with discrete or continuous random variables. For a discrete random variable, we have that $f_X(x)$ is called the **probability mass function** and is defined as[4]

$$f_X(x) = \begin{cases} P(X = x) & \text{if } X(s) = x \text{ for some } s \in \mathbb{S} \\ 0 & \text{otherwise} \end{cases}, \qquad (2.1)$$

and $F_X(x)$ is called the **cumulative distribution function (cdf)** and is defined as

$$F_X(x) = P(X \leq x) = \sum_{x_i \leq x} f_X(x) = \sum_{x_i \leq x} P(X = x_i). \qquad (2.2)$$

For a continuous random variable, we first define the cumulative distribution function, $F_X(x)$, as

$$F_X(x) = P(X \leq x). \qquad (2.3)$$

Then, the **probability density function (pdf)**, $f_X(x)$, is defined as

$$f_X(x) = \frac{dF_X(x)}{dx}. \qquad (2.4)$$

In practice, it is the probability density function that is first defined and then integrated to obtain the cumulative distribution function, that is,

---

[4] Alternatively, we could define this function using indicator (or with a slight abuse of notation Dirac δ-) functions. This would then unify discrete and continuous variables when it comes to the computation of further properties.

$$P(X \leq a) = \int_{-\infty}^{a} f(x)dx. \tag{2.5}$$

It can be noted that the subscript on the probability density function is often not written unless it is desired to emphasis the associated random variable.

By Kolmogorov's axiom that $P(\mathbb{S}) = 1$, the probability density function has the following property:

$$\int_{-\infty}^{\infty} f(x)dx = 1. \tag{2.6}$$

Furthermore, by Kolmogorov's axiom that $P(\mathbb{E}) \geq 0$,

$$f(x) \geq 0 \text{ for all } x. \tag{2.7}$$

The properties given by Eqs. (2.6) and (2.7) are useful for determining if a given candidate function is in fact a probability density function or if the result obtained is indeed correct.

In order to describe the resulting space for the random variable, it is useful to consider two terms previously introduced that will now be formally defined: **mean** and **variance**. For a discrete function, the mean, μ, is defined as

$$\mu = \sum_{\substack{\forall x: X(s) = x \\ \text{for some } s \in \mathbb{S}}} x P(X = x). \tag{2.8}$$

The **variance**, σ^2, is defined as

$$\sigma^2 = \sum_{\substack{\forall x: X(s) = x \\ \text{for some } s \in \mathbb{S}}} (x - \mu)^2 P(X = x). \tag{2.9}$$

The standard deviation would be defined as the square root of the variance.

For continuous functions, the mean would be computed as

$$\mu = \int_{-\infty}^{\infty} x f(x)dx, \tag{2.10}$$

while the variance would be computed as

$$\text{var}(x) = \sigma^2 = \int_{-\infty}^{\infty} (x - \mu)^2 f(x)dx.$$

$$= \int_{-\infty}^{\infty} x^2 f(x)dx - \mu^2 \tag{2.11}$$

The variance can either be denoted by σ^2 or by var. It is common to use var when it is desired to treat the variance as an operator and perform additional manipulations with it.

Finally, the ith **uncentred moment** of the probability density function $f(x)$, written as m_i, is

$$m_i = \int_{-\infty}^{\infty} x^i f(x)dx. \tag{2.12}$$

It can be noted that the first moment is equivalent to the mean. In certain cases, centred moments are preferred. In such cases, the ith **centred moment** for the probability density function $f(x)$, written as $\overline{m}_i$ is

$$\overline{m}_i = \int_{-\infty}^{\infty} (x - \mu)^i f(x)dx. \tag{2.13}$$

The second centred moment, $\overline{m}_2$, is equivalent to the variance.

For a random variable, we denote the underlying probability space using a tilde, for example, $X \sim \mathfrak{N}(0, 1)$ means that the random variable X is based on a normal distribution with a mean of zero and a variance of 1. Formally, we can state this as

$$P(X \leq a) = \int_{-\infty}^{a} f_N(x)dx, \tag{2.14}$$

where f_N is the probability density function for the normal distribution.

Example 2.2: *Determining Acceptable Probability Density Functions*

Determine if the following functions can be used as probability density functions:

(1) $f(x) = x^2$ for $0 \leq x \leq 2$, zero otherwise;
(2) $f(x) = N^{-1}$ for $0 \leq x \leq N$, zero otherwise; N a positive number.

If a candidate function cannot be used, suggest how to make it a valid probability density function.

Solution

The graph of the function given by $f(x) = x^2$ for $0 \leq x \leq 2$, zero otherwise is shown in Fig. 2.1. From Fig. 2.1, it can be seen that the function satisfies the condition $f(x) \geq 0$ for all x (Eq. 2.7). In order to determine, if the other constraint given by Eq. (2.6) is satisfied, it is necessary to integrate the given function, that is,

$$\int_{-\infty}^{\infty} f(x)dx = \int_{-\infty}^{0} 0\,dx + \int_{0}^{2} x^2\,dx + \int_{2}^{\infty} 0\,dx.$$

$$= \frac{1}{3}x^3 \Big|_{0}^{2} = \frac{1}{3}\left(2^3 0^3\right) = \frac{8}{3} \neq 1$$

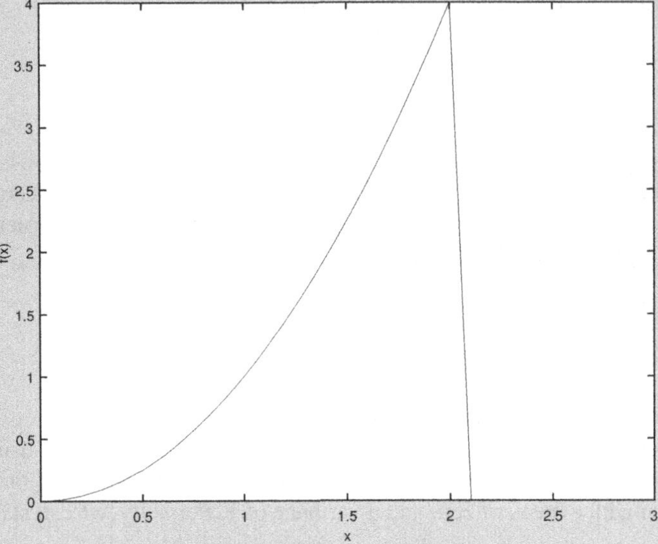

Fig. 2.1 Plot of the probability density function 1 in Example 2.2

Since the integral does not equal 1, the given function is not a probability density function. In order to make it one, it is necessary to divide the function by the inverse of what was obtained above, that is,

$$f(x) = \frac{3}{8}x^3 \text{ for } 0 \leq x \leq 2, \text{ zero otherwise.}$$

In general, if the value of the integral is $K (\neq 0)$ for a candidate function f, then $K^{-1}f$ will be a probability density function assuming that Eq. (2.7) is satisfied. The procedure is commonly used and is called **normalisation**.

For the function given by $f(x) = N^{-1}$ for $0 \leq x \leq N$, zero otherwise, with N a positive number, it is easy to verify that Eq. (2.7) holds. To verify Eq. (2.6), integration gives

$$\int_{-\infty}^{\infty} f(x)dx = \int_{-\infty}^{0} 0\,dx + \int_{0}^{N} N^{-1}\,dx + \int_{N}^{\infty} 0\,dx.$$

$$= \frac{1}{N}x\Big|_0^N = 1$$

Since the integral equals 1, this implies that Eq. (2.6) is satisfied. Since both conditions are satisfied, this is a candidate probability density function.

Example 2.3: *Computing Mean and Variance from the Probability Density Function*

For the corrected probability density function from Example 2.2, compute the mean, standard deviation, and variance.

Solution

Since we have that $f(x) = 3x^2 / 8$ for $0 \leq x \leq 2$, the mean can be found as follows:

$$\mu = \int_{-\infty}^{\infty} xf(x)dx = \frac{3}{8}\int_0^2 xx^2dx = \frac{3}{(8)4}x^4\Big|_0^2 = \frac{3}{(8)4}2^4 = 1.5.$$

The variance can be found as

$$\sigma^2 = \int_{-\infty}^{\infty} x^2 f(x)dx - \mu^2 = \frac{3}{8}\int_0^2 x^2x^2dx - 1.5^2$$

$$= \frac{3}{(8)5}x^5\Big|_0^2 - 1.5^2 = \frac{3}{(8)5}2^5 - 1.5^2 = 0.15$$

The standard deviation can be written as $\sigma = \sqrt{0.15}$.

Therefore, for the given probability density function, the mean is 1.5, the variance is 0.15, and the standard deviation is $\sqrt{0.15}$.

2.2 Expectation Operator

In order to simplify some of the mathematical operations that are required in manipulating statistical properties, the concept of the **expectation operator**, E, needs to be introduced. The expectation operator determines the mean, or expected, value, of a distribution and is defined as

$$E(X) = \int_{-\infty}^{\infty} xf(x)dx = \mu. \tag{2.15}$$

For the discrete case, the expectation operator can be computed as

$$E(X) = \sum_{x \in \mathbb{S}} x P(X = x) = \mu. \tag{2.16}$$

For any two random variable X and Y, and $c \in \mathbb{R}$, the expectation operator has the following properties:

(1) $E(X + c) = E(X) + c$;
(2) $E(cX) = cE(X)$;
(3) $E(X + Y) = E(X) + E(Y)$;
(4) $E(g(X)) = \int_{-\infty}^{\infty} g(x) f(x) dx$;
(5) $E(XY) = E(X)E(Y) + \text{covar}(X, Y)$, where *covar* is the covariance (or the degree of relationship between) of X and Y. Note that if two variables are **independent** of each other, then their covariance is zero. The covariance of the same variable, that is $\text{cov}(X, X)$ or the autocovariance, is by definition equal to the variance of the variable, that is, σ_X^2.

Properties 1, 2, and 3 show that the expectation operator is a linear operator.

Example 2.4: *Using the Expectation Operator*
   Consider two variables X ($\mu = 5$ and $\sigma^2 = 4$) and Y ($\mu = 2$ and $\sigma^2 = 2$), with covariance, $\text{covar}(X, Y) = 2$, compute the following values:

1) $E(2X - 5)$;
2) $E(3XY)$.

Solution

For part 1, we have that

$$E(2X - 5) = 2E(X) - 5 = 2(5) - 5 = 5$$

and for part 2, we have that

$$E(3XY) = 3E(XY) = 3(E(X)E(Y) + \text{covar}(X, Y)) = 3((5)(2) + 2) = 36.$$

Both results are obtained by a straightforward application of the rules for the expectation operator.

2.3 Multivariate Statistics

So far it has been assumed that there is only a single variable that governs the behaviour of the probability space. However, in many cases, it is useful to deal with multivariate probability spaces where multiple variables determine the outcome. In general, all of the univariate results generalise straightforwardly to the multivariate case. In order to simplify the presentation, all results will be derived first for the bivariate ($n = 2$) situation. The extension to an arbitrary n simply requires adding additional integrations.

Assume that $\vec{X} = \langle X_1, X_2 \rangle$ and $\vec{x}$ is similarly defined. Let the **joint probability density function**, $f_{\vec{X}}(\vec{x})$, be defined over the region $\mathbb{R}^2$. The joint probability density function satisfies the following three properties:

(1) $f_{\vec{X}}(\vec{x}) \geq 0$, for all $\vec{x} \in \vec{X}$;

(2) $\int_{-\infty}^{\infty} \int_{-\infty}^{\infty} f_{\vec{X}}(\vec{x}) d\vec{x} = 1$; and

(3) $P\left(\vec{X} \in \mathbb{H}\right) = \int_{-\infty}^{x_1} \int_{-\infty}^{x_2} f_{\vec{X}}(\vec{x}) d\vec{x}$, where $\mathbb{H}$ is some subspace (region) of the $\mathbb{R}^2$ space.

The **marginal probability density function** represents the probability for a subset of random variables in the original joint probability density function. The subset considered is used as the subscript for the function, for example, $f_X(x)$ would be the marginal probability density function for X. The marginal probability density function is obtained by integrating out all the remaining variables, that is,

$$f_{X_1}(x_1) = \int_{-\infty}^{\infty} f_{\vec{X}}(\vec{x}) dx_2. \tag{2.17}$$

The process of removing some subset of random variables is called **marginalisation** and the removed variables (x_2 in this case) are said to be **marginalised out**.

The **conditional probability** represents the probability, given information about some of the other variables, for example, if there are two variables X and Y, then once the value of X is determined, what is the probability of obtaining a given Y. Using probability notation, the conditional probability of Y given X is defined as

$$P(Y|X) = \frac{P(X \cap Y)}{P(X)}. \tag{2.18}$$

The **conditional probability density function** for Y given $X = x$ is written as

$$f_{Y|x}(y) = \frac{f_{\vec{X}}(\vec{x})}{f_X(x)}. \tag{2.19}$$

Since this is a probability density function, it will satisfy all the usual properties. The mean value of X_1 can be obtained as

$$\mu_{X_1} = E(X_1) = \int_{-\infty}^{\infty} x_1 f_{X_1}(x_1) dx_1. \tag{2.20}$$

Similarly, the **conditional mean** of X_2 given $X_1 = x_1$ can be written as

$$\mu_{X_2|x_1} = E(X_2|x_1) = \int_{-\infty}^{\infty} x_2 f_{X_2|x_1}(x_2) dx_2. \tag{2.21}$$

The variance of each variable would be defined similarly to the means. The **covariance**, σ_{XY} or **cov(X, Y)**, represents the degree of linear relationship between two variables and is defined as

$$\sigma_{X_1 X_2} = \int_{-\infty}^{\infty} \int_{-\infty}^{\infty} x_1 x_2 f_{X_1 X_2}(x_1, x_2) dx_1 dx_2 - \mu_{X_1} \mu_{X_2}. \tag{2.22}$$

The **covariance matrix**, Σ, is defined as the $n \times n$ matrix whose elements are the covariances of the given variables, that is, $\Sigma_{ij} = \text{cov}(X_i, X_j)$. It should be noted that σ_{XX} is equal to the variance of X. Formally, the covariance matrix is defined as

$$\Sigma = E\left(\vec{X}^T \vec{X}\right) - E\left(\vec{X}^T\right) E\left(\vec{X}\right), \tag{2.23}$$

where $\vec{X}$ is the $1 \times n$ vector of random variables. The **correlation** is the normalised covariance and is defined as

$$\text{corr}(X_1, X_2) = \rho_{X_1 X_2} = \frac{\sigma_{X_1 X_2}}{\sigma_{X_1} \sigma_{X_2}} = \frac{\sigma_{X_1 X_2}}{\sqrt{\sigma_{X_1}^2 \sigma_{X_2}^2}}. \tag{2.24}$$

The correlation is bounded between -1 and 1.

Two (or more) variables are said to be **independent** if the following hold:

(1) $f_{\vec{X}}(\vec{x}) = \prod_{i=1}^{n} f_{X_i}(x_i)$ for all x_i;

(2) $f_{X_j|x_i}(x_j) = f_{X_j}(x_j)$ for all x_i and x_j with $f_{X_j}(x_j) > 0$; and

(3) $P(X_1 \in \mathbb{B}_1, X_2 \in \mathbb{B}_2, \ldots, X_n \in \mathbb{B}_n) = \prod_{i=1}^{n} P(X_i \in \mathbb{B}_i)$ for any sets $\mathbb{B}_i$ in the range of x_i.

When dealing with a multivariate distribution, the computation of the marginal and conditional probabilities is more complex. Let $\vec{D} = \langle D_1, D_2, \ldots, D_m \rangle$ be an m-dimensional subset of the n-dimensional vector $\vec{X}$, and let $\vec{d}$ be defined similarly. Let $\vec{X}_r$ be defined such that it contains all the variables in $\vec{X}$ that are not in $\vec{D}$ and let $\vec{x}_r$ be defined similarly. Let $\mathbb{D}$ be the subset of $\mathbb{R}^n$ for the $\vec{D}$ vector. The marginal probability density function can then be written as

$$f_{\vec{D}}\left(\vec{d}\right) = \int_{\mathbb{R}^n \setminus \mathbb{D}} f_{\vec{X}}(\vec{x}) d\vec{x}_r, \tag{2.25}$$

where the integration is performed on the interval $]-\infty, \infty[$ for all the variables not in $\vec{D}$.

The conditional probability for $\vec{D}$ given $\vec{D}'$, such that there are no variables in common with both vectors, is defined as

$$f_{\vec{D}|\vec{d}}\left(\vec{d}\right) = \frac{f_{\vec{D}\vec{D}'}\left(\vec{d}, \vec{d}\right)}{f_{\vec{D}'}\left(\vec{d}\right)}. \tag{2.26}$$

Equations (2.25) and (2.26) reduce to the bivariate examples provided previously.

Example 2.5: *Dealing with a Multivariate Distribution*
Consider the following multivariate distribution

$$f_{XYZ}(x, y, z) = 40e^{-5x-2y-4z} \quad x > 0, y > 0, z > 0$$

and compute the following information:

(1) $f_X(x)$.
(2) $f_{YZ}(y, z)$.
(3) $f_{YZ|x}(y, z)$. Does the value depend on x?
(4) Compute the covariance between Y and Z.
(5) Determine if the variables are independent.

Solution

For (1), let $\vec{D} = \langle X \rangle$, $\vec{d} = \langle x \rangle$, $\vec{X}_r = \langle Y, Z \rangle$, and $\vec{x}_r = \langle y, z \rangle$. $\mathbb{D}$ will be $]0, \infty[$. Equation (2.25) can then be rewritten as

$$f_X(x) = \int_0^\infty \int_0^\infty f_{XYZ}(x, y, z) dy dz = \int_0^\infty \int_0^\infty 40e^{-5x-2y-4z} dy dz$$

$$= 40e^{-5x} \int_0^\infty e^{-2y} dy \int_0^\infty e^{-4z} dz = 40e^{-5x}(0.5)(0.25).$$

$$= 5e^{-5x}$$

It should be noted that $\int_0^\infty e^{-cz} dz = c^{-1}$.

For (2), let $\vec{D} = \langle Y, Z \rangle$, $\vec{d} = \langle y, z \rangle$, $\vec{X}_r = \langle X \rangle$, and $\vec{x}_r = \langle x \rangle$. $\mathbb{D}$ will be $]0, \infty[\times]0, \infty[$. Equation (2.25) can then be rewritten as

$$f_{YZ}(y, z) = \int_0^\infty f_{XYZ}(x, y, z)dx = \int_0^\infty 40e^{-5x-2y-4z}dx.$$
$$= 8e^{-2y-4z}$$

For (3), let $\vec{D}' = \langle X \rangle, \vec{d} = \langle x \rangle, \vec{D} = \langle Y, Z \rangle$, and $\vec{d} = \langle y, z \rangle$. Equation (2.26) can then be rewritten as

$$f_{YZ|x}(y, z) = \frac{f_{XYZ}(x, y, z)}{f_X(x)} = \frac{40e^{-5x-2y-4z}}{5e^{-5x}} = 8e^{-2y-4z}.$$

It can be seen that the conditional probability does not depend on x.

For (4), first compute the mean value of both Y and Z:

$$\mu_Y = \int_0^\infty y f_Y(y)dy = \int_0^\infty 2y e^{-2y}dy = 2 \times \frac{1}{4} = 0.5,$$

$$\mu_Z = \int_0^\infty z f_Z(z)dy = \int_0^\infty 4z e^{-4z}dz = 4 \times \frac{1}{16} = 0.25.$$

Note that $\int_0^\infty z e^{-cz}dz = c^{-2}$. Then, the covariance between Y and Z can be determined as

$$\sigma_{YZ}^2 = \int_0^\infty \int_0^\infty yz f_{YZ}(y, z)dydz - (0.5)(0.25) = \int_0^\infty \int_0^\infty yz \left(8e^{-2y-4z}\right)dydz - 0.125$$
$$= 8 \int_0^\infty y e^{-2y}dy \int_0^\infty z e^{-4z}dz - 0.125 = 8 \times \frac{1}{4} \times \frac{1}{16} - 0.125.$$
$$= 0$$

For (5), note that $f_{XYZ} = f_X f_Y f_Z$, which corresponds to the first statement for independence. Since it is satisfied, we can conclude that all three variables are independent of each other. Note that f_{YZ} and $f_{YZ|x}$ are the same, which should be the case if the variables are independent.

When computing the variance and covariance matrices using vectors, there can be some confusion about the order in which the transposes are placed (on the first or second member). Part of the confusion stems from the fact that depending on the particular application, the same vector can be used to obtain two different values: either a scalar variance value or a matrix covariance value. The following rules can be used to resolve any potential issues:

(1) **Scalar column vector rule**: Consider $\vec{X}$ to be a $1 \times n$ vector of random variables, for which it is desired to compute the second moment, m_2. In such a case, the

correct formula is given as $m_2 = E\left(\vec{X}\vec{X}^T\right)$, where T is the transpose operator. It is easy to verify that this will give a scalar value.

(2) **Scalar row vector rule**: Consider $\vec{X}$ to be an $n\times 1$ vector of random variables, for which it is desired to compute the second moment, m_2. In such a case, the correct formula is given as $m_2 = E\left(\vec{X}^T\vec{X}\right)$, where T is the transpose operator. It is easy to verify that this will give a scalar value.

(3) **Matrix column vector rule**: Consider $\vec{X}$ to be a $1\times n$ vector of random variables, for which it is desired to compute the covariance matrix, Σ. In such a case, the correct formula is given as $\Sigma = E\left(\vec{X}^T\vec{X}\right)$, where T is the transpose operator. It is easy to verify that this will give a matrix value.

(4) **Matrix row vector rule**: Consider $\vec{X}$ to be a $n\times 1$ vector of random variables, for which it is desired to compute the covariance matrix, Σ. In such a case, the correct formula is given as $\Sigma = E\left(\vec{X}\vec{X}^T\right)$, where T is the transpose operator. It is easy to verify that this will give a matrix value.

(5) **Product of vector rule**: Consider two vectors $\vec{X}_1$ and $\vec{X}_2$ that are multiplied together. It is desired to determine the second moment for the product of these two vectors $\vec{X}_1\vec{X}_2$. In such a case, the first vector in the product will determine which of the two rules apply. For example, if $\vec{X}_1$ is a column vector and $\vec{X}_2$ the corresponding row vector, then the second moment would be computed as follows, $m_2 = E\left(\vec{X}\vec{X}^T\right) = E\left(\vec{X}_1\vec{X}_2\left(\vec{X}_1\vec{X}_2\right)^T\right) = E\left(\vec{X}_1\vec{X}_2\vec{X}_2^T\vec{X}_1^T\right)$, where the last manipulation results from the property of transpose and inverse for matrices, that is,

$$(AB)^T = B^T A^T. \tag{2.27}$$

where A and B are two appropriately sized matrices. *Mutatis mutandis* for the inverse.

2.4 Common Statistical Distributions

In statistics, a commonly encountered generalised probability space, given by either actual probabilities or a probability density function, is called a **distribution**. Such distributions show how a variable is distributed among all possible available values. In this section, the following common distributions will be considered: **normal**, **Student's t-**, χ^2-, **F-**, **binomial**, and **Poisson** distributions. Except for the last two distributions which are discrete, all the other distributions are continuous.

The swing dash (~) is used to denote that a random variable follows a given distribution, for example, $y \sim \mathfrak{N}(0, \sigma^2)$ means that the random variable y follows a normal distribution with mean zero and variance σ^2. In general, a capital Fraktur letter will be used to refer to a specific distribution. Occasionally, due to historical precedent, a different symbol may be used. In many applications, it is desired to find the x-value corresponding to a specific probability for the given distribution. In general, this will be denoted using the same symbol as before, but with the addition of a subscript probability value, that is, $\mathfrak{N}_{p,\mu,\sigma^2}$ would represent the x-value of the normal probability density function with parameters μ and σ^2, such that the area under the curve equals p, that is, finding the x-value for the cumulative probability function for the normal distribution with parameters μ and σ^2 given a probability of p. In general, this problem does not have a closed-form solution.

2.4.1 Normal Distribution

The **normal** (or **Gaussian**) distribution, denoted by $\mathfrak{N}(\mu, \sigma^2)$, is the most common distribution in statistics. It can be fully described by the mean, μ, and standard deviation, σ. It is a symmetric distribution centred on the mean. The spread of values is determined by the standard deviation. The larger the standard deviation, the more spread out the values are. The following are some useful rules of thumb regarding a normal distribution:

(1)   68% of all values drawn from a normal distribution lie within the interval $\mu\pm\sigma$.
(2)   95% of all values drawn from a normal distribution lie within the interval $\mu\pm2\sigma$.
(3)   99.7% of all values drawn from a normal distribution lie within the interval $\mu\pm3\sigma$.
(4)   999,999 out of one million values drawn from a normal distribution lie within the interval $\mu\pm6\sigma$. This observation is the origin for 6σ process control and analysis.

Common properties of the normal distribution are summarised in Table 2.1, while Fig. 2.2 gives a probability plot of the normal distribution.

The **standard normal distribution**, denoted by Z, is defined as a normal distribution with $\mu = 0$ and $\sigma = 1$. A variable following any normal distribution can be standardised as follows:

$$Z = \frac{x - \mu}{\sigma} \tag{2.28}$$

where Z is often called the Z-score for the given variable. This normalisation allows for easier comparison between variables drawn from different distributions. The cumulative distribution function for the standard normal distribution, denoted by

Table 2.1 Useful properties of the normal distribution

Property	Value		
Notation	$\mathfrak{N}(\mu, \sigma^2)$ $\mathfrak{N}(0, 1)$ is denoted by Z		
Probability density function	$f(x) = \frac{1}{\sigma\sqrt{2\pi}} e^{-\frac{(x-\mu)^2}{2\sigma^2}}$		
Mean	μ		
Variance	σ^2		
Probability density function	MATLAB®	`normpdf(x, μ, σ)`	
	Excel®	`norm.dist (x, μ, σ, false)`	
Generate Numbers Drawn From Given Distribution	MATLAB®	`normrnd(n, p, Size)`	
Inverse cumulative distribution function	MATLAB®	`norminv (p, μ, σ)`	
	Excel®	`norm.inv (p, μ, σ)`	

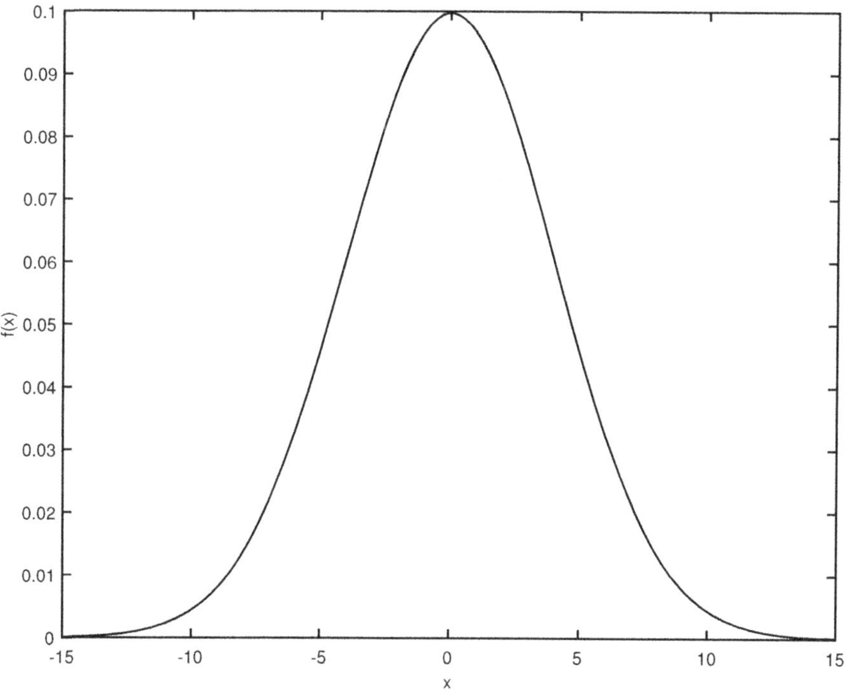

Fig. 2.2 Probability density function for the normal distribution where $\mu = 0$ and $\sigma = 4$

$\Phi(z)$, is

$$\Phi(z) = P(X \leq z) = \frac{1}{\sqrt{2\pi}} \int_{-\infty}^{z} e^{\frac{-x^2}{2}} dx. \tag{2.29}$$

The normal distribution is commonly encountered when performing regression analysis, system identification, or analysing systems in process control. As well, due to the behaviour of large numbers and means, it is used to describe the distribution of many common parameters, including heights and weights of people, grades, and machine errors.

In some cases, the multivariate normal distribution is useful. For a vector of random variables, $\vec{X}$, with a mean vector $\vec{\mu}$ and a covariance matrix Σ, then the multivariate normal distribution is

$$f(\vec{x}) = (2\pi)^{-\frac{n}{2}} |\Sigma|^{-0.5} e^{-\frac{1}{2}(\vec{x}-\vec{\mu})^T \Sigma (\vec{x}-\vec{\mu})}, \tag{2.30}$$

where $|\cdot|$ is the determinant function. The shape and behaviour of the multivariate normal distribution is the same as the univariate normal distribution.

2.4.2 Student's t-Distribution

The **Student's t-distribution**, denoted as $t(\nu)$ or more commonly as t_ν, is a statistical distribution that is used for dealing with the estimation of the mean of a normal distribution when the sample size is small and the population standard deviation is unknown. It approaches the normal distribution as the number of degrees of freedom, ν, approaches infinity. In general, the Student's t-distribution has larger tails than the normal distribution. Useful properties of the Student's t-distribution are summarised in Table 2.2, while Fig. 2.3 compares the Student's t-distribution with the normal distribution.

2.4.3 χ^2-Distribution

The χ^2-**distribution**, denoted as $\chi^2(\nu)$ or χ_ν^2, is a distribution that can be used to model the distribution of standard deviations. It depends on the number of degrees of freedom, ν, for the given set of observations. Useful properties of the χ^2-distribution are summarised in Table 2.3, while Fig. 2.4 gives a plot of the probability density function for the χ^2-distribution for different values of ν.

Table 2.2 Useful properties of the Student's t-distribution

Property	Value	
Notation	$t(\nu)$ or t_ν	
Probability density function	$f(x) = \dfrac{\Gamma\left(\frac{\nu+1}{2}\right)}{\Gamma\left(\frac{\nu}{2}\right)\sqrt{\nu\pi}}\left(1 + \frac{x^2}{\nu}\right)^{-\frac{\nu+1}{2}},$ where Γ is the gamma function	
Mean	0 for $\nu > 1$	
Variance	$\nu/(\nu - 2)$ for $\nu > 2$	
Probability density function	MATLAB®	tpdf (x, ν)
	Excel®	t.dist $(x, \nu,$ false$)$
Generate numbers drawn from given distribution	MATLAB®	trnd$(\nu,$ Size$)$
Inverse cumulative distribution function	MATLAB®	tinv (p, ν)
	Excel®	t.inv (p, ν)

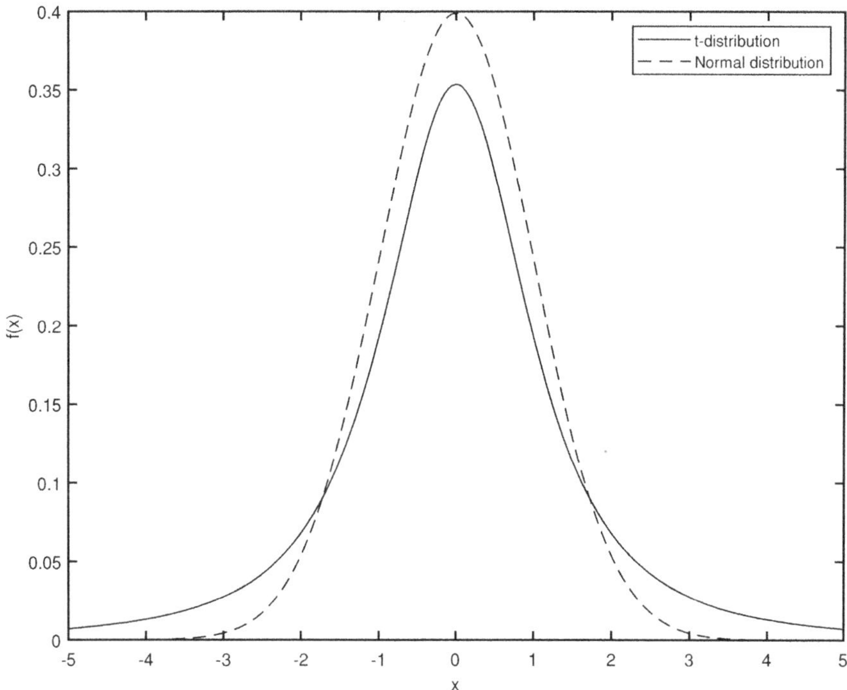

Fig. 2.3 Comparison between the t-distribution with two degrees of freedom and the standardised normal distribution

Table 2.3 Useful properties of the χ^2-distribution

Property	Value	
Notation	$\chi^2(\nu)$ or χ^2_ν	
Probability density function	$f(x) = \dfrac{1}{\Gamma(\frac{\nu}{2})2^{\frac{\nu}{2}}} x^{\frac{\nu}{2}-1} e^{-\frac{x}{2}}$, where Γ is the gamma function	
Mean	ν	
Variance	2ν	
Probability density function	MATLAB®	`chi2pdf (x, ν)`
	Excel®	`chisq.dist(x, ν, false)`
Generate Numbers Drawn From Given Distribution	MATLAB®	`chi2rnd (ν, Size)`
Inverse cumulative distribution function	MATLAB®	`chi2inv(p, ν)`
	Excel®	`chisq.inv(p, ν)`

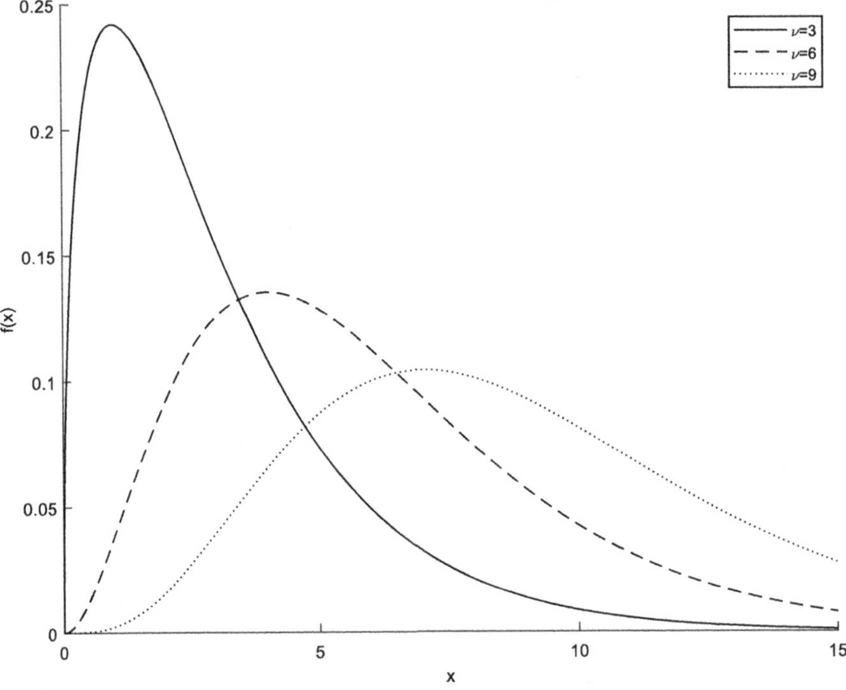

Fig. 2.4 Probability density function for the χ^2-distribution as a function of the degrees of freedom

Table 2.4 Useful properties of the F-distribution

Property	Value		
Notation	$\mathfrak{F}(\nu_1, \nu_2)$ or F_{ν_1, ν_2}		
Probability density function	$$f(x) = \dfrac{\sqrt{\dfrac{(\nu_1 x)^{\nu_1} \nu_2^{\nu_2}}{(\nu_1 x + \nu_2)^{\nu_1 + \nu_2}}}}{x B\left(\frac{\nu_1}{2}, \frac{\nu_2}{2}\right)},$$ where B is the beta function		
Mean	$\nu_2 / (\nu_2 - 2)$ for $\nu_2 > 2$		
Variance	$\dfrac{2\nu_2^2(\nu_1 + \nu_2 - 2)}{\nu_1(\nu_2 - 2)^2(\nu_2 - 4)}$ for $\nu_2 > 4$		
Probability density function	MATLAB®	fpdf (x, ν_1, ν_2)	
	Excel®	f.dist $(x, \nu_1, \nu_2,$ false$)$	
Generate Numbers Drawn From Given Distribution	MATLAB®	frnd $(\nu_1, \nu_2, Size)$	
Inverse cumulative distribution function	MATLAB®	finv (p, ν_1, ν_2)	
	Excel®	f.inv (p, ν_1, ν_2)	

2.4.4 F-Distribution

The **F-distribution**, denoted as $\mathfrak{F}(\nu_1, \nu_2)$ or F_{ν_1, ν_2}, is a distribution that can be used to model the distribution of ratios. Its shape depends on the number of degrees of freedom for both the numerator, ν_1, and denominator, ν_2, of the ratio. Useful properties of the F-distribution are summarised in Table 2.4, while Fig. 2.5 gives a plot of the probability density function for the F-distribution.

2.4.5 Binomial Distribution

The **binomial distribution**, denoted by $\mathfrak{B}(n, q)$, is a discrete distribution used to model the outcome of a series of binary (0 and 1 or *yes* and *no*) events. For each trial or realisation, the value 1 can occur with probability q and the value 0 with probability $1 - q$. It is assumed that there are n trials and the number of 1's is k. The order in which the events occur is not important, only their total number, for example, $\{1, 0, 0, 1\}$ and $\{1, 1, 0, 0\}$ would be equivalent, since each has cases of 1 and two cases of 0. The meaning assigned to 1 and 0 can be arbitrary as long as the outcome is binary. For example, 1 could represent success and 0 failure; or 1 heads and 0 tails. Useful properties of the binomial distribution are summarised in Table 2.5.

Setting $n = 1$, that is, only a single trial occurs, we get the **Bernoulli distribution**. It models the probability of a single trial given two possibilities. Historically, this distribution was first proposed and then generalised to the binomial distribution.

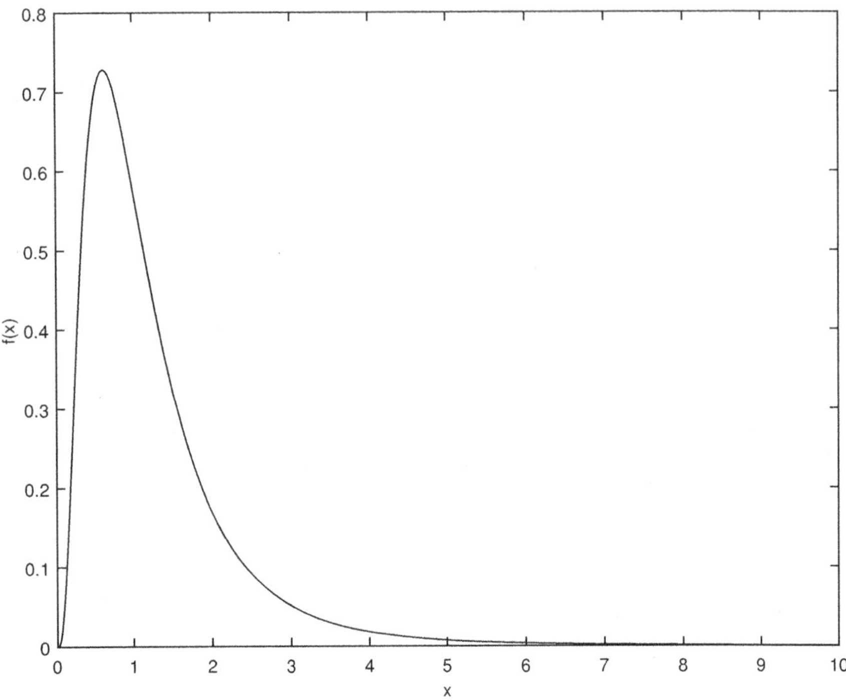

Fig. 2.5 Probability density function for the F-distribution for $v_1 = 8$ and $v_2 = 10$

Table 2.5 Useful properties of the binomial distribution

Property	Value	
Notation	$\mathfrak{B}(n, q)$	
Probability measure function	$P(X = k) = \dbinom{n}{k} q^k (1 - q)^{n-k},$ where n is the total number of trials and k is the number of trials with outcome 1	
Mean	nq	
Variance	$nq(1 - q)$	
Probability measure function	MATLAB®	`binopdf (k, n, q)`
	Excel®	`binom.dist(k, n, q, false)`
Generate Numbers Drawn From Given Distribution	MATLAB®	`binornd (n, q, Size)`
Inverse	MATLAB®	`binoinv (p, n, q)`
	Excel®	`binom.inv(n, q, p)`

Finally, it should be noted that it is possible to approximate the binomial distribution using the standard normal distribution if $np > 5$ and $n(1 - p) > 5$. In this case,

$$Z = \frac{X - np}{\sqrt{np(1 - p)}} \tag{2.31}$$

is approximately a standard normal variable. When computing probabilities, it is normal to add a correction factor to deal with the fact that the binomial distribution is discrete, while the normal distribution is continuous. The correction can be written as

$$P(X < x) \approx P\left(Z < \frac{x + 0.5 - np}{\sqrt{np(1 - p)}}\right)$$
$$P(X > x) \approx P\left(Z > \frac{x - 0.5 - np}{\sqrt{np(1 - p)}}\right) \tag{2.32}$$

It can be noted that in both cases the final probability will be larger than if no correction factor had been used.

2.4.6 *Poisson Distribution*

The **Poisson distribution**, denoted by $\mathfrak{p}(\lambda)$,[5] is a discrete distribution used to model the occurrence of independent events in a given time interval or space. It is the result of taking the binomial distribution and extending the number of trials to infinity. The Poisson distribution is encountered in reliability engineering to model the time occurrences of failure and used in queuing theory to model the behaviour of a queue. Useful properties of the Posson distribution are summarised in Table 2.6.

It should be noted that, as for the binomial distribution from which it is derived, it is possible to approximate the Poisson distribution using the standard normal distribution. If $\lambda > 5$,

$$Z = \frac{X - \lambda}{\sqrt{\lambda}} \tag{2.33}$$

is approximately a standard normal variable.

---

[5] Note a lowercase Fraktur p is used as the symbol here.

Table 2.6 Useful properties of the Poisson distribution

Property	Value		
Notation	$p(\lambda)$		
Probability Measure Function	$P(X = k) = \frac{\lambda^k}{k!}e^{-\lambda}$, where $k \geq 0$ is an integer		
Mean	λ		
Variance	λ		
Probability Measure Function	MATLAB®	`poisspdf(k, λ)`	
	Excel®	`poisson.dist(k, λ, false)`	
Generate Numbers Drawn From Given Distribution	MATLAB®	`poissrnd(λ, Size)`	
Inverse	MATLAB®	`poissinv(p, λ)`	

2.5 Parameter Estimation

Parameter estimation is the name given to the procedure used to estimate, or approximate, the true population parameters, based on a sample of the population. This is commonly encountered when, after running some experiment or other data collection method, it is necessary to obtain estimated values for the parameters. Since the true parameter values are not known, a method needs to be developed for estimating the values so that they are as close to the true values as possible.

In general, the estimated parameter is denoted by placing a circumflex ($\hat{\ }$) over the theoretical or population parameter, for example, $\hat{\mu}$ is the estimated value for the mean μ. Very often, an arbitrary parameter is denoted by θ and its estimated value is given as $\hat{\theta}$.

2.5.1 Considerations for Parameter Estimation

When estimating the value of a parameter, the following points should be borne in mind:

(1) **Bias**: A good estimate of a parameter should be **unbiased**, that is, $E(\hat{\theta}) = \theta$. The bias, δ, in a parameter estimate is defined as:

$$\delta = E\left(\hat{\theta}\right) - \theta. \tag{2.34}$$

(2) **Variance**: The variance of the parameter should be as small as possible. A parameter estimate with the smallest variance over all possible parameter estimates is called the **minimum variance estimator** (MVE) for that parameter.

(3) **Mean-squared error (MSE)**: The mean-squared error of a parameter is defined as

$$\text{MSE}\left(\hat{\theta}\right) = \text{E}\left(\left(\hat{\theta} - \theta\right)^2\right) = \sigma_{\hat{\theta}}^2 + \delta^2, \tag{2.35}$$

where $\sigma_{\hat{\theta}}^2$ is the variance of the parameter estimate. If the estimate is unbiased, then the mean-squared error is equal to the variance. A minimum mean-squared-error estimate for a parameter need not be equal to its minimum variance estimator.

(4) **Consistency**: This says that as the number of samples used to estimate the parameter goes to infinity, then the estimate of the parameter goes to the true value of the parameter with probability one, that is,

$$\plim_{n \to \infty} \hat{\theta}_n = \theta, \tag{2.36}$$

where $\hat{\theta}_n$ is the parameter estimate using n data points and *plim* denotes convergence in a probabilistic manner, that is, given a sequence $\{X_n\}$, it is said to converge in a probabilistic manner to the random variable X, if for any ε,

$$\lim_{n \to \infty} P(|X_n - X| \geq \varepsilon) = 0. \tag{2.37}$$

The **Cramér–Rao lower bound** for a parameter estimate provides a bound on how low the variance of the estimated parameters can be. Achieving the lower bound implies that we have a minimum variance estimate. The Cramér–Rao lower bound is defined as

$$\sigma_{\hat{\theta}}^2 \geq \mathcal{F}(\theta)^{-1}, \tag{2.38}$$

where $\mathcal{F}$ is the Fisher information matrix defined as

$$\mathcal{F}(\theta) = -E\left(\frac{\partial^2 \log(L(x, \theta))}{\partial \theta^2}\right) \tag{2.39}$$

and $L(\theta|x)$ is the likelihood function for the parameter estimates. The Cramér–Rao lower bound can be used to define the **efficiency** of a parameter estimate, that is, how close does the parameter estimate come to the lower bound. Let the efficiency of a parameter estimate, $g_{\hat{\theta}}$, be defined as

$$g_{\hat{\theta}} = \frac{\mathcal{F}(\theta)^{-1}}{\sigma_{\hat{\theta}}^2} \leq 1. \tag{2.40}$$

A parameter estimator achieving the Cramér–Rao lower bound will have an efficiency of 1.

2.5.2 Methods of Parameter Estimation

In general, there are three different methods that can be used to estimate a parameter:

(1) **Method of Moments**, where the parameter estimates are obtained by calculating the moments of the sample and comparing them with the theoretical moments.
(2) **Maximum Likelihood Method**, where the likelihood function given the data set is optimised to determine the optimal parameter values.
(3) **Regression**, where the error between the estimated and actual data points is minimised to determine the parameter estimates. Since this method is so commonly used, the following chapter (Chap. 3: Regression) is devoted to examining this concept further.

2.5.2.1 Method of Moments

In the **method of moments** approach, theoretical expressions for the moments of the distribution are determined. Using the sample obtained, the sample moments are then computed and compared with the theoretical expressions. Solving the resulting system of equations for the unknown parameters will give the method of moments parameter estimates. If there are m parameters that describe a given distribution, then at least the first m moments will need to be computed. The sample moments can be computed as follows:

$$\hat{m}_k = \frac{1}{n} \sum_{i=1}^{n} x_i^k. \tag{2.41}$$

The method of moments can provide easy to obtain parameter estimates for a distribution. However, the parameter estimates may be biased. As well, solving the system of equations can be difficult.

Example 2.6: *Method of Moments for a Normal Distribution*

Consider estimating the mean and standard deviation for a normal distribution using the method of moments. Are the parameter estimates biased?

Solution

The theoretical expressions for the two moments are

$$m_1 = \mu, \; m_2 = \sigma^2 + \mu^2.$$

Let the sample moments be given as

$$\hat{m}_1 = \frac{1}{n} \sum_{i=1}^{n} x_i, \; \hat{m}_2 = \frac{1}{n} \sum_{i=1}^{n} x_i^2.$$

Equating the corresponding moments gives

$$\hat{m}_1 = \frac{1}{n} \sum_{i=1}^{n} x_i = \mu \Rightarrow \hat{\mu} = \frac{1}{n} \sum_{i=1}^{n} x_i$$

$$\hat{m}_2 = \frac{1}{n} \sum_{i=1}^{n} x_i^2 = \sigma^2 + \mu^2 \Rightarrow \hat{\sigma}^2 = \frac{1}{n} \sum_{i=1}^{n} x_i^2 - \left(\frac{1}{n} \sum_{i=1}^{n} x_i \right)^2.$$

To determine the **bias** of the parameter estimates, take the expected value of the equations obtained above, namely,

$$E(\hat{\mu}) = E\left(\frac{1}{n} \sum_{i=1}^{n} x_i \right) = \frac{1}{n} \sum_{i=1}^{n} E(x_i) = \frac{1}{n} \sum_{i=1}^{n} \mu = \mu \quad \text{(unbiased!)}.$$

$$E(\hat{\sigma}^2) = E\left(\frac{1}{n} \sum_{i=1}^{n} x_i^2 - \left(\frac{1}{n} \sum_{i=1}^{n} x_i \right)^2 \right) = \frac{1}{n} E\left(\sum_{i=1}^{n} x_i^2 - \frac{1}{n} \left(\sum_{i=1}^{n} x_i \right)^2 \right)$$

$$= \frac{1}{n} \sum_{i=1}^{n} E\left(x_i^2 - \frac{2}{n} x_i \sum_{j=1}^{n} x_j + \frac{1}{n^2} \sum_{j=1}^{n} x_j \sum_{k=1}^{n} x_k \right)$$

$$= \frac{1}{n} \sum_{i=1}^{n} \frac{n-2}{n} E(x_i^2) - \frac{2}{n} \sum_{j \neq i}^{n} E(x_i x_j) + \frac{1}{n^2} \sum_{j=1}^{n} \sum_{k \neq j}^{n} E(x_j x_k) + \frac{1}{n^2} \sum_{j=1}^{n} E(x_j^2)$$

$$= \frac{1}{n} \sum_{i=1}^{n} \frac{n-2}{n} (\sigma^2 + \mu^2) - \frac{2(n-1)}{n} \mu^2 + \frac{n-1}{n} \mu^2 + \frac{1}{n} (\sigma^2 + \mu^2)$$

$$= \frac{1}{n} \sum_{i=1}^{n} \frac{n-2}{n} (\sigma^2) + \frac{1}{n} (\sigma^2)$$

$$= \frac{n-1}{n}\sigma^2 \quad \text{(biased!)}$$

Therefore, the estimate for the mean is unbiased, while for the variance the estimate is biased.

2.5.2.2 Maximum Likelihood Method

The **maximum likelihood method** seeks to maximise the likelihood function for the parameters given the data set. The **likelihood function** for the parameters θ given the data set $\vec{x}$, $L(\theta|\vec{x})$, can be written as[6]

$$L(\theta|\vec{x}) = f(\vec{x}, \theta), \tag{2.42}$$

where $f(\vec{x}, \theta)$ is the assumed probability density function from which the data set came. If it can be assumed that the individual data points are independent of each other, then Eq. (2.42) reduces to

$$L(\theta|\vec{x}) = \prod_{i=1}^{n} f(x_i, \theta), \tag{2.43}$$

where $f(x, \theta)$ is the corresponding univariate probability density function and n is the number of samples or data points. Since it is difficult to deal with a product, the logarithm of the likelihood function is most often used for optimisation. The **log-likelihood function**, $\ell(\theta|\vec{x})$, is defined as

$$\boxed{\ell(\theta|\vec{x}) = \log L(\theta|\vec{x}) = \sum_{i=1}^{n} \log f(x_i, \theta).} \tag{2.44}$$

Maximisation is performed by setting the derivatives of ℓ with respect to the parameters to zero and solving the resulting system of equations, that is,

$$\boxed{\hat{\theta}_{MLE} = \arg\max_{\theta} \ell(\theta|\vec{x}).} \tag{2.45}$$

The maximum likelihood method has the following asymptotic properties:

---

[6] The likelihood function is similar in form to a probability density function, but the relationship between the parameters and data points is reversed, that is, the probability density function assumes the parameters and seeks the data points, while the likelihood function assumes the data points and seeks the parameters.

(1) $E(\hat{\theta}) \approx \theta$;
(2)   The variance of the estimate attains the Cramér–Rao lower bound;
(3)   The estimate is consistent;
(4)   The estimate is efficient; and
(5) $\hat{\theta}$ is approximately normally distributed.

Asymptotic means that these properties are attained as the number of samples approaches infinity ($n \to \infty$). Furthermore, the maximum likelihood estimate has the advantage that a function of a parameter estimate is also a maximum likelihood estimate of the function of the true values, that is, if $\hat{\theta}$ is the maximum likelihood estimate for θ, then $g(\hat{\theta})$ is the corresponding estimate for $g(\theta)$. This is called the transformative property.

The main problem with the maximum likelihood method is that a closed-form solution cannot always be obtained for the parameter estimates. Solving such a system numerically may not always be easy.

Example 2.7: *Maximum Likelihood Estimates for a Normal Distribution*

Consider the problem of finding the maximum likelihood estimates given X_1, $X_2,\ldots, X_n$ random, independent samples drawn from a normal distribution. Determine the maximum likelihood parameter estimates.

Solution

The probability density function of interest can be written as

$$f(x, \mu, \sigma^2) = \frac{1}{\sigma\sqrt{2\pi}} e^{\frac{-(x-\mu)^2}{2\sigma^2}}.$$

Therefore, given n samples the log-likelihood function can be written as

$$\ell(\theta|\vec{x}) = \sum_{i=1}^{n} \log\left(\frac{1}{\sigma\sqrt{2\pi}} e^{\frac{-(x_i-\mu)^2}{2\sigma^2}}\right) = \sum_{i=1}^{n} \log\left(\frac{1}{\sigma\sqrt{2\pi}}\right) - \frac{(x_i - \mu)^2}{2\sigma^2}.$$

$$= -\frac{n}{2}\log 2\pi - n\log\sigma - \frac{\sum_{i=1}^{n}(x_i - \mu)^2}{2\sigma^2}$$

Taking the derivative of the log-likelihood function with respect to μ gives

$$\frac{\partial \ell}{\partial \mu} = \frac{\partial\left(-\frac{n}{2}\log 2\pi - n\log\sigma - \frac{\sum_{i=1}^{n}(x_i-\mu)^2}{2\sigma^2}\right)}{\partial \mu} = -\frac{\sum_{i=1}^{n}(x_i - \mu)}{\sigma^2}.$$

Equating to zero and solving for μ gives

$$-\frac{\sum_{i=1}^{n} x_i - n\mu}{\sigma^2} = 0 \Rightarrow \hat{\mu} = \frac{\sum_{i=1}^{n} x_i}{n}.$$

Taking the derivative with respect to σ, equating to zero, and solving the resulting equation for σ gives

$$\frac{\partial \ell}{\partial \sigma} = \frac{\partial\left(-\frac{n}{2}\log 2\pi - n\log\sigma - \frac{\sum_{i=1}^{n}(x_i-\mu)^2}{2\sigma^2}\right)}{\partial \sigma} = -\frac{n}{\sigma} + \frac{\sum_{i=1}^{n}(x_i-\mu)^2}{\sigma^3} = 0$$

$$\Rightarrow \hat{\sigma}^2 = \frac{\sum_{i=1}^{n}(x_i - \hat{\mu})^2}{n}.$$

It can be noted that both parameter estimates are equivalent to the method of moment estimator. This is not necessarily true in general. Finally, note that the estimate for the variance is biased. An exploration of the meaning of this is given in Sect. 2.5.3.

2.5.3 Remarks on Estimating the Mean, Variance, and Standard Deviation

As can be seen from the above results for the normal distribution, the estimate of the mean is unbiased irrespective of the method selected.

Unfortunately, the same cannot be said about the variance, which is biased. The bias in the variance is

$$\delta = E\left(\hat{\theta}\right) - \theta = \frac{n-1}{n}\sigma^2 - \sigma^2 = -\frac{\sigma^2}{n}. \tag{2.46}$$

It can thus be seen that the variance will be underestimated using the maximum likelihood or method of moment values. Furthermore, note that as $n \to \infty$, $n^{-1} \to 0$, so that the bias will decrease to zero. This clearly shows that the estimate is an asymptotically unbiased estimator. In order to obtain an unbiased estimator, the formula needs to be changed using Bessel's correction. The unbiased estimate of the variance is then

$$\sigma^2 = \frac{\sum_{i=1}^{n}(x_i - \hat{\mu})^2}{n-1}, \tag{2.47}$$

where the n in the original denominator is replaced by $n-1$. One way to explain this change is to note that since the true mean is not known, one degree of freedom is used to compute its value. Therefore, the variance does not have n, but rather $n-1$

degrees of freedom. This correction should only be used if both the population mean and variance are being estimated from the same data set. If the population mean is known, then there is no need to use the correction.

Since the standard deviation is equal to the square root of the variance, it follows that the corrected standard deviation would be

$$\sigma = \sqrt{\frac{\sum_{i=1}^{n}\left(x_i - \hat{\mu}\right)^2}{n - 1}}. \tag{2.48}$$

However, unlike the variance, this estimate will be biased. In fact, since the standard deviation is equal to the square root of the variance, the bias will only be corrected by using a different estimator for standard deviation. It can be noted that asymptotically the bias will go to zero as given by the transformation property of the maximum likelihood method.

2.6 Central Limit Theorem

The **central limit theorem** is an important result concerning the behaviour of the mean of any distribution computed from multiple different samplings of the original distribution.

Theorem 2.1 *Given* $\{X_1, X_2,\ldots,X_n\}$, *a set of random variables that are independent and identical, with mean* μ *and finite variance* $\sigma^2 \neq 0$. *Let* $S_n = X_1 + X_2 + \ldots + X_n$, *then*

$$\lim_{n \to \infty} P\left(\frac{S_n - n\mu}{\sigma\sqrt{n}} \leq z\right) = \Phi(z) \tag{2.49}$$

that is the probability density function of S_n *converges to the standard normal distribution.*

The result of the central limit theorem explains why many observations can be treated as coming from a normal distribution. Specifically, for the mean, which is computed as the sum of all observations divided by the number, n, we get

$$\lim_{n \to \infty} P\left(\frac{\hat{\mu} - \mu}{\sigma/\sqrt{n}} \leq z\right) = \Phi(z). \tag{2.50}$$

This can be obtained from Eq. (2.49) by dividing both the numerator and denominator by n and noting that $S_n / n = \hat{\mu}$. This implies that the estimated mean comes from a normal distribution with $\mu = \hat{\mu}$ and $\sigma^2 = \hat{\sigma}^2 / n$.

2.7   Hypothesis Testing and Confidence Intervals

One of the most common applications of statistics is to test different "questions" about the relationship between the true (or assumed) value and the estimated value obtained after sampling some population. Hypothesis testing always consists of two parts: the **null hypothesis**, H_0, which represents the default position and the **alternative hypothesis**, H_1, which represents the other option. If it is assumed that the true parameter is θ and the corresponding estimated parameter value is $\hat{\theta}$, then the null hypothesis can be written as

$$H_0 : \hat{\theta} = \theta. \tag{2.51}$$

There are three different possibilities for the alternative hypothesis:

(1)   **Case 1**: $H_1 : \hat{\theta} \neq \theta$;
(2)   **Case 2**: $H_1 : \hat{\theta} < \theta$; and
(3)   **Case 3**: $H_1 : \hat{\theta} > \theta$.

Case 1 is often referred to as a **two-sided** or **two-tailed** test, while Cases 2 and 3 are often called **single-sided** or **single-tailed** tests. The hypothesis test is performed at some confidence level $100(1 - \alpha)\%$, where α is the **α-error, Type I error,** or **false positive rate**, which is the fraction of times with which the null hypothesis will be rejected even though the estimated parameter value did indeed come from the sample space. The opposite situation of accepting the null hypothesis, even though the alternative hypothesis is correct, is called a **Type II error, β-error,** or **false negative rate**. The confidence level typically determines what critical value should be used to determine which hypothesis better describes the obtained estimate. It should be noted that if the alternative hypothesis is **rejected** based on the test statistic and the critical value, then the null hypothesis is more likely, and it is said that the null hypothesis **may be** correct, since some other untested hypothesis may be an even better fit with the relationship between the true and estimated parameters. On the other hand, if the null hypothesis is **rejected**, then it is said that the alternative hypothesis is **accepted**.

Consider the probability density functions shown in Fig. 2.6, where the solid, black curve is the probability density function for the null hypothesis, and the dashed, green curve is the probability density function for the alternative hypothesis. The black line represents the selected critical value (r_{critical}). The area to the right of this line and under the solid, black curve represents the likelihood of rejecting the null hypothesis even if it is true. This region is called the α-error (or Type I error) region. On the other hand, the area to the left of this line and under the dashed, green curve represents the likelihood of rejecting the alternative hypothesis even if it is true. This region is called the β-error (or Type II error) region. Ideally, it would be nice to reduce the size of both regions simultaneously. However, this is not practical, as decreasing the α-error, by shifting the critical value (denoted by the black line) to the right, will

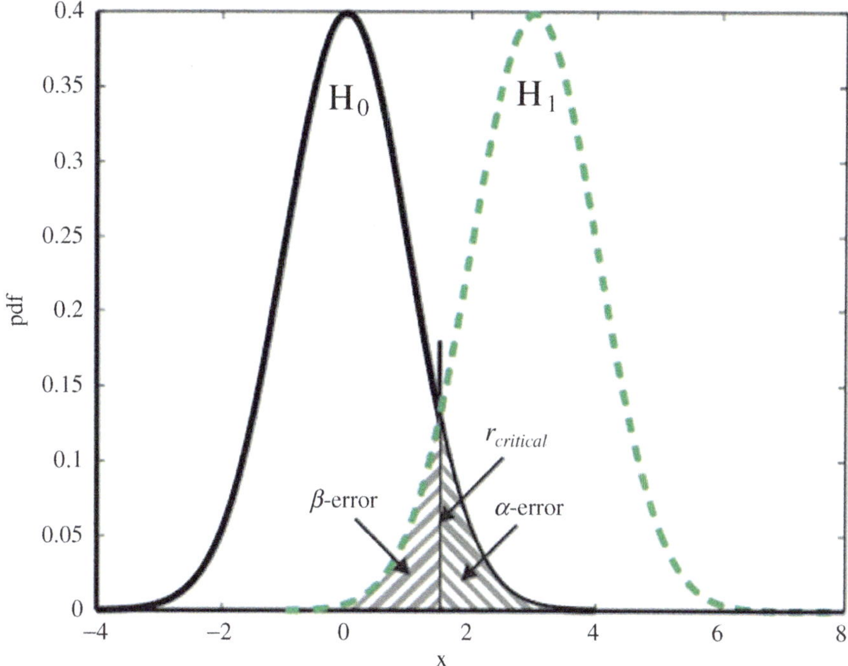

Fig. 2.6 Probability densities for the two hypotheses

lead to an increase in the β-error. Therefore, there is a trade-off between the two objectives.

To illustrate this trade-off, Fig. 2.7 shows three different distributions and how they overlap with each other. From this figure, it can be seen that if we take the solid curve as the basis (or null hypothesis) and compare it with the dashed curves, we see that only the dash-dot curve with $\mu = 10$ is substantially different from the null hypothesis. This shows the importance of the variance and mean on the tests. If the mean changes substantially, then, even if the variance is large, the difference will be clearer. On the other hand, to detect small changes in the process requires that the parameter variance also be small. One way in which the parameter variance can be decreased is to increase the number of data points used to estimate the given value.

The general procedure for hypothesis testing can be written as

(1)   Determine an appropriate **test statistic** for the problem at hand. The test statistic is some transformation of the available data that leads to a parameter that follows a known distribution.

(2)   Compute the value of the test statistic using the available data.

(3)   Determine the critical value based on the value of α, the number of data points, and any other relevant criteria.

(4)   Compare the test statistic with the critical value.

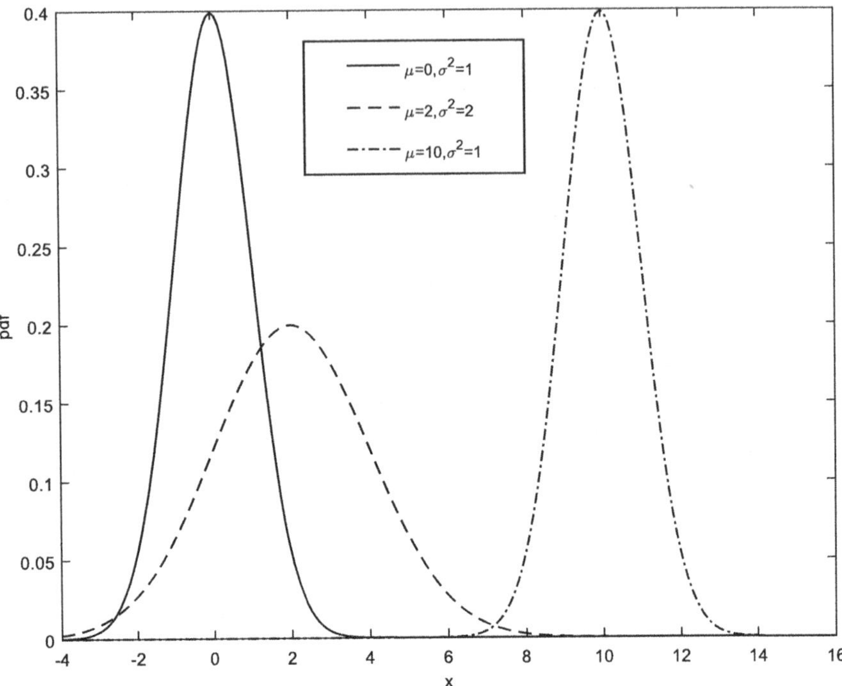

Fig. 2.7 Three different distributions and their overlap

(5)   Draw the appropriate conclusion given the relationship between the test statistic, the critical value, and the hypotheses being considered.

A **100(1-α)% confidence interval** for a parameter represents the region in which it would be expected to find the true parameter value. Confidence intervals exist for all three cases, of which the confidence interval obtained from the first case is the most common. For case 1, the corresponding confidence interval is

$$\hat{\theta} - r_{lower}\sigma_\theta \leq \theta \leq \hat{\theta} + r_{upper}\sigma_\theta, \tag{2.52}$$

where r_{lower} is the lower bound critical value, r_{upper} is the upper bound critical value, and σ_θ is the standard deviation of the parameter estimate. For case 2, the corresponding confidence interval is

$$\hat{\theta} - r_{lower}\sigma_\theta \leq \theta. \tag{2.53}$$

For case 3, the corresponding confidence interval is

$$\theta \leq \hat{\theta} + r_{upper}\sigma_\theta. \tag{2.54}$$

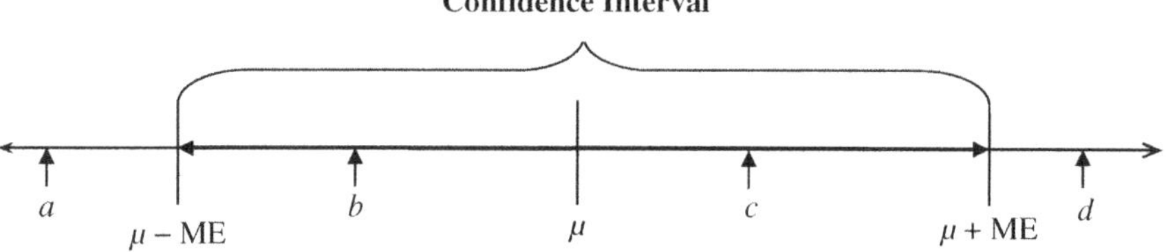

Fig. 2.8 Confidence intervals and covering a value (ME $= r_{crit}\sigma_\theta$)

A confidence interval is said to **cover** a value at the given confidence value α, if the given value lies within the given confidence bounds. Consider the confidence interval shown in Fig. 2.8. Points b and c are inside the confidence interval and hence it would be said that the confidence interval covers the given values. On the other hand, points a and d are outside the confidence interval, which implies that the points are not covered by the confidence interval. A point inside the confidence interval can be considered to be equal to the value expressed by the confidence interval. For example, if the true value is 5 and the estimated value is 6±3 (95% confidence interval), then it can be concluded that the estimated value covers the true value, and hence, it is likely that the values are the same. On the other hand, if we had 10±2 (95% confidence interval) and the same true value, then we can conclude that the true value and the estimated value are different.

2.7.1 Computing the Critical Value

In order to compute the critical values for the test statistics, there is a need to understand the difference between left and right probabilities. Define the **left probability, p_l,** to be

$$p_l = \int_{-\infty}^{r_{critical}} f(x)dx, \tag{2.55}$$

where $r_{critical}$ is the desired critical value. Define the **right probability, p_r,** to be

$$p_r = \int_{r_{critical}}^{\infty} f(x)dx. \tag{2.56}$$

In all cases, the following relationship holds

$$p_l + p_r = 1. \tag{2.57}$$

Figure 2.9 shows the difference between left and right probabilities.

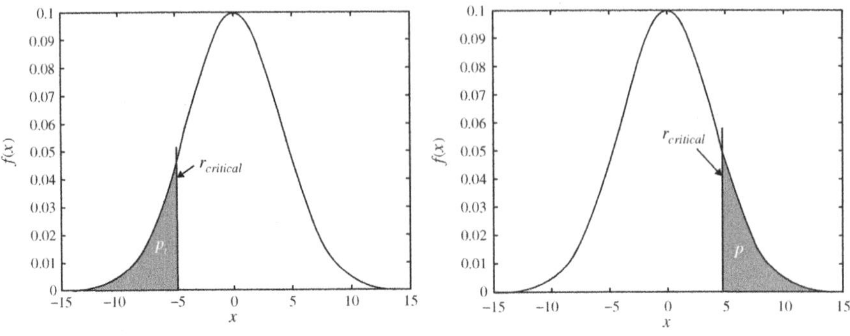

Fig. 2.9 Difference between (**left**) left and (**right**) right probabilities

Table 2.7 Different software and the probability values they return

Software	Left Probability	Right Probability
MATLAB®	`finv`, `tinv`, `chi2inv`, `norminv`	—
Excel®*	`t.inv`, `f.inv`, `norm.inv`, `chi2.inv`	—
Statistical Tables	—	Always give right probabilities

*The old Excel® functions, such as `finv`, `chiinv`, and `tinv`, give right-side probabilities with unusual definitions, for example, `tinv` is designed to be used to solve two-tailed problems.

The need to distinguish between left and right probabilities arises from the way different software and books tabulate the relationship between α (a probability) and the critical value. Table 2.7 summarises the different software and the location returned. It should be noted that this textbook strives to consistently use left probabilities in defining all relationships.

2.7.2 Converting Confidence Intervals

It can happen that the variable for which the confidence intervals have been computed is not the variable that is actually desired. This can happen often when there is a need to transform one variable (or set of variables) into another in order to obtain a better statistical result. Such transformations are often encountered in regression analysis.

Consider the case, where the parameter for which the confidence interval has been computed, β, is some function of the desired parameter, α, for example, $\beta = f(\alpha)$. If the confidence interval is obtained as $\beta_L \leq \beta \leq \beta_U$, then the following three different cases can be distinguished for converting the given confidence interval into the desired confidence interval:

(1) If $f(\alpha)$ is a **one-to-one, monotonic function**, then the desired confidence
interval can be given as $f^{-1}(\beta_L) \leq \alpha \leq f^{-1}(\beta_U)$, where $f^{-1}(\beta)$ is the inverse
function for $f(\alpha)$. In this context, a one-to-one function is simply a function that
contains a unique inverse. For example, $y = x^2$ is not one-to-one, while $y = x$ is.
A **monotonic function** is a function that on a given interval is either constant
and increasing or constant and decreasing. For example, $y = x$ is monotonic
everywhere, while $y = x^2$ is monotonic over the regions $[0, +\infty[$ and $]-\infty, 0]$,
but not over $]-\infty, +\infty[$.

(2) If $f(\alpha)$ is a **one-to-one function, but not necessarily monotonic**, then the
maximum and minimum values of the function need to be determined over the
given region, that is, the following three steps need to be performed:

   a. Find the values of $f^{-1}(\beta_L)$ and $f^{-1}(\beta_U)$.

   b. Determine whether $\frac{df^{-1}(\beta_i)}{d\beta_i} = 0$ over the interval $\beta_L \leq \beta \leq \beta_U$ and
determine the value of the function at these points.

   c. Take the maximum and minimum of the values for α obtained in the above
steps to be the confidence intervals for the parameter, α.

(3) If none of the above holds, then the following method can be used to obtain an
estimate for the converted confidence interval. This method can also be used if
the function depends on more than one of the parameters. The general formula
is given as

$$\sigma_\alpha^2 = \sum_{i=1}^{l} \left(\left| \frac{df^{-1}\left(\alpha, \vec{\beta}\right)}{d\beta_i} \right| \hat{\sigma}_i^2 \right), \tag{2.58}$$

where $\hat{\sigma}_i^2$ is the variance associated with the ith parameter, and σ_α^2 is the variance
associated with the parameter of interest.

2.7.3 Testing the Mean

For testing hypotheses involving the mean, the test statistic is

$$t_{computed} = \frac{\hat{\mu} - \mu}{\hat{\sigma}/\sqrt{n}}, \tag{2.59}$$

where n is the number of data points. Invariably, the correct critical value can be
obtained from the Student's t-distribution. If the true population standard deviation is
known or $n > 30$, then the normal distribution can be used instead. Strictly speaking,
this test only applies to samples drawn from a normal distribution. However, by

Table 2.8 Summary of the required critical values, bounds, and confidence intervals for testing hypotheses about the mean

Case	H_1	Probability (p)		Test to reject H_0	Confidence Intervals
		Left	Right		
1	$\hat{\mu} \neq \mu$	$1 - \alpha/2$	$\alpha/2$	$\lvert t_{computed} \rvert > t_{p,\,n-1}$	$\hat{\mu} \pm t_{p,n-1}\frac{\hat{\sigma}}{\sqrt{n}}$
2	$\hat{\mu} < \mu$	α	$1 - \alpha$	$t_{computed} < t_{p,\,n-1}$	$\hat{\mu} - t_{p,n-1}\frac{\hat{\sigma}}{\sqrt{n}}$
3	$\hat{\mu} > \mu$	$1 - \alpha$	α	$t_{computed} > t_{p,\,n-1}$	$\hat{\mu} + t_{p,n-1}\frac{\hat{\sigma}}{\sqrt{n}}$

the central limit theorem, this result can be used even for nonnormal distributions provided that a relatively large number of samples is available (say $n > 30$).

Table 2.8 summarises the appropriate probabilities, critical values, testing conditions, and corresponding confidence intervals for the different hypothesis cases. In all cases, the t-score can be replaced by the Z-score if the true population variance is known (in which it should be used in lieu of $\hat{\sigma}$) or $n > 30$.

Example 2.8: *Testing the Mean—Computing a Confidence Interval*

Consider the following data that are claimed to come from a normal distribution with a mean of 1 and a standard deviation of 1. Compute a 95% confidence interval and determine if it covers the true mean.

$$\mathbb{X} = \{2.16, 2.71, 1.09, 0.40, 1.47, 1.13, 1.97\}$$

Solution

The mean is

$$\hat{\mu} = \frac{2.16 + 2.71 + 1.09 + 0.40 + 1.47 + 1.13 + 1.97}{7}$$

$$= \frac{10.93}{7}$$

$$= 1.56$$

Since the population standard deviation is known, $Z_{1-\alpha/2}$ will be used, which in this case is 1.96 (useful number to memorise). Therefore, the confidence interval is

$$\hat{\mu} \pm Z_p \frac{\sigma}{\sqrt{n}}$$

$$1.56 \pm 1.96 \frac{1}{\sqrt{7}}.$$

$$1.56 \pm 0.741$$

Since the confidence interval includes the true value of 1, it can be concluded that the true value is covered by the mean.

Example 2.9: *Testing the Mean—Hypothesis Testing*
   Consider the same data and set-up as in Example 2.8, but now perform the following hypothesis test:

$$H_0 : \hat{\mu} = \mu$$
$$H_1 : \hat{\mu} \neq \mu.$$

Solution

Since we are dealing with case 1, a two-tailed test will be performed. Since the population standard deviation is known, the Z-test will be used. The computed statistic can be written as

$$Z_{computed} = \frac{1.56 - 1}{\frac{1}{\sqrt{7}}} = 1.48.$$

The critical value of Z_{crit} is 1.96. Since $|Z_{computed}| < Z_{crit}$, the null hypothesis cannot be rejected. Note that the same conclusion was reached with the confidence interval.

Example 2.10: *Testing the Mean—Unknown Variances*
   Consider the same data and set-up as in Example 2.8, but now assume that the standard deviation is unknown. Perform the following hypothesis test:

$$H_0 : \hat{\mu} = \mu$$
$$H_1 : \hat{\mu} \neq \mu.$$

Solution

Since we are dealing with case 1, a two-tailed test will be performed. Since the sample space standard deviation is unknown, the t-test will be used. The estimated standard deviation is 0.78. The computed statistic can be written as

$$t_{computed} = \frac{1.56 - 1}{\frac{0.78}{\sqrt{7}}} = 1.90.$$

The critical value of t_{crit} with $7 - 1 = 6$ degrees of freedom is 2.97. Since $|t_{computed}| < t_{crit}$, the null hypothesis cannot be rejected. Note that the value of the t-score is much larger than the corresponding Z-score, since we are dealing with a very small sample.

Example 2.11: *Testing the Mean—Detailed Example*

As a plant engineer of a bitumen processing plant, you have specified in the supplier's contract that the delivered processed bitumen must have a purity of at least 99.0% at a 95% confidence level. You obtain a shipment of bitumen which you test. With 200 samples, you obtain a mean purity of 98.6% with a standard deviation of 0.5%. The company claims that this is sufficiently close that the shipment should be accepted. Should the shipment be rejected at the 95% confidence level?

Solution

The statistical set-up is

$$H_0 : \hat{\mu} = \mu$$
$$H_1 : \hat{\mu} < \mu$$

which implies that we are dealing with Case 2. Since there are more than 30 samples (precisely 200 samples), the Z-score can be used. The computed value is given as

$$Z_{computed} = \frac{98.6 - 99.0}{\frac{0.5}{\sqrt{200}}} = -11.3.$$

Since the value of $Z_{critical}$ is -1.64 (single-tail, left probability with a p-value of 0.05) and $Z_{computed} < Z_{critical}$, it can be concluded that the mean is indeed less than the specified value and the shipment should be rejected.

2.7.4 Testing the Variance

For testing hypotheses involving the variance, the test statistic is

$$\chi^2_{computed} = \frac{(n-1)\hat{\sigma}^2}{\sigma^2}, \tag{2.60}$$

where n is the number of data points. Invariably, the correct critical value can be obtained from the χ^2-distribution. It should be noted that this assumes that the underlying distribution is normal.

Table 2.9 summarises the appropriate probabilities, critical values, testing conditions, and corresponding confidence intervals for the different hypothesis cases.

Example 2.12: *Testing the Variance*

You have been hired to design a new control strategy for a temperature controller. In order to show that your controller is better, you need to show that the variance of the new control strategy has been decreased. Historically speaking, the previous controller had a variance of 2 K^2. Upon implementing the new control strategy for 100 samples, the variance is 1.2 K^2. At an $\alpha = 0.05$, has the new control strategy decreased the variance of the temperature controller?

Solution

The statistical set-up is

$$H_0 : \hat{\sigma}^2_u = \sigma^2$$
$$H_1 : \hat{\sigma}^2 < \sigma^2$$

which implies that we are dealing with Case 2. The correct test statistic is the χ^2-test. The computed value is

Table 2.9 Summary of the required critical values, bounds, and confidence intervals for testing hypotheses about the variance

Case	H_1	Probability (p)		Test to Reject H_0	Confidence Interval
		Left	Right		
1	$\hat{\sigma}^2 \neq \sigma^2$	$p_1 = 1 - 0.5\alpha$ $p_2 = 0.5\alpha$	$p_1 = 0.5\alpha$ $p_2 = 1 - 0.5\alpha$	$\chi^2_{computed} > \chi^2_{p_1,n-1}$ **OR** $\chi^2_{computed} < \chi^2_{p_2,n-1}$	$\frac{(n-1)\hat{\sigma}^2}{\chi^2_{p_1,n-1}} \leq \sigma^2 \leq \frac{(n-1)\hat{\sigma}^2}{\chi^2_{p_2,n-1}}$
2	$\hat{\sigma}^2 < \sigma^2$	α	$1 - \alpha$	$\chi^2_{computed} < \chi^2_{p,n-1}$	$\frac{(n-1)\hat{\sigma}^2}{\chi^2_{p,n-1}}$
3	$\hat{\sigma}^2 > \sigma^2$	$1 - \alpha$	α	$\chi^2_{computed} > \chi^2_{p,n-1}$	$\frac{(n-1)\hat{\sigma}^2}{\chi^2_{p,n-1}}$

$$\chi^2_{computed} = \frac{(n-1)\hat{\sigma}^2}{\sigma^2}$$
$$= \frac{(100-1)(1.2)}{2}$$
$$= 59.4.$$

The critical value of the χ^2-test with $100 - 1 = 99$ degrees of freedom at a p-value of 0.05 is 77.0. Comparing the critical value with the computed value it can be seen that $59.4 < 77.0$, which implies that the null hypothesis should be rejected and the alternative hypothesis accepted. Thus, it can be concluded that the variance of the process has improved (decreased).

2.7.5 Testing a Ratio or Proportion

For testing hypotheses involving binomial proportions, r, then the appropriate test statistic is

$$Z_{computed} = \frac{x - nr}{\sqrt{nr(1-r)}} = \frac{\hat{r} - r}{\sqrt{n^{-1}r(1-r)}}, \tag{2.61}$$

where n is the number of data points and x is the number of successes. In order to apply the normal approximation to the binomial distribution, $nr \geq 5$ and $n(1-r) \geq 5$ should hold.

Table 2.10 summarises the appropriate probabilities, critical values, testing conditions, and corresponding confidence intervals for the different hypothesis cases.

Table 2.10 Summary of the required critical values, bounds, and confidence intervals for testing hypotheses about a ratio

Case	H_1	Probability (p)		Test to reject H_0	Confidence interval
		Left	Right		
1	$\hat{r} \neq r$	$1 - \alpha/2$	$\alpha/2$	$|Z_{computed}| > Z_p$	$\hat{r} \pm Z_p\sqrt{\frac{\hat{r}(1-\hat{r})}{n}}$
2	$\hat{r} < r$	α	$1 - \alpha$	$Z_{computed} < Z_p$	$\hat{r} - Z_p\sqrt{\frac{\hat{r}(1-\hat{r})}{n}}$
3	$\hat{r} < r$	$1 - \alpha$	α	$Z_{computed} > Z_p$	$\hat{r} + Z_p\sqrt{\frac{\hat{r}(1-\hat{r})}{n}}$

Example 2.13: *Testing a Ratio*

As the plant engineer, you have been monitoring the incidences of faults in your plant. Historically, you have observed that the hourly rate of faults is 30%. After implementing a new process management technique, you noticed that the hourly rate of faults has decreased to 25% over a period of 200 h, that is, there were some type of faults during 40 h of operation. You have been asked by your boss to determine if the rate of incidence of faults has decreased?

Solution

The statistical set-up is

$$H_0 : \hat{r} = r$$
$$H_1 : \hat{r} < r$$

which implies that we are dealing with case 2. The correct test statistic is the Z-test. The computed value is

$$Z_{computed} = \frac{\hat{r} - r}{\sqrt{n^{-1}r(1 - r)}}$$
$$= \frac{0.25 - 0.30}{\sqrt{200^{-1}(0.3)(1 - 0.3)}}.$$
$$= -1.54$$

The critical value the Z-test at a p-value of 0.05 is -1.64. Comparing the critical value with the computed value it can be seen that $-1.64 < -1.54$, which implies that the null hypothesis cannot be rejected. Thus, it can be concluded that the new process management software has not necessarily improved the hourly fault rate.

2.7.6 Testing Two Samples

So far all the examples considered have examined testing some value against some reference or known benchmark value. However, in many cases, it is interesting to compare two estimated values against each other to determine if they are different. This section will consider the tests required to obtain such conclusions.

2.7.6.1   Testing the Mean

When comparing two sample means, there is a need to carefully consider not only the available information, but also the experimental set-up. In all cases, it is assumed that the underlying distribution is normal for both samples and that the two samples are distinct and mutually independent of each other. Five different cases can be distinguished:

(I)   The true (population) variances for both samples are known.
(II)   The two variances are unknown, but a large number of data points was used for both samples (n_1 and n_2 both greater than about 30). This case can be treated as if the variances were known for both samples and analysed using the first approach.
(III)   The two variances are unknown but can be assumed to be equal.
(IV)   The two variances are unknown and cannot be assumed to be equal.
(V)   The mean has been obtained by taking the difference between two samples, that is, paired data are being used.

The null hypothesis for cases I–IV can be stated as H_0: $\mu_1 - \mu_2 = \Delta$, while for case V, it can be stated as H_0: $\mu_D = 0$, where μ_D is the mean value of all the differences. Paired tests are useful when the individual characteristics of an object may vary from sample to sample and it is desired to determine the overall effect on the system.

For case I, where the true variances, are known, or case II, where the variances are unknown, but the sample sizes are large, the test statistic can be written as

$$Z_{computed} = \frac{\hat{\mu}_1 - \hat{\mu}_2 - \Delta}{\sqrt{\frac{\sigma_1^2}{n_1} + \frac{\sigma_2^2}{n_2}}}, \tag{2.62}$$

where the subscripts refer to the two data sets being compared. The required critical values and testing conditions are shown in Table 2.11.

For case III, where the true variances are unknown but equal, the test statistic can be written as

Table 2.11 Summary of the required critical values and bounds for testing hypotheses about a difference when the true variances are known

Case	H_1	Probability (p)		Test to reject H_0
		Left	Right	
1	$\hat{\Delta} \neq \Delta$	$1 - \alpha / 2$	$\alpha / 2$	$|Z_{computed}| > Z_p$
2	$\hat{\Delta} < \Delta$	α	$1 - \alpha$	$Z_{computed} < Z_p$
3	$\hat{\Delta} > \Delta$	$1 - \alpha$	α	$Z_{computed} > Z_p$

Table 2.12 Summary of the required critical values and bounds for testing hypotheses about a difference when the true variances are unknown, but assumed equal

Case	H_1	Probability (p)		Test to reject H_0 $v = n_1 + n_2 - 2$
		Left	Right	
1	$\hat{\Delta} \neq \Delta$	$1 - \alpha/2$	$\alpha/2$	$|t_{computed}| > t_{p,v}$
2	$\hat{\Delta} < \Delta$	α	$1 - \alpha$	$t_{computed} < t_{p,v}$
3	$\hat{\Delta} > \Delta$	$1 - \alpha$	α	$t_{computed} > t_{p,v}$

$$t_{computed} = \frac{\hat{\mu}_1 - \hat{\mu}_2 - \Delta}{\hat{\sigma}_p \sqrt{\frac{1}{n_1} + \frac{1}{n_2}}}, \tag{2.63}$$

where the subscripts refer to the two data sets being compared and $\hat{\sigma}_p$ is the pooled standard deviation computed from

$$\hat{\sigma}_p^2 = \frac{(n_1 - 1)\hat{\sigma}_1^2 + (n_2 - 1)\hat{\sigma}_2^2}{n_1 + n_2 - 2}. \tag{2.64}$$

The degrees of freedom (v) for this test are equal to $n_1 + n_2 - 2$. The required critical values and testing conditions are shown in Table 2.12.

For case IV, where the true variances are unknown and cannot be assumed equal, the test statistic can be written as

$$t_{computed} = \frac{\hat{\mu}_1 - \hat{\mu}_2 - \Delta}{\sqrt{\frac{\hat{\sigma}_1^2}{n_1} + \frac{\hat{\sigma}_2^2}{n_2}}}, \tag{2.65}$$

where the subscripts refer to the two data sets being compared. The degrees of freedom (v) for this test are computed from

$$v = \left\lfloor \frac{\left(\frac{\hat{\sigma}_1^2}{n_1} + \frac{\hat{\sigma}_2^2}{n_2}\right)^2}{\frac{\hat{\sigma}_1^4}{n_1^2(n_1-1)} + \frac{\hat{\sigma}_2^4}{n_2^2(n_2-1)}} \right\rfloor, \tag{2.66}$$

where $\lfloor \cdot \rfloor$ is the floor or round down function, that is, $\lfloor -1.23 \rfloor$ would be -2 and $\lfloor 1.86 \rfloor$ would be 1. The required critical values and testing conditions are the same as those shown in Table 2.12, except that v would be computed using Eq. (2.66).

Finally, for case V, where a paired difference is being considered, the test statistic can be written as

$$t_{computed} = \frac{\hat{\mu}_D - \mu_D}{\hat{\sigma}_D / \sqrt{n}}, \tag{2.67}$$

Table 2.13 Summary of the required critical values, bounds, and confidence intervals for testing hypotheses about a paired mean value

Case	H_1	Probability (p)		Test to reject H_0	Confidence interval
		Left	Right		
1	$\hat{\mu}_D \neq \mu_D$	$1 - \alpha/2$	$\alpha/2$	$|t_{computed}| > t_{p,\,n-1}$	$\hat{\mu}_D \pm t_{p,n-1}\dfrac{\hat{\sigma}_D}{\sqrt{n}}$
2	$\hat{\mu}_D < \mu_D$	α	$1 - \alpha$	$t_{computed} < t_{p,\,n-1}$	$\hat{\mu}_D - t_{p,n-1}\dfrac{\hat{\sigma}_D}{\sqrt{n}}$
3	$\hat{\mu}_D > \mu_D$	$1 - \alpha$	α	$t_{computed} > t_{p,\,n-1}$	$\hat{\mu}_D + t_{p,n-1}\dfrac{\hat{\sigma}_D}{\sqrt{n}}$

where $\hat{\sigma}_D$ is the standard deviation of the differences and n is the number of differences available. In most cases, $\mu_D = 0$, since we are testing whether the pair differences are significant. The required critical values, testing conditions, and corresponding confidence intervals are shown in Table 2.13.

Example 2.14: *Testing Differences in Means—Variances Known.*
Two separate samples from the same batch were taken and tested by two different operators. As the plant engineer, you are interested in knowing if the two operators provide similar test results. Historically, it has been determined that the variability of the test results is 4.0. It can be assumed that the variability will be the same for both operators as the samples are drawn from the same batch. The first operator performed ten tests and obtained the following values: 9.6, 9.9, 8.9, 12.0, 10.4, 13.8, 10.8, 10.3, 9.2, and 9.3. The second operator performed ten tests and obtained: 10.8, 11.0, 11.1, 11.2, 7.6, 9.8, 14.0, 7.8, 8.6, and 8.4. Are the two operators similar? Assume $\alpha = 0.05$.

Solution

In order to answer the question, we will need to compute the mean value for the two operators:

$$\hat{\mu}_1 = \frac{9.6 + 9.9 + 8.9 + 12.0 + 10.4 + 13.8 + 10.8 + 10.3 + 9.2 + 9.3}{10} = 10.4,$$

$$\hat{\mu}_2 = \frac{10.8 + 11.0 + 11.1 + 11.2 + 7.6 + 9.8 + 14.0 + 7.8 + 8.6 + 8.4}{10} = 10.0.$$

Since we are testing to determine if there is a difference between the two operators, $\Delta = 0$. The statistical set-up is

$$H_0 : \hat{\Delta} = \Delta$$

$$H_1 : \hat{\Delta} \neq \Delta.$$

Since the variance is known, this is case I and the test statistic can be computed
as

$$Z_{computed} = \frac{10.4 - 10.0 - 0}{\sqrt{\frac{4.0}{10} + \frac{4.0}{10}}} = 0.4602.$$

The critical value is $Z_{crit} = 1.96$. Comparing the critical and computed values
gives, $|Z_{computed}| < Z_{crit}$, which suggests that the null hypothesis cannot be
rejected. Therefore, it can be concluded that the two operators give similar
results.

Example 2.15: *Testing the Difference in Means—Unknown, Common Variance*

As a plant engineer, you are testing a new drying procedure for the plant.
Two options exist: A and B. Option A is currently in place, while option B
is a faster new method. It is desired to determine whether option B should
be implemented. The criterion for implementation is that the quality of the
material being dried should not decrease by more than 0.1 at an $\alpha = 0.05$.
This value has been determined based on a cost–benefit analysis of the costs
of drying compared with the costs of production. It will be assumed that the
variance for both options is the same.

Option A gave the following product quality: 95.6, 97.3, 95.6, 95.4, 99.4,
97.2, 92.2, 92.8, 94.3, and 92.6. Option B gave the following product quality:
89.2, 94.2, 93.9, 93.2, 94.7, 91.7, 93.2, 92.4, 91.8, and 91.5.

Solution

Before performing the tests, compute the mean and variance for both options.

$$\hat{\mu}_A = \frac{95.6 + 97.3 + 95.6 + 95.4 + 99.4 + 97.2 + 92.2 + 92.8 + 94.3 + 92.6}{10}$$

$$= 95.24,$$

$$\hat{\sigma}_A^2 = \frac{90,755.3 - (10)95.24^2}{10 - 1} = 5.41$$

$$\hat{\mu}_B = \frac{89.2 + 94.2 + 93.9 + 93.2 + 94.7 + 91.7 + 93.2 + 92.4 + 91.8 + 91.5}{10}$$

$$= 92.58.$$

$$\hat{\sigma}_B^2 = \frac{85,734.2 - (10)92.58^2}{10 - 1} = 2.626.$$

Since we are testing to determine if the observed difference is significant to warrant a change in operating conditions, $\mu_A - \mu_B = \Delta = 0.1$. The statistical set-up is

$$H_0 : \hat{\Delta} = \Delta$$

$$H_1 : \hat{\Delta} > \Delta.$$

This is a single-sided test since we only care about degradation in process quality. Should Option B increase product quality then it is another reason to implement it. Note that it is very important to carefully state all the definitions at this point as there are multiple equivalent approaches that one can take. In this particular example, it has been assumed that the difference is defined as $\mu_A - \mu_B$, which implies that if this difference is greater than 0.1, then it can be assumed that the new drying option is not good. If the difference had been defined as $\mu_B - \mu_A$ then the alternative hypothesis would have been defined as $H_1 : \hat{\Delta} < \Delta$, with $\Delta = -0.1$.

Since the true variances are not known, this is case III of the options for testing sample means. Thus, we need to compute the pooled standard deviation. The formula gives

$$\hat{\sigma}_p^2 = \frac{(n_A - 1)\hat{\sigma}_A^2 + (n_B - 1)\hat{\sigma}_B^2}{n_A + n_B - 2} = \frac{(10 - 1)(5.41) + (10 - 1)(2.626)}{10 + 10 - 2}$$

$$= 4.018.$$

The test statistic can be computed as

$$t_{computed} = \frac{95.24 - 92.58 - 0.1}{\sqrt{4.015}\sqrt{\frac{1}{10} + \frac{1}{10}}} = 2.857.$$

The degrees of freedom are $10 + 10 - 2 = 18$. The critical value is $t_{crit, 18} = 1.73$. Comparing the critical and computed values gives $t_{computed} > t_{crit}$, which suggests that the null hypothesis can be rejected. Therefore, option B decreases product quality and hence should not be implemented.

Example 2.16: *Testing Two Means—Unknown Variance*

Consider the same data as in Example 2.15. This time assume that the variances cannot be assumed to be equal. Determine if option B should be implemented.

Solution

Since the required mean and variance have already been computed, we can proceed to the next step. The statistical set-up will be identical to before, that is,

$$H_0 : \hat{\Delta} = \Delta$$

$$H_1 : \hat{\Delta} > \Delta.$$

This is a single-sided test for case IV. The test statistic can be written as

$$t_{computed} = \frac{95.24 - 92.58 - 0.1}{\sqrt{\frac{5.41}{10} + \frac{2.626}{10}}} = \frac{2.56}{\sqrt{0.08036}} = 2.856.$$

The degrees of freedom can be computed as

$$\nu = \left\lfloor \frac{\left(\frac{5.41}{10} + \frac{2.626}{10}\right)^2}{\frac{(5.41)^2}{(10)^2(10-1)} + \frac{(2.626)^2}{10^2(10-1)}} \right\rfloor = \left\lfloor \frac{0.64577}{0.04018} \right\rfloor = \lfloor 16.071 \rfloor = 16.$$

The critical value $t_{crit, 16} = 1.75$. Comparing the critical and computed values gives $t_{computed} > t_{crit}$, which suggests that the null hypothesis can be rejected. Therefore, option B decreases product quality and hence should not be implemented. Notice that the same conclusion has been reached irrespective of whether the variances were assumed to be equal. In practice, the equality of the two sample variances could be tested using the method presented in Sect. 2.7.6.2.

Example 2.17: *Testing a Paired Mean*

You have been asked to test whether two different approaches to testing a sample of wood to determine heat capacity are the same. Each wood sample is divided in two and tested with each approach. Since the sample is burnt afterwards, a new sample is required to repeat the procedure. However, the individual properties of each sample could be sufficiently different that the values of the heat capacity obtained are different. Therefore, it is desired to use the difference between each sample to determine whether the methods are the same at a level of $\alpha = 0.05$. The data are provided below.

Sample	Approach 1 ($J \cdot g^{-1} \cdot K^{-1}$)	Approach 2 ($J \cdot g^{-1} \cdot K^{-1}$)	Difference ($J \cdot g^{-1} \cdot K^{-1}$)
1	1.75	1.73	0.02
2	1.65	1.64	0.01
3	1.67	1.67	0.00
4	1.53	1.51	0.02
5	1.76	1.74	0.02
6	1.72	1.70	0.02
7	1.78	1.77	0.01
8	1.68	1.65	0.03

Solution

In order to solve the question, it is first necessary to compute the mean and variance of the differences. Doing so, gives

$$\hat{\mu} = \frac{0.02 + 0.01 + 0.00 + 0.02 + 0.02 + 0.02 + 0.01 + 0.03}{8} = 0.01625$$

$$\hat{\sigma}^2 = \frac{0.0027 - (8)0.01625^2}{8 - 1} = 8.39 \times 10^{-5}.$$

Since we are testing a paired difference, this is case V. The statistical set-up will be

$$H_0 : \hat{\mu}_D = \mu_D = 0.$$
$$H_1 : \hat{H}_D \neq \mu_D$$

This is a two-sided test for case V. The test statistic can be written as

$$t_{computed} = \frac{0.01625 - 0}{\sqrt{8.39 \times 10^{-5}}/\sqrt{8}} = 5.017.$$

The degrees of freedom, ν, is $8 - 1 = 7$. The corresponding critical value is $t_{crit, 7} = 2.36$ (at $p = 0.975$, since this is a two-sided test). Since $|t_{computed}| > t_{crit}$, it can be concluded that the null hypothesis should be rejected and the alternative hypothesis accepted. Therefore, it can be concluded that there is a (statistical) difference between the two approaches to determining the heat capacity of the wood samples.

2.7.6.2 Testing Two Variances

In order to compare the variance of two normally distributed samples, it is useful to form a ratio of the two variances and compare them against 1. If the ratio is close to 1, then it can be concluded that the two variances are the same; otherwise there is a difference. The null hypothesis in this case is

$$H_0 : \sigma_1^2 = \sigma_2^2.$$

The test statistic can be written as

$$F_{computed} = \frac{\hat{\sigma}_1^2}{\hat{\sigma}_2^2}. \tag{2.68}$$

Table 2.14 summarises the appropriate probabilities, critical values, and testing conditions for the different hypothesis cases. Testing two sample variances forms the foundation for various methods that can be used to compare different data subsets to determine their significance.

Example 2.18: *Testing Two Sample Variances*
    Consider the same data as in Example 2.15 and test to determine if the variances could be assumed to be the same.

Solution

The statistical set-up will be

$$H_0 : \sigma_1 = \sigma_2$$
$$H_1 : \sigma_1 \neq \sigma_2.$$

This is a two-sided test. The test statistic can be written as

Table 2.14 Summary of the required critical values and bounds for testing hypotheses about the two variances

Case	H_1	Probability (p)		Test to reject H_0
		Left	Right	
1	$\sigma_1^2 \neq \sigma_2^2$	$p_1 = 1 - \alpha/2$ $p_2 = \alpha/2$	$p_1 = \alpha/2$ $p_2 = 1 - \alpha/2$	$F_{computed} > F_{p_1, n_1-1, n_2-1}$ **OR** $F_{computed} < F_{p_2, n_1-1, n_2-1}$
2	$\sigma_1^2 > \sigma_2^2$	$1 - \alpha$	α	$F_{computed} > F_{p, n_1-1, n_2-1}$

$$F_{computed} = \frac{\hat{\sigma}_1^2}{\hat{\sigma}_2^2} = \frac{5.41}{2.626} = 2.06.$$

The two required degrees of freedom are $v_1 = 10 - 1 = 9$ and $v_2 = 9$. The critical values are $F_{1, 9, 9} = 4.026$ and $F_{2, 9, 9} = 0.2484$. Since $F_2 < F_{computed} < F_1$, the null hypothesis cannot be rejected. This suggests that the two variances are actually the same. This result agrees with the previously observed results that showed that assuming that the variances were the same gave very similar results to assuming that they were not.

2.7.6.3 Testing Two Proportions

In order to test two sample proportions, assume that a large number of samples has been taken, so that the normal approximation to the binomial distribution can be used. The null hypothesis can be written as.

$$H_0 : r_1 = r_2.$$

The appropriate test statistic can then be written as

$$Z_{computed} = \frac{\hat{r}_1 - \hat{r}_2}{\sqrt{\hat{r}_p(1 - \hat{r}_p)(n_1^{-1} + n_2^{-1})}}, \tag{2.69}$$

where $\hat{r}_p$ is the common or pooled proportion computed as follows:

$$\hat{r}_p = \frac{n_1 \hat{r}_1 + n_2 \hat{r}_2}{n_1 + n_2}. \tag{2.70}$$

Table 2.15 summarises the appropriate probabilities, critical values, and testing conditions for the different hypothesis cases.

Table 2.15 Summary of the required critical values and bounds for testing hypotheses about two proportions

Case	H_1	Probability (p)		Test to Reject H_0
		Left	Right	
1	$r_1 \neq r_2$	$1 - \alpha/2$	$\alpha/2$	$|Z_{computed}| > Z_p$
2	$r_1 < r_2$	α	$1 - \alpha$	$Z_{computed} < Z_p$
3	$r_1 > r_2$	$1 - \alpha$	α	$Z_{computed} > Z_p$

Example 2.19: *Testing Two Proportions.*

A test for defects was performed on two different batches of drugs. In the first batch, out of fifty samples, there were twenty defects. In the second batch, out of forty samples, there were fifteen defects. Is the proportion of defects in the two batches different? Assume $\alpha = 0.10$.

Solution

The statistical set-up for this problem can be written as

$$H_0 : r_1 = r_2$$
$$H_1 : r_1 \neq r_2.$$

The pooled proportion is

$$\hat{r}_p = \frac{n_1 \hat{r}_1 + n_2 \hat{r}_2}{n_1 + n_2} = \frac{20 + 15}{50 + 40} = \frac{7}{18} \approx 0.389.$$

The respective proportions are $\hat{r}_1 = 20 / 50 = 0.40$ and $\hat{r}_2 = 15 / 40 = 0.375$. The test statistic can be written as

$$Z_{computed} = \frac{0.40 - 0.375}{\sqrt{\frac{7}{18}\left(1 - \frac{7}{18}\right)\left(50^{-1} + 40^{-1}\right)}} = \frac{0.025}{\sqrt{\frac{77}{7200}}} = 0.2417.$$

The two-tailed critical value is $Z_{crit,\,0.95} = 1.65$. Since $|Z_{computed}| < Z_{crit}$, we cannot reject the null hypothesis and conclude that the defects in the two batches are indeed the same.

2.8 Further Reading

The following are references that provide additional information about the topic:

(1) **Statistical Theory**:

    a. Varberg, D. E. (1963). The development of modern statistics. *The Mathematics Teacher*, 56(4), 252–257.

    b. Kallenberg, O. (2002). *Foundations of Modern Probability*. New York, New York, United States of America: Springer-Verlag.

    c. Ogunnaike, B. A. (2010). *Random Phenomena: Fundamentals of Probability and Statistics for Engineers*. Boca Raton, Florida, United States of America: CRC Press.

(2)   **Introduction to Statistics**:

    a.   Box, G. E., Hunter, W. G., & Hunter, J. S. (1978). *Statistics for Experimenters: An Introduction to Deisgn, Data Analysis, and Model Building.* New York, New York, United States of America: John Wiley & Sons, Inc.

    b.   Johnson, R. A., & Wichern, D. W. (2007). *Applied Multivariate Statistical Analysis* (6th ed.). Upper Saddle River, New Jersey, United States of America: Prentice Hall.

    c.   Montgomery, D. C., & Runger, G. C. (2007). *Applied Statistics and Probability for Engineers* (4th ed.). Hoboken, New Jersey, United States of America: John Wiley & Sons, Inc.

    d.   Sokal, R. R., & Rohlf, F. J. (1995). *Biometry: The Principles and Practice of Statisics in Biological Research* (3rd ed.). New York, New York, United States of America: W. H. Freeman and Company.

    e.   Steel, R. G., & Torrie, J. H. (1980). *Principles and Procedures of Statistics: A Biometrical Approach* (2nd ed.). Tokyo, Japan: McGraw-Hill Kogakusha, Ltd.

2.9   Chapter Problems

Problems at the end of the chapter consist of three different types: (a) Basic Concepts (True/False), which seek to test the reader's comprehension of the key concepts in the chapter; (b) Short Exercises, which seek to test the reader's ability to compute the required parameters for a simple data set using simple or no technological aids. This section also includes proofs of theorems; and (c) Computational Exercises, which require not only a solid comprehension of the basic material, but also the use of appropriate software to easily manipulate the given data sets.

2.9.1   Basic Concepts

Determine if the following statements are true or false and state why this is the case.

(1)   If the covariance of two variables X and Y is zero, then they are independent.

(2)   The function $f(x) = \sin(x)$ is a probability density function.

(3)   The function $f(x) = 2x$, $0 \leq x \leq 1$, zero otherwise is a probability density function.

(4)   The function $f(x) = 0.25|\cos(x)|$ for $0 \leq x \leq 2\pi$, zero otherwise, is a probability density function.

(5)   If X has a mean value of 5, then $E(2X) = 10$.

(6)   If $E(X) = 1$ and $E(X^2) = 2$, then the variance is 2.

(7)   If two variables are independent, then their joint probability density function is equal to the product of their individual probability density functions.

(8) If $f_{XY|z} = f_{XY}$ for all z, then it can be concluded that variables X and Y are independent of Z.

(9) A two-tailed test is necessary when testing the following alternative hypothesis: $H_1: \mu \geq \hat{\mu}$.

(10) If $E(\hat{a}) = a + 1$, then it can be said that the parameter estimate of a is unbiased.

(11) If $E(\hat{g}) = g$, then it is said that the parameter estimator for g is unbiased.

(12) If ten samples were taken and a mean calculated, then the appropriate test statistic is the t-test with nine degrees of freedom.

(13) An unbiased, minimum variance estimator is also a minimum mean-squared-error estimator.

(14) In order to compute the confidence intervals for variance, the χ^2-distribution is required.

(15) When testing the hypothesis $H_1: \hat{\mu} > 0$ with ten samples at a 90% confidence level, the appropriate test statistic is the t-score with nine degrees of freedom at a value of 0.05.

(16) If the 95% confidence interval is [0.95, 4], then at a 95% confidence level it can be concluded that the data set comes from a population with a mean of 3.

(17) If a parameter and its 95% confidence interval can be written as 5 ± 6.2 m, then this parameter does not equal zero at a 95% confidence level.

2.9.2 Short Exercises

These questions should be solved using only a simple, nonprogrammable, nongraphical calculator combined with pen and paper.

(18) Consider a game of Three Card Monte, where three cards, Q♥, Q♠, and Q♣, are shown and it is desired to select, after shuffling, the Queen of Hearts (Q♥). Answer the following questions:

    a.   What are $\mathbb{S}$, $\mathbb{F}$, and P?

    b.   What is the probability of winning?

    c.   If each win is worth \$2 and each loss is worth $-\$1$ (you lose money), what is the expected (or mean) payout? What is the variance of this payout?

(19) Consider the case of flipping a single biased coin where heads will turn up 70% of the time. Answer the following questions for this case:

    a.   What are $\mathbb{S}$, $\mathbb{F}$, and P?

    b.   If heads are worth \$1 and tails are worth $-\$2$ (you lose money), what is the expected (or mean) payout? Will you make money in the long run?

    c.   What is the break-even point, that is, the point at which the expected value is zero?

(20)   Consider the case of flipping two coins A and B. Coin A is biased and will turn up heads 60% of the time, while Coin B is unbiased and will turn up heads 50% of the time. Answer the following questions for this case:

     a.   What are $\mathbb{S}$, $\mathbb{F}$, and P?

     b.   If heads are worth \$2 and tails are worth \$1, what is the expected (or mean) payout?

         Hint: The possible outcomes can be denoted as HH (for heads A, heads B), HT (for heads A, tails B), and so on.

(21)   Consider the following potential probability density functions. Determine if they can be used as probability density functions and if so compute the mean, variance, and $E(|x|)$:

     a. $f(x) = \cos(x)$ for $-\pi \le x \le \pi$, zero otherwise.

     b. $f(x) = \sin(x)$ for $0 \le x \le \pi$, zero otherwise.

     c. $f(x) = N^{-1}$ for $0 \le x \le N$, zero otherwise.

(22)   Consider a variable Y whose probability density function is

$$f(y) = \frac{\sqrt{2}}{\sigma\sqrt{\pi}} \exp\left(-\frac{y^2}{2\sigma^2}\right), \quad y \ge 0 \tag{2.71}$$

     Answer the following questions:

     a.   Compute $\mu = E(Y)$.

     b.   Compute the variance.

     c.   Consider a normal variable X drawn from a zero-mean normal distribution and compute $E(|X|)$. (*Hint: Split the integral into two parts, one from $-\infty$ to 0 and the other from 0 to $+\infty$. Note that $|x|$ can be written as $-x$ if $x \le 0$ and x if $x \ge 0$.*)

     d.   Explain how the result from c) compares with that from a).

(23)   Consider two independent variables X ($\mu = 1$ and $\sigma^2 = 2$) and Y ($\mu = 2$ and $\sigma^2 = 1$), compute the following values: $E(2X - 4Y)$, $E(3XY)$, and $E(X^2)$.

(24)   Consider the gamma distribution

$$f(x) = \frac{x^{\alpha-1}e^{-x/\beta}}{\beta^{\alpha}\,\Gamma(\alpha)}, \, x > 0, \tag{2.72}$$

     with the following moments

$$E(X) = \alpha\beta$$
$$E(X^2) = \beta^2\alpha(1+\alpha). \tag{2.73}$$

Derive the method of moment estimators for α and β.

(25) Devise a method for estimating the mean of n samples drawn from an arbitrary distribution so that the parameter estimate:

    a.   Is biased but consistent.

    b.   Is unbiased but not consistent.

(26) Consider the production of glass sheets in a plant. Every day, twenty samples of these sheets are randomly taken and tested for quality. Based on historical considerations, the distribution of quality is normally distributed with a mean 0.85 and a variance of 0.2. Assume that the sample is an independent random variable.

    a.   What is the probability that quality is below 0.75 for five samples in a row?

    b.   What is the probability that the quality is above 0.90 for all twenty samples?

(27) Consider estimating the mean of a variable drawn from a population with mean μ and standard deviation σ, using the following formulae:

$$\hat{\mu}_1 = x_1$$

$$\hat{\mu}_2 = \frac{x_1 + x_2}{2}$$

$$\hat{\mu}_3 = \frac{x_1 + 2x_2}{2}$$

$$\hat{\mu}_4 = \frac{x_1 + x_2}{3}$$

$$\hat{\mu}_5 = \frac{\sum_{i=1}^{n} x_i}{n}$$

Assume that in all cases the individual estimates x_i are independent of each other. Determine the bias, variance, and mean-squared error for each of the estimators. Which ones are biased?

(28) You are an engineer working in the process control department of a petro-chemical plant. You have hired George, a postdoctoral fellow, to improve the performance of your control system. George has proposed a new control algorithm that involves the use of soft sensors. It gives a mean of 0.9855 for the purity with a standard deviation of 0.005 on a test of 1,000 samples. Historically, the process has been run at 0.985 with a standard deviation of 0.01. Answer the following questions:

   a.   Has the performance of the process been degraded by introducing this new control algorithm? Performance would be degraded if the mean value is less than the historical one. Formulate an appropriate hypothesis and test it with an $\alpha = 0.05$.

   b.   Has the performance of the process improved? Formulate appropriate hypotheses and test them with an $\alpha = 0.05$.

   c.   Is the new variance significantly different from historical one? Compute the 95% confidence interval for the sample variance.

   d.   Would you implement the new control algorithm?

(29)   For the following experiments, compute the 95% confidence intervals and determine whether the data comes from the stated population.

   a. $\mathbb{X} = \{3, 2.3, 4.5, 1.2, 5.6, 2.3, 4.5\}$, $\mu = 3$.

   b. $\hat{\mu} = 4.2$, $\hat{\sigma} = 1.2$, $n = 100$, $\mu = 3$.

   c. $\hat{\mu} = 0.2$, $\hat{\sigma} = 5$, $n = 10$, $\mu = 3$.

   d. $\hat{\sigma} = 1.5$, $n = 10$, $\sigma = 5$.

(30)   Given the data, resolve the following hypotheses with $\alpha = 0.05$.

   a. $H_1 : \mu > \hat{\mu}$, for $\hat{\mu} = -5$, $\hat{\sigma} = 1.2$, $\mu = 2$, $n = 100$.

   b. $H_1 : \mu \neq \hat{\mu}$, for $\hat{\mu} = -5$, $\hat{\sigma} = 1.2$, $\mu = 2$, $n = 100$.

2.9.3 Computational Exercises

The following problems should be solved with the help of a computer and appropriate software packages, such as MATLAB® *or* Excel®.

(31)   Consider the following distributions and their key parameters. Plot each distribution on a single plot and compare how the distribution changes the given parameter is changed.

   a.   **Normal distribution**: Change the standard deviation from 0.5 to 5 in increments of at least 0.5.

   b.   **Student's t-distribution**: Change the value of ν from 2 to 50 in increments of 2. Compare against each other as well as the standard normal distribution. How do the two distributions compare as $\nu \to \infty$.

   c.   **χ^2-distribution**: Change the value of ν from 3 to 20 in increments of at least 2.

   d.   **F-distribution**: Change of the value of ν_1 from 1 to 20, keeping ν_2 fixed at 5. Repeat with ν_2 between 1 and 20 and ν_1 fixed at 5. Compare the results. What happens to the two graphs?

(32)   Verify the central limit theorem for the following distributions: normal, χ^2-, and F-distributions. Compute the mean value for multiple samplings of these distributions. Do they converge to the normal distribution?

(33)   Compute the maximum likelihood estimators for the gamma distribution (see Question 24), Eq. (2.72) for the probability density function). Using a numerical solver, obtain parameter estimates for the following data set: 8.1864, 8.7553, 18.0286, 22.4389, 8.0564, 2.1472, 9.1224, 4.1870, 8.3551, 12.4235, 6.9026, 11.3712, 9.4377, 8.8809, 5.3927, 9.9001, 6.5891, 6.8874, 6.5011, and 7.3799.

(34)   Consider the following probability density function

$$f(x) = \begin{cases} 0.5 \sin(x) & 0 \le x \le \pi \\ 0 & \text{otherwise} \end{cases}$$

and address the following questions:

  a.   Show that $f(x)$ is indeed a probability density function.
  b.   Compute the mean value of this distribution.
  c.   If a parameter ψ is well-described by this distribution, compute the 95% confidence interval for ψ given $\hat{\psi} = 1$ and $\hat{\sigma}_{\hat{\psi}} = 1.0$. Could the true value of ψ be 0.5?
  d.   After sampling the above distribution 200 times, the mean value was obtained to be 1.1 with a variance of 0.5. Formally, test at the 95% level ($\alpha = 0.05$) whether this result equals the true value obtained in b). Note that you should clearly state what the null and alternative hypotheses are and what test needs to be performed.
  e.   Compute $P(0.25 < x < 0.5)$ for this probability density function.

Appendix A2: A Brief Review of Set Theory and Notation

In mathematics, sets are defined as a collection of objects that share some type of property. A set is delimited using curly brackets "{ }" and often denoted using double struck letters ($\mathbb{A}, \mathbb{B}, \mathbb{C},\ldots$). Common sets include:

(1) $\mathbb{R}$ (U + 211D),[7] which is the set of all real numbers;
(2) $\mathbb{N}$ (U + 2215), which is the set of all natural numbers and defined as $\mathbb{N} = \{0, 1, 2,\ldots\}$;
(3) $\mathbb{Z}$ (U + 2124), which is the set of all integers;
(4) $\mathbb{C}$ (U + 2102), the set of complex numbers; and
(5)   { } or $\emptyset$ (U + 2205), the empty set. This is used to represent a set that contains no members or elements.

---

[7] The values in brackets are the Unicode code points that will allow the given character to be easily entered on the keyboard.

The **element** operator $\in$ (U + 2208) states that a given variable is a member or **element** of the set, for example, $1 \in \mathbb{N}$, states that 1 belongs to (or is an element of) the set of natural numbers. The **exclusion** operator \ states that some given set is to be excluded, for example $\mathbb{N}\backslash\{0\}$ is the set of natural numbers excluding zero.

There are two common set operations: **union** and **intersection**. The **union** of two sets, denoted as $\cup$ (U + 222A), is the set that contains all elements found in both sets, while the **intersection** of two sets, denoted as $\cap$ (U + 2229), is the set that contains only those elements that are common (found) in both sets. For example, if $\mathbb{A} = \{1, 2, 3, 4\}$ and $\mathbb{B} = \{4, 5, 6, 7\}$, then $\mathbb{A} \cup \mathbb{B} = \{1, 2, 3, 4, 5, 6, 7\}$, while $\mathbb{A} \cap \mathbb{B} = \{4\}$.

Chapter 3
Regression

Regression is the method by which a set of data is fit to a relationship with unknown parameter values by minimising some criterion. The most common criterion is least squares, where the sum of the square of the residuals, that is the difference between the measured and predicted values, is minimised. Regression can be subdivided into two broad categories: **linear regression**, where the model of interest can be written as the sum of linear functions, and **nonlinear regression**, where this is not the case. In turn, there are two main methods to deal with regression: **ordinary, least-squares** and **weighted, least-squares** analysis.

3.1 Regression Analysis Framework

The regression analysis framework, shown in Fig. 3.1, is an iterative procedure that seeks to determine the best model for the data. Before the procedure can be started, three things must be determined:

(1) **Data**: what information is available about the system of interest and which variables can be measured? If no data sets are readily available, then it may be necessary to perform experiments to obtain the required data (see Chap. 4: Design of Experiments for further information).

(2) **Model**: what model (relationship) will be fit? How many parameters will be considered? The selection of an appropriate model is neither a trivial nor easy task. Luckily in many processes applications there may be some idea of what the model should like based on a theoretical analysis of the system.

(3) **Regression Method**: which method will be used to determine the model parameters? The selection of the correct method will impact the validity of the model obtained and what type of analysis can be performed.

Once these three aspects have been determined, the regression framework consists of three steps:

© The Author(s), under exclusive license to Springer Nature Switzerland AG 2022, corrected publication 2023
Y. A. W. Shardt, *Statistics for Chemical and Process Engineers*,
https://doi.org/10.1007/978-3-030-83190-5_3

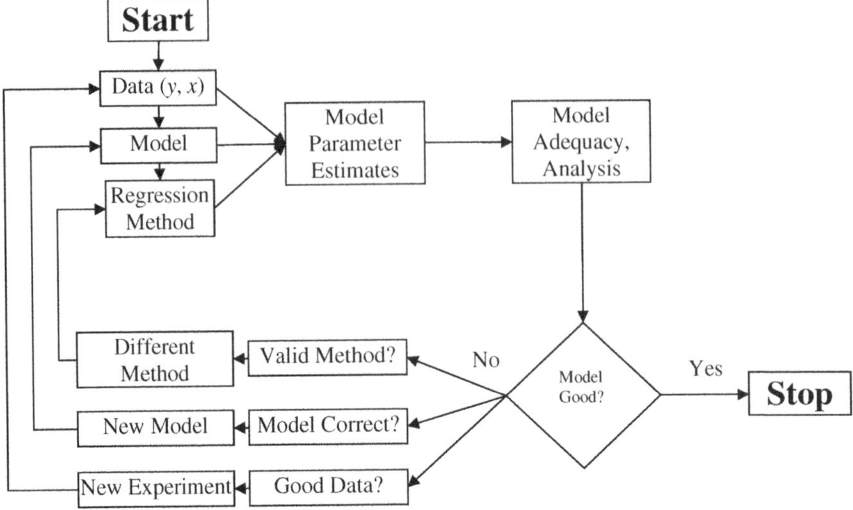

Fig. 3.1 Flowchart for regression analysis

(1) **Parameter Determination**: In this step, the model parameters are determined using an appropriate regression method. Values required for further analysis can also be computed at this point.

(2) **Model Validation**: In this step, the adequacy of the model is determined by considering two different aspects: the assumptions of the regression model and the overall fit. It is important to verify that the resulting regression satisfies the assumptions. Failure to satisfy the assumptions means that subsequent analysis may be incorrect. Assessing the overall fit seeks to determine how well the model fits the data.

(3) **Decision**: Based on the desired use of the model, the time available, and any other relevant factors, a decision needs to be made whether to accept the model as it currently is or to improve the model. Improving the model considers changing any of the three initial inputs: data, model, and regression method. For example, additional experiments could be performed to provide more data and help make a better decision or a more complex model could be selected in order to improve model performance.

3.2 Regression Models

Consider the following generalised representation of the regression problem

$$y = g\left(\vec{\beta}, \vec{x}, \varepsilon\right)$$

(3.1)

where.

(1) y is the **output** (or dependent) variable. This is the variable that we seek to model or describe. Often in process engineering, it will be a variable of importance such as concentration or quality.

(2) $\vec{x}$ is an $l \times 1$ vector containing the **regressor** (or independent) variables. These are the variables that can be used to describe how the output will behave. In many cases, irrelevant variables may be considered in order to be certain that a complete model has been obtained. The regressors need not have any physical relationship with the output; rather, they may solely be correlated with the output variable. If they are correlated, then the quality of the model may suffer once the correlation no longer holds.

(3) $\vec{\beta}$ is an $n \times 1$ vector containing the **parameters**, which are model constants whose values will be determined during regression. The parameters will most often be treated as statistical variables coming from a normal distribution, so that the significance of the results can be obtained.

(4) ε is the **error**. In practical situations, the output values obtained are subject to variability, that is, under the same conditions, it is unlikely that the same output values will be obtained due to measurement error, analysis errors, or the like. These errors mean that the predicted value given by g will be different from the measured values. The error is often assumed to be a stochastic, that is, random, variable whose values follow the normal distribution.

(5) g is the complete model that describes the relationship between the regressors, parameters, errors, and the output. The form of the model determines what type of analysis can be performed.

In order to simplify calculation, the complete model, g, is often split into two components. The first component, denoted as $f\left(\vec{\beta}_x, \vec{x}\right)$, and called the **regression model**, deals solely with the relationship between the inputs and outputs. For this reason, it is often called the **deterministic component** of the model. The second component, denoted as $k\left(\vec{\beta}_\varepsilon, \varepsilon\right)$, and called the **error structure or model**, deals solely with the relationship between the error and output. For this reason, it is often called the **stochastic component** of the model. There are two main ways to relate these two components to the complete model:

(1) **Additive Approach**: In the additive approach, it is assumed that $g = f + k$. This implies that the error is added solely to the output. In regression analysis, this is the only acceptable model.

(2) **Multiplicative Approach**: In the multiplicative approach, it is assumed that $g = fk$. This implies that the magnitude of the error depends on the value of the deterministic component. In regression analysis, this type of model must be re-arranged into an additive model in order to obtain a solution.

Although the stochastic component is important for analysing the results obtained using regression, it is useless from the perspective of using the model, as the exact value of the stochastic component can never be known. Therefore, although specifying the stochastic component is important for obtaining and analysing model parameters, it is often ignored when using the model to predict future or new values. In many applications, the two components can be modelled separately. The methods for modelling the stochastic component using the residuals will be presented in Chap. 5: Modelling Stochastic Processes. An application to the identification of a process for application in process control will be presented in Chap. 6: Modelling Dynamic Process. As well, some cases where both components need to be simultaneously modelled will be considered in that chapter.

Finally, assume that $m \geq n$ experiments were performed. Thus, there are m data points to be fit with n unknown parameter values. Once again, estimated values will be denoted using a circumflex ($\hat{o}$).

3.2.1 *Linear and Nonlinear Regression Functions*

A model is said to be **linear** if the first derivative of the regression model with respect to the parameters $\vec{\beta}$ is not a function of any of the parameters. A regression model whose first derivatives with respect to the parameters depend on the parameters is called a **nonlinear** model. For example, $y = \beta_0 + \beta_1 x + \beta_2 x^2$ is a linear model in terms of the parameters, since none of the derivatives with respect to the parameters depend on the parameters. On the other hand, $y = \beta_0 e^{-\beta x}$ is nonlinear since the derivatives with respect to the parameters depend on the parameter estimates (*e.g.* $\frac{\partial y}{\partial \beta_0} = e^{-\beta x}$).

A nonlinear model can occasionally be converted into a linear model by transforming the model to obtain a linear model. However, it should be noted that transforming the model introduces problems into the model by changing the error structure, which may imply that the assumptions of linear regression are not satisfied. In this case, the parameter estimates obtained from a linearised model can then be used as the initial guess for the nonlinear regression case.

Example 3.1 Linearising Nonlinear Models

Consider the following nonlinear models. Linearise them so that linear regression methods can be applied. Explain in which cases the error structure will be that of the standard, linear model (additive error) and where it will not be.

(1) **Arrhenius's Equation I:** $K = Ae^{\varepsilon}e^{-\frac{E_a}{RT}}$, where K is the reaction rate, A the reaction constant, E_a the activation energy, R the universal gas constant, and T the temperature.

(2) **Arrhenius's Equation II:** $K = Ae^{-\frac{E_a}{RT}} + \varepsilon$, where K is the reaction rate, A the reaction constant, E_a the activation energy, R the universal gas constant, and T the temperature.

(3) **Michaelis–Menten Equation**: $K = \frac{K_{\max} S}{K_m + S} + \varepsilon$, where $K_{\max}$ is the maximum reaction rate, S the substrate concentration, and K_m a reaction constant.

(4) $y = A \varepsilon x^b$, where y is the output, A a constant, and b the exponent. Show that this is a nonlinear equation.

Solution

(1) **Arrhenius's Equation I**: For the first equation, taking the natural logarithm of this equation will give

$$\ln K = \ln\left(A e^{\varepsilon} e^{-\frac{E_a}{RT}}\right) = \ln A + \ln e^{\varepsilon} + \ln e^{-\frac{E_a}{RT}} = \ln A - \frac{E_a}{RT} + \varepsilon$$

which by defining

$$\vec{y} = \langle \ln K \rangle, \, \vec{\beta} = \left(\ln A \; -\frac{E_a}{R} \right)^T, \, \vec{x} = \left(1 \; T^{-1} \right)$$

can be written in the standard linear regression format. Therefore, if we assume that the original error structure is lognormal, as in this example, then the system will be linearisable, and the linear regression methods will completely apply.

(2) **Arrhenius's Equation II**: The second equation, which is the same as the first equation, except that the error structure is different now: it is additive rather than multiplicative. Therefore, when the logarithm is taken, the error term will remain inside the logarithm, namely,

$$\ln K = \ln\left(A e^{-\frac{E_a}{RT}} + \varepsilon\right)$$

and the complete sum cannot be simplified in order to obtain a solution that includes the error. In such cases, it is common to simply ignore the error structure and proceed to linearise the equation. The final result will be the same as was obtained for (1).

(3) **Michaelis–Menten Equation**: Since this is a very common equation in biochemical engineering, different forms of linearisation have been devised. In this example, two different linearisation will be considered.

Form 1: The following steps can be taken to linearise the equation:

$$K = \frac{K_{\max} S}{K_m + S} + \varepsilon$$

$$K S = K_{\max} - K K_m + (K_m + S)\varepsilon.$$

Let

$$\vec{y} = \langle K\,S \rangle, \vec{\beta} = \begin{pmatrix} K_{\max} & K_m \end{pmatrix}^T, \vec{x} = \begin{pmatrix} 1 & -K \end{pmatrix}.$$

Note that the error structure, although additive, has a variance that depends on the value of the unknown parameter K_m.

Form 2: Take the following steps to linearise the equation

$$\frac{1}{K} = \frac{K_m + S}{K_{\max} S + (K_m + S)\varepsilon}$$

$$\frac{1}{K} = \frac{K_m}{K_{\max} S + (K_m + S)\varepsilon} + \frac{1}{K_{\max} + \frac{(K_m + S)}{S}\varepsilon}.$$

If we ignore the fact that the error is in the denominator and proceed by dropping it, we can see that

$$\vec{y} = \begin{pmatrix} K^{-1} \end{pmatrix}, \vec{\beta} = \begin{pmatrix} K_{\max}^{-1} & K_m K_{\max}^{-1} \end{pmatrix}^T, \vec{x} = \begin{pmatrix} 1 & S^{-1} \end{pmatrix}.$$

This approach is called the Lineweaver–Burk plot. It can be seen that the error structure is not additive and there is a need to take the inverse of potentially small numbers which can introduce further errors. In general, none of the methods is very good at obtaining accurate parameter values, and nonlinear regression is better.

(4) $y = A\varepsilon x^b$: For this equation, take the following steps to linearise it

$$\log y = \log A + \log \varepsilon + \log x^b$$
$$= \log A + \log \varepsilon + b \log x.$$

Let

$$\vec{y} = \langle \log y \rangle, \vec{\beta} = \begin{pmatrix} \log A & b \end{pmatrix}^T, \vec{x} = \begin{pmatrix} 1 & \log x \end{pmatrix}.$$

Note that the error structure, although additive, has a different form. It is no longer a normal distribution. As well, there can be issues with negative numbers, as although the original equation can deal with them, the linearised version cannot.

To show that the given equation is nonlinear, compute the derivatives of the function with respect to the parameters, that is,

$$\frac{\partial y}{\partial A} = \varepsilon x^b$$

$$\frac{\partial y}{\partial b} = Ab\varepsilon x^{b-1}.$$

From these two equations, it is obvious that the derivatives depend on the parameters, and so the resulting system is nonlinear.

3.3 Linear Regression

In linear regression, it is assumed that the function of interest can be written as:

$$y = \sum_{i=1}^{n} \beta_i f_i(\vec{x}) + \varepsilon = \vec{a}\vec{\beta} + \varepsilon \tag{3.2}$$

where the individual functions f_i are all known, and $\vec{a}$ is the vector containing the values of the functions, that is, $\vec{a} = \langle f_1(\vec{x}), f_2(\vec{x}), \ldots, f_n(\vec{x}) \rangle$. Assume that there are $m \geq n$ data points that are to be fit to the model given by Eq. (3.2). The data points are given as $\vec{x}_j$, where $j = 1, 2, 3, \ldots, m$, and the individual parameters are given as β_i, where $i = 1, 2, 3, \ldots, n$.

3.3.1 Ordinary, Least-Squares Regression

This is the most common type of linear regression. In ordinary, least-squares regression, the objective function to be optimised is given as

$$\min_{\vec{\beta}} \left(\left(\vec{y} - \mathcal{A}\vec{\beta} \right)^T \left(\vec{y} - \mathcal{A}\vec{\beta} \right) \right) \tag{3.3}$$

where $\mathcal{A}$, denoted as the **regression matrix**, is defined as

$$\mathcal{A} = \begin{bmatrix} f_1(\vec{x}_1) & f_2(\vec{x}_1) & \cdots & f_n(\vec{x}_1) \\ f_1(\vec{x}_2) & f_2(\vec{x}_2) & \cdots & f_n(\vec{x}_2) \\ \vdots & & \ddots & \vdots \\ f_1(\vec{x}_m) & f_2(\vec{x}_m) & \cdots & f_n(\vec{x}_m) \end{bmatrix} \tag{3.4}$$

and

$$\vec{\beta} = \langle \beta_1, \beta_2, \ldots, \beta_n \rangle^T \tag{3.5}$$

$$\vec{y} = \langle y_1, y_2, ..., y_m \rangle^T \tag{3.6}$$

Theorem 3.1 *The solution for Eq. (3.3) is*

$$\hat{\vec{\beta}} = \left(A^T A\right)^{-1} A^T \vec{y}. \tag{3.7}$$

Proof In order to find the solution to the optimisation problem given by Eq. (3.3), it is necessary to determine the points at which the derivative of the objective function is zero and then solve for the desired unknown values.

Therefore, taking the derivative of Eq. (3.3) with respect to the unknown parameters, $\vec{\beta}$, gives

$$\frac{\partial}{\partial\vec{\beta}}\left(\vec{y} - A\vec{\beta}\right)^T\left(\vec{y} - A\vec{\beta}\right) = \frac{\partial}{\partial\vec{\beta}}\left(\vec{y}^T\vec{y} - \vec{y}^T A\vec{\beta} - \vec{\beta}^T A^T\vec{y} + \vec{\beta}^T A^T A\vec{\beta}\right)$$
$$= -2A^T\vec{y} + 2A^T A\vec{\beta}. \tag{3.8}$$

Setting Eq. (3.8) equal to zero and assuming that $A^T A$ is full-rank, that is, invertible, gives

$$-2A^T\vec{y} + 2A^T A\vec{\beta} = 0$$
$$A^T A\vec{\beta} = A^T\vec{y}$$
$$\vec{\beta} = \left(A^T A\right)^{-1} A^T\vec{y}, \tag{3.9}$$

which is equivalent to Eq. (3.7).

$$Q.E.D.$$

In order to analyse the resulting parameter estimates, it is necessary to make four assumptions about the underlying error structure:

(1)   The errors have a mean of zero.
(2)   The errors are independent.
(3)   The errors are homoscedastic, that is, they have the same variance.
(4)   The errors are normally distributed.

The first three assumptions are required to obtain an understanding of the properties of the least-squares estimate, while the last assumption allows for hypothesis testing to performed on the estimates, as well as making the regression estimates equal to the maximum likelihood parameter estimates.

Theorem 3.2 *Under the assumptions stated above, the regression parameter estimates are unbiased.*

Proof To show that the parameter estimates are unbiased, $E(\hat{\theta}) = \theta$ must be satisfied. Substituting the parameter estimates given by Eq. (3.7) into this equation gives

$$E\left((A^T A)^{-1} A^T \vec{y}\right). \tag{3.10}$$

Substituting the true model $\vec{y} = A\vec{\beta} + \varepsilon$ into Eq. (3.10) and simplifying gives

$$E\left((A^T A)^{-1} A^T \left(A\vec{\beta} + \varepsilon\right)\right) = E\left((A^T A)^{-1} A^T A\vec{\beta} + (A^T A)^{-1} A^T \varepsilon\right)$$
$$= \vec{\beta} + E\left((A^T A)^{-1} A^T \varepsilon\right). \tag{3.11}$$

Since it has been assumed that $E(\varepsilon) = 0$, Eq. (3.11) becomes $\vec{\beta}$, which is the true parameter value. This shows that the parameter estimates are unbiased.
Q.E.D.

Theorem 3.3 *Under the assumptions stated above, the covariance matrix of the regression parameter estimates is*

$$\sigma_{\hat{\vec{\beta}}}^2 = \sigma^2 (A^T A)^{-1} \tag{3.12}$$

where σ^2 is the variance of the error.

Proof From Eq. (2.23), the variance of the parameters can be written as

$$\sigma_{\hat{\vec{\beta}}}^2 = E\left(\hat{\vec{\beta}}\hat{\vec{\beta}}^T\right) - E\left(\hat{\vec{\beta}}\right) E\left(\hat{\vec{\beta}}\right)^T. \tag{3.13}$$

Making the same substitutions as in Theorem 3.2 for $\hat{\vec{\beta}}$, and noting that from that theorem we have the value for $E\left(\hat{\vec{\beta}}\right)$ gives

$$\sigma_{\hat{\vec{\beta}}}^2 = E\left((A^T A)^{-1} A^T \left(A\vec{\beta} + \varepsilon\right)\left((A^T A)^{-1} A^T \left(A\vec{\beta} + \varepsilon\right)\right)^T\right) - \vec{\beta}\vec{\beta}^T. \tag{3.14}$$

Multiplying out and simplifying gives

$$\sigma_{\hat{\vec{\beta}}}^2 = E\left((A^T A)^{-1} A^T \left(A\vec{\beta} + \varepsilon\right)\left(A\vec{\beta} + \varepsilon\right)^T A\left((A^T A)^{-1}\right)^T\right) - \vec{\beta}\vec{\beta}^T$$

$$\sigma_{\hat{\vec{\beta}}}^2 = E\left(\begin{array}{l} \vec{\beta}\vec{\beta}^T + \vec{\beta}\varepsilon^T A(A^T A)^{-1} \\ +(A^T A)^{-1} A^T \varepsilon\vec{\beta}^T \\ +(A^T A)^{-1} A^T \varepsilon\varepsilon^T A(A^T A)^{-1} \end{array}\right) - \vec{\beta}\vec{\beta}^T. \tag{3.15}$$

Applying the linearity of the expectation operator gives

$$\sigma_{\hat{\vec{\beta}}}^2 = E\left(\vec{\beta}\vec{\beta}^T\right) + E\left(\vec{\beta}\varepsilon^T A(A^T A)^{-1}\right) + E\left((A^T A)^{-1} A^T \varepsilon \vec{\beta}^T\right)$$
$$+ E\left((A^T A)^{-1} A^T \varepsilon \varepsilon^T A(A^T A)^{-1}\right) - \vec{\beta}\vec{\beta}^T. \tag{3.16}$$

Since $E(\varepsilon) = 0$, the above equation simplifies to

$$\sigma_{\hat{\vec{\beta}}}^2 = E\left((A^T A)^{-1} A^T \varepsilon \varepsilon^T A(A^T A)^{-1}\right). \tag{3.17}$$

Since it is assumed that the individual errors (residuals) are independent, have a mean of zero, and have the same error covariance, the error covariance can be rewritten as

$$\sigma_{\varepsilon}^2 = E\left(\varepsilon \varepsilon^T\right) = \sigma^2 \mathcal{I}. \tag{3.18}$$

Therefore, the covariance of the parameters can be rewritten as

$$\sigma_{\hat{\vec{\beta}}}^2 = \sigma_{\varepsilon}^2 (A^T A)^{-1} A^T A(A^T A)^{-1}$$
$$= \sigma^2 (A^T A)^{-1}. \tag{3.19}$$

$$Q.E.D.$$

Further, if it is assumed that the errors follow a normal distribution, Theorem 3.3 allows the confidence intervals for the parameter estimates to be established, that is,

$$\hat{\beta}_i \pm t_{1-\frac{\alpha}{2}, m-n} \hat{\sigma} \sqrt{(A^T A)_{ii}^{-1}} \tag{3.20}$$

where i refers to the ith parameter estimate, $(A^T A)_{ii}^{-1}$ represents the entry in the ith column and row of the $(A^T A)^{-1}$ matrix, and $\hat{\sigma}$ is the estimated value of the error standard deviation.

Having established the properties of the parameter estimates, it will now be useful to look at the predictive properties of the overall model: what are the values and confidence intervals for a given prediction. Given a new data point, $\vec{x}_d$, there are two different measures for answering this question: the **mean response** and the **predicted response confidence intervals**. The mean response represents the average value at a given point given the available data, that is, $E(y|\vec{x}_d)$. Essentially, this represents the value that would be obtained if multiple experiments are repeated at similar conditions with the same input values. The predicted response represents the value obtainable from a single new experiment. In all cases, the confidence interval for a predicted response will be larger than that of the mean response, since there is additional uncertainty in the predicted value.

Theorem 3.4 *Under the assumptions stated above, the mean response and its confidence interval for the point $\vec{x}_d$ are given as.*

$$\vec{a}_{\vec{x}_d}\hat{\vec{\beta}} \pm t_{1-\frac{\alpha}{2},m-n}\hat{\sigma}\sqrt{\vec{a}_{\vec{x}_d}(A^T A)^{-1}\vec{a}_{\vec{x}_d}^T} \tag{3.21}$$

where $\vec{a}_{\vec{x}_d} =< f_1(\vec{x}_d), f_2(\vec{x}_d), \ldots, f_n(\vec{x}_d) >$.

Proof This requires showing two things: the mean response value and its associated confidence interval. First, consider the mean response value given by $E(y|\vec{x}_d) = \mu_{Y|\vec{x}_d}$. The best estimate for this value is the mean value obtained from the regression equation, that is,

$$E\left(y|\vec{a}_{\vec{x}_d}\right) = \vec{a}_{\vec{x}_d}\hat{\vec{\beta}}. \tag{3.22}$$

This is an unbiased estimate for the mean response since from Theorem 3.2 we have that $\hat{\vec{\beta}}$ is an unbiased estimate of $\vec{\beta}$. This establishes the best estimate of the mean response.

Next, consider the confidence interval. In order to do this, we need to determine the variance of the mean response estimate, that is, $\sigma^2_{\mu_{Y|\vec{x}_d}}$,

$$\sigma^2_{\mu_{Y|\vec{x}_d}} = E\left(\vec{a}_{\vec{x}_d}\hat{\vec{\beta}}\left(\vec{a}_{\vec{x}_d}\hat{\vec{\beta}}\right)^T\right) - E\left(\vec{a}_{\vec{x}_d}\hat{\vec{\beta}}\right)E\left(\vec{a}_{\vec{x}_d}\hat{\vec{\beta}}\right)^T. \tag{3.23}$$

Simplifying Eq. (3.23) by noting that $\vec{a}_{\vec{x}_d}$ is a constant gives

$$\sigma^2_{\mu_{Y|\vec{x}_d}} = \vec{a}_{\vec{x}_d}E\left(\hat{\vec{\beta}}\hat{\vec{\beta}}^T\right)\vec{a}_{\vec{x}_d}^T - \vec{a}_{\vec{x}_d}E\left(\hat{\vec{\beta}}\right)E\left(\hat{\vec{\beta}}\right)^T\vec{a}_{\vec{x}_d}^T. \tag{3.24}$$

Rewriting Eq. (3.24) by noting that the variance of the parameters can be written as $\sigma^2_{\hat{\vec{\beta}}} = E\left(\hat{\vec{\beta}}\hat{\vec{\beta}}^T\right) - E\left(\hat{\vec{\beta}}\right)E\left(\hat{\vec{\beta}}\right)^T$ gives

$$\sigma^2_{\mu_{Y|\vec{x}_d}} = \vec{a}_{\vec{x}_d}\sigma^2_{\hat{\vec{\beta}}}\vec{a}_{\vec{x}_d}^T. \tag{3.25}$$

Theorem 3.3, which gives the parameter variance in terms of the variance of the errors, allows Eq. (3.25) to be rewritten as

$$\sigma^2_{\mu_{Y|\vec{x}_d}} = \sigma^2\vec{a}_{\vec{x}_d}(A^T A)^{-1}\vec{a}_{\vec{x}_d}^T. \tag{3.26}$$

This establishes the variance of the mean response. In practice, σ^2 is not known and must be replaced by its estimated value, $\hat{\sigma}^2$. It can be noted that since the parameter estimates themselves are normally distributed (under the assumption that the errors themselves are normally distributed) then the confidence interval can be written as

$$\vec{a}_{\vec{x}_d}\hat{\beta} \pm t_{1-\frac{\alpha}{2},m-n}\hat{\sigma}\sqrt{\vec{a}_{\vec{x}_d}(A^T A)^{-1}\vec{a}_{\vec{x}_d}^T}. \tag{3.27}$$

Q.E.D.

Theorem 3.5 *Under the assumptions stated above, the predicted response and its confidence interval for the point $\vec{x}_d$ is given as*

$$\vec{a}_{\vec{x}_d}\hat{\beta} \pm t_{1-\frac{\alpha}{2},m-n}\hat{\sigma}\sqrt{1 + \vec{a}_{\vec{x}_d}(A^T A)^{-1}\vec{a}_{\vec{x}_d}^T} \tag{3.28}$$

where $\vec{a}_{\vec{x}_d} =< f_1(\vec{x}_d), f_2(\vec{x}_d), \ldots, f_n(\vec{x}_d) >$.

Proof This requires showing two things: the predicted response value and its associated confidence interval. First, the predicted response value will be the same as for the mean response value and so will be unbiased.

For the confidence interval, consider the difference between the true value, y, and the predicted value, $\hat{y}$, at the point, that is,

$$\varepsilon_p = y - \hat{y}. \tag{3.29}$$

The variance of ε will determine the variance of the predicted value estimate. The required variance can be obtained as follows

$$\begin{aligned}
\sigma_{\varepsilon_p}^2 &= E\left((y - \hat{y})(y - \hat{y})^T\right) - E(y - \hat{y})E(y - \hat{y})^T \\
&= \sigma_y^2 + \sigma_{\hat{y}}^2.
\end{aligned} \tag{3.30}$$

From Theorem 3.4, the value of $\sigma_{\hat{y}}^2$ will be equal to that given by Eq. (3.26). The only remaining term will be the variance due to the true value given by σ_y^2. This can be computed as follows:

$$\begin{aligned}
\sigma_y^2 &= E\left(yy^T\right) - E(y)E(y)^T \\
&= E\left((\vec{a}\vec{\beta} + \varepsilon)(\vec{a}\vec{\beta} + \varepsilon)^T\right) - E(\vec{a}\vec{\beta} + \varepsilon)E(\vec{a}\vec{\beta} + \varepsilon)^T \\
&= E\left(\vec{a}\vec{\beta}\vec{\beta}^T\vec{a}^T + \varepsilon\vec{\beta}^T\vec{a}^T + \vec{a}\vec{\beta}\varepsilon^T + \varepsilon\varepsilon^T\right) - E(\vec{a}\vec{\beta} + \varepsilon)E(\vec{a}\vec{\beta} + \varepsilon)^T.
\end{aligned} \tag{3.31}$$

It can be noted that the terms that do not contain ε are all constant. Furthermore, by assumption, ε has a mean value of 0. Therefore, Eq. (3.31) can be rewritten as

$$\sigma_y^2 = \vec{a}\vec{\beta}\vec{\beta}^T\vec{a}^T + \varepsilon\varepsilon^T - \vec{a}\vec{\beta}\vec{\beta}^T\vec{a}^T$$
$$= \varepsilon\varepsilon^T = \sigma^2. \tag{3.32}$$

Thus, the variance of the predicted response value can be written as

$$\sigma_{\varepsilon_p}^2 = \sigma^2 + \sigma^2 \vec{a}_{\vec{x}_d}\left(A^T A\right)^{-1}\vec{a}_{\vec{x}_d}^T = \sigma^2\left(1 + \vec{a}_{\vec{x}_d}\left(A^T A\right)^{-1}\vec{a}_{\vec{x}_d}^T\right). \tag{3.33}$$

In practice, σ^2 is not known and must be replaced by its estimated value, $\hat{\sigma}^2$. Similarly to the mean response, noting that both the parameter estimates and the errors are normally distributed, then the confidence interval can be written as

$$\vec{a}_{\vec{x}_d}\hat{\vec{\beta}} \pm t_{1-\frac{\alpha}{2},m-n}\hat{\sigma}\sqrt{1 + \vec{a}_{\vec{x}_d}\left(A^T A\right)^{-1}\vec{a}_{\vec{x}_d}^T}. \tag{3.34}$$

Q.E.D.

The **residual** is defined as the difference between the predicted and measured values, that is,

$$\vec{\varepsilon} = \vec{y} - \hat{\vec{y}} = \vec{y} - A\hat{\vec{\beta}} = \vec{y} - A\left(A^T A\right)^{-1}A^T\vec{y} = \left(\mathcal{I} - A\left(A^T A\right)^{-1}A^T\right)\vec{y} \tag{3.35}$$

where $\mathcal{I}$ is the appropriately sized identity matrix.

3.3.2 Analysis of Variance of the Regression Model

Analysis of variance is an approach to determining the significance and validity of a regression model using variances obtained from the data and model. The goal is to simultaneously test multiple means in order to determine the overall significance. Unfortunately, the naïve approach of simply comparing pairwise each mean can easily lead to too large an overall α-error. For this reason, the analysis of variance approach seeks to decompose the total variability in the data into various orthogonal components that can then be independently analysed. For the purposes of analysing the regression, let the **total sum of squares**, denoted by *TSS*, be defined as

$$TSS = \sum (y_i - \bar{y})^2 = \left\|\vec{y} - \bar{y}\vec{1}\right\|^2 \tag{3.36}$$

where $\vec{1}$ is the unit column vector, that is, $\vec{1} = \langle 1, 1, 1, \ldots, 1\rangle^T$ and $\|\cdot\|$ is the standard, vector two-norm. Let the **sum of squares due to regression**, *SSR*, be defined as

$$SSR = \sum \left(\hat{y}_i - \bar{y}\right)^2 = \left\|\hat{\vec{y}} - \bar{y}\vec{1}\right\|^2. \tag{3.37}$$

Let the **sum of squares due to the error**, *SSE*, be defined as

$$SSE = \sum (y_i - \hat{y}_i)^2 = \|\vec{\varepsilon}\|^2 = \left\| \vec{y} - \hat{\vec{y}} \right\|^2. \tag{3.38}$$

Theorem 3.6 *SSR and SSE are an orthogonal decomposition of TSS.*

Proof In order to simplify the computations, the proof will be performed using suitable vector manipulations. By definition, we can rewrite the total sum of squares as

$$
\begin{aligned}
TSS &= \left\| \vec{y} - \bar{y}\vec{1} \right\|^2 \\
&= \left\| \vec{y} - \bar{y}\vec{1} + \hat{\vec{y}} - \hat{\vec{y}} \right\|^2 \\
&= \left\| \left(\vec{y} - \hat{\vec{y}} \right) + \left(\hat{\vec{y}} - \bar{y}\vec{1} \right) \right\|^2.
\end{aligned} \tag{3.39}
$$

By the polarisation identity between norms and dot products $\|a + b\|^2 = \|a\|^2 + \|b\|^2 + 2b^T a$,[1] Eq. (3.39) can be rewritten as

$$TSS = \left\| \left(\vec{y} - \hat{\vec{y}} \right) \right\|^2 + \left\| \left(\hat{\vec{y}} - \bar{y}\vec{1} \right) \right\|^2 + 2\left(\vec{y} - \hat{\vec{y}} \right)^T \left(\hat{\vec{y}} - \bar{y}\vec{1} \right). \tag{3.40}$$

It is now necessary to show that the last term in Eq. (3.40) equals zero. This can be accomplished by writing the last term as

$$\left(\vec{y} - \hat{\vec{y}} \right)^T \left(\hat{\vec{y}} - \bar{y}\vec{1} \right) = \left(\vec{y} - \hat{\vec{y}} \right)^T \hat{\vec{y}} + \left(\vec{y} - \hat{\vec{y}} \right)^T \bar{y}\vec{1}. \tag{3.41}$$

Since $\hat{\vec{y}} = A\hat{\vec{\beta}}$, the first part of the last term simplifies to

$$
\begin{aligned}
\left(\vec{y} - \hat{\vec{y}} \right)^T A\hat{\vec{\beta}} &= \left(\left(I - (A^T A)^{-1} A^T \right) \vec{y} \right)^T A\hat{\vec{\beta}} = \vec{y}^T \left(I - A(A^T A)^{-1} \right) A\hat{\vec{\beta}} \\
&= \vec{y}^T \left(A - A(A^T A)^{-1} A^T A \right) \hat{\vec{\beta}} \\
&= \vec{y}^T (A - A)\hat{\vec{\beta}} = 0.
\end{aligned} \tag{3.42}
$$

For the second part of the last term, it can be noted that $\sum \varepsilon_i = \left(\vec{y} - \hat{\vec{y}} \right)^T \vec{1} = 0$, by the assumptions of linear regression. This shows that the second part is also equal to zero. Thus, the last term in Eq. (3.40) is equal to zero. This shows that

---

[1] For two column vectors a and b, the dot product $a \cdot b$ can be defined as the matrix multiplication $a^T b$ or $b^T a$.

$$TSS = \left\| \left(\vec{y} - \hat{\vec{y}} \right) \right\|^2 + \left\| \left(\hat{\vec{y}} - \overline{y}\vec{1} \right) \right\|^2 = SSE + SSR. \qquad (3.43)$$

Q.E.D.

It can be noted that the sum of squares due to regression can be further partitioned if an orthogonal basis is used to define the regression parameters. This will be explored in greater detail in Chap. 4, including how to define an orthogonal basis for regression.

In the analysis of variance approach, the **F-statistic** can be calculated as follows:

$$\boxed{F = \frac{SSR/k}{SSE/m - n}} \qquad (3.44)$$

where k is defined as follows. If there is a function, such that $f_i(x) = 1$ for all values of x, then $k = n-1$; otherwise, $k = n$. This value should be compared to the critical F-value computed as $F_{1 - \alpha, k, m-n}$, where α is the α-error. The F-statistic computed using Eq. (3.44) should be greater than the critical value to conclude that the parameter estimates are significant compared to the noise in the system. Basically, the analysis of variance seeks to determine which of the two components, the regression model or the noise variance, is greater. A good regression model should give a larger SSR value than a bad regression model.

Another useful measure of the regression model is **Pearson's coefficient of regression**, R^2. It can be calculated as follows:

$$\boxed{R^2 = \frac{SSR}{TSS} = 1 - \frac{SSE}{TSS}.} \qquad (3.45)$$

The closer the value is to 1, the better the regression model is. Furthermore, R^2 measures the fraction of the total variance in the model explained by the regression with the given variables. The value of R^2 lies between $[0, 1]$. Unfortunately, this parameter has the tendency that, as the number of parameters is increased, the value of R^2 approaches 1. This would suggest that the model is improved by an increase in the number of parameters. However, if there are exactly n data points, then any n-parameter model will fit the data quite closely. A possible approach to fix this problem is to calculate the adjusted R^2, which is given as

$$\boxed{R^2_{adj} = 1 - \frac{SSE/m - n}{TSS/m - 1} = 1 - \left(1 - R^2\right)\left(\frac{m - 1}{m - n}\right).} \qquad (3.46)$$

It should be noted that the adjusted R^2 is not constrained to lie in the interval $[0, 1]$. Thus, negative values can be expected using this measure.

3.3.3   Useful, Formulae for Ordinary, Least-Squares Regression

This section provides a convenient summary of all the equations required for ordinary, least-squares regression.

For the regression model given by

$$y = \sum_{i=1}^{n} \beta_i f_i(\vec{x}) + \varepsilon = \vec{a}\vec{\beta} + \varepsilon \tag{3.47}$$

the solution can be obtained by solving the following equation for the unknown, estimated coefficients, $\hat{\vec{\beta}}$:

$$A^T A \hat{\vec{\beta}} = A^T \vec{y} \tag{3.48}$$

where

$$A = \begin{bmatrix} f_1(\vec{x}_1) & f_2(\vec{x}_1) & \cdots & f_n(\vec{x}_1) \\ f_1(\vec{x}_2) & f_2(\vec{x}_2) & \cdots & f_n(\vec{x}_2) \\ \vdots & & \ddots & \vdots \\ f_1(\vec{x}_m) & f_2(\vec{x}_m) & \cdots & f_n(\vec{x}_m) \end{bmatrix} \tag{3.49}$$

$$\hat{\vec{\beta}} = < \beta_1, \beta_2, ..., \beta_n >^T \tag{3.50}$$

$$\vec{y} = < y_1, y_2, \ldots, y_m >^T . \tag{3.51}$$

The **standard deviation** for this model is given by

$$\hat{\sigma} = \sqrt{\frac{\vec{y}^T \vec{y} - \hat{\vec{\beta}}^T A^T \vec{y}}{m - n}} = \sqrt{\frac{\vec{\varepsilon}^T \vec{\varepsilon}}{m - n}} \tag{3.52}$$

where $\vec{\varepsilon}$ are the residuals given as

$$\vec{\varepsilon} = \vec{y} - \hat{\vec{y}} = \vec{y} - A\hat{\vec{\beta}} = \vec{y} - A\left(A^T A\right)^{-1} A^T \vec{y} = \left(I - A\left(A^T A\right)^{-1} A^T\right)\vec{y}. \tag{3.53}$$

The **100(1−α)% Confidence Interval for** β_i is given by

$$\hat{\beta}_i \pm t_{1-\frac{\alpha}{2},m-n}\hat{\sigma}\sqrt{(\mathcal{A}^T\mathcal{A})_{ii}^{-1}} \qquad (3.54)$$

where $(\mathcal{A}^T\mathcal{A})_{ii}^{-1}$ represents the value located at (i, i) in the matrix $(\mathcal{A}^T\mathcal{A})^{-1}$.

The **100(1−α)% Mean Response Confidence Intervals**, that is, the region within which the mean value of repeated measurements will lie $100(1-\alpha)\%$ of the time, for the point $\vec{x}_d$ is given by

$$\vec{a}_{\vec{x}_d}\hat{\vec{\beta}} \pm t_{1-\frac{\alpha}{2},m-n}\hat{\sigma}\sqrt{\vec{a}_{\vec{x}_d}(\mathcal{A}^T\mathcal{A})^{-1}\vec{a}_{\vec{x}_d}^T} \qquad (3.55)$$

where $\vec{a}_{\vec{x}_d} =< f_1(\vec{x}_d), f_2(\vec{x}_d), \ldots, f_n(\vec{x}_d) >$. It can be noted that if the parameter estimates are obtained using some transformed equation, then the mean response confidence interval cannot be converted into the original units.

The **100(1−α)% Predictive Confidence Intervals**, that is, the region within which the actual value will lie $100(1-\alpha)\%$ of the time, for the point $\vec{x}_d$ is given by

$$\vec{a}_{\vec{x}_d}\hat{\vec{\beta}} \pm t_{1-\frac{\alpha}{2},m-n}\hat{\sigma}\sqrt{1 + \vec{a}_{\vec{x}_d}(\mathcal{A}^T\mathcal{A})^{-1}\vec{a}_{\vec{x}_d}^T}. \qquad (3.56)$$

The **sum of squares due to regression**, *SSR*, can be calculated as follows:

$$SSR = \sum (\hat{y}_i - \bar{y})^2 = \left\|\hat{\vec{y}} - \bar{y}\vec{1}\right\|^2 = \hat{\vec{\beta}}^T\mathcal{A}^T\vec{y} - \frac{1}{m}\left(\vec{y}^T\vec{1}\vec{1}^T\vec{y}\right) \qquad (3.57)$$

where $\vec{1}$ is the unit column vector of size $m\times 1$, that is, $\vec{1} = \langle 1, 1, 1, \ldots, 1\rangle^T$.

The **sum of squares due to the errors**, *SSE*, can be calculated as follows:

$$SSE = \sum (y_i - \hat{y}_i)^2 = \left\|\vec{y} - \hat{\vec{y}}\right\|^2 = \vec{y}^T\vec{y} - \hat{\vec{\beta}}^T\mathcal{A}^T\vec{y}. \qquad (3.58)$$

The **total sum of squares**, *TSS*, can be calculated as follows:

$$TSS = \sum (y_i - \bar{y})^2 = \left\|\vec{y} - \bar{y}\vec{1}\right\|^2 = \vec{y}^T\vec{y} - \frac{1}{m}\left(\vec{y}^T\vec{1}\vec{1}^T\vec{y}\right). \qquad (3.59)$$

Appendix A3.1 presents a traditional, nonmatrix, approach to performing univariate (one variable), linear, least-squares regression.

3.3.4 Computational Example Part I: Determining the Model Parameters

Consider the problem of trying to estimate the yield strength of a new steel compound based on the addition of compounds A and B. A series of experiments were performed for different amounts of compounds A and B and the yield strength of the resulting compound was measured. Based on previous knowledge, it is known that the yield strength can be modelled as

$$\sigma_y = \beta_0 + \beta_1 x_A + \beta_2 x_B \tag{3.60}$$

where σ_y is the yield strength, x_A the amount of compound A in mg that was added, and x_B the amount of compound B in mg that was added. The data are provided in Table 3.1. Using these data, determine the parameter estimates and their confidence intervals, the standard deviation of the model, for the point, $x_A = 5.00$ mg and $x_B = 10.00$ mg, determine the mean and predictive confidence intervals for the yield, obtain SSR, TSS, SSE, R^2, and the F-statistic, and provide the model significance at $\alpha = 0.05$.

Solution

The first step in obtaining a solution is to convert the given problem into its corresponding mathematical description using the standard notation, which gives

$$y = \sigma_y$$
$$\vec{\beta} = \langle \beta_0, \beta_1, \beta_2 \rangle^T$$
$$\vec{a} = \langle 1, x_A, x_B \rangle. \tag{3.61}$$

Therefore, writing the matrices for the ten data points gives

Table 3.1 Yield strength data

Yield strength, σ_y (MPa)	Compound A, x_A (mg)	Compound B, x_B (mg)
105.3	1.00	0.25
104.8	3.00	2.25
104.2	4.00	4.00
101.3	6.00	9.00
99.5	7.00	12.20
97.2	8.00	16.00
91.5	10.00	25.00
88.1	11.00	30.20
84.2	12.00	36.00
75.4	14.00	49.00

$$\vec{y} = \begin{bmatrix} 105.3 & 104.8 & 104.2 & 101.3 & 99.5 & 97.2 & 91.5 & 88.1 & 84.2 & 75.4 \end{bmatrix}^T$$

$$\mathcal{A} = \begin{bmatrix} 1 & 1.00 & 0.25 \\ 1 & 3.00 & 2.25 \\ 1 & 4.00 & 4.00 \\ 1 & 6.00 & 9.00 \\ 1 & 7.00 & 12.20 \\ 1 & 8.00 & 16.00 \\ 1 & 10.00 & 25.00 \\ 1 & 11.00 & 30.20 \\ 1 & 12.00 & 36.00 \\ 1 & 14.00 & 49.00 \end{bmatrix}. \tag{3.62}$$

The value of m is 10 (as there are ten data points) and n is 3 (as there are three parameters of interest).

The least-squares parameter estimates are given by

$$\hat{\vec{\beta}} = \left(\mathcal{A}^T \mathcal{A} \right)^{-1} \mathcal{A}^T \vec{y}$$

$$= \left(\begin{bmatrix} 1 & 1.00 & 0.25 \\ 1 & 3.00 & 2.25 \\ 1 & 4.00 & 4.00 \\ 1 & 6.00 & 9.00 \\ 1 & 7.00 & 12.20 \\ 1 & 8.00 & 16.00 \\ 1 & 10.00 & 25.00 \\ 1 & 11.00 & 30.20 \\ 1 & 12.00 & 36.00 \\ 1 & 14.00 & 49.00 \end{bmatrix}^T \begin{bmatrix} 1 & 1.00 & 0.25 \\ 1 & 3.00 & 2.25 \\ 1 & 4.00 & 4.00 \\ 1 & 6.00 & 9.00 \\ 1 & 7.00 & 12.20 \\ 1 & 8.00 & 16.00 \\ 1 & 10.00 & 25.00 \\ 1 & 11.00 & 30.20 \\ 1 & 12.00 & 36.00 \\ 1 & 14.00 & 49.00 \end{bmatrix} \right)^{-1} \begin{bmatrix} 1 & 1.00 & 0.25 \\ 1 & 3.00 & 2.25 \\ 1 & 4.00 & 4.00 \\ 1 & 6.00 & 9.00 \\ 1 & 7.00 & 12.20 \\ 1 & 8.00 & 16.00 \\ 1 & 10.00 & 25.00 \\ 1 & 11.00 & 30.20 \\ 1 & 12.00 & 36.00 \\ 1 & 14.00 & 49.00 \end{bmatrix}^T \begin{bmatrix} 105.3 \\ 104.8 \\ 104.2 \\ 101.3 \\ 99.5 \\ 97.2 \\ 91.5 \\ 88.1 \\ 84.2 \\ 75.4 \end{bmatrix}$$

$$= \begin{bmatrix} 1.191 & -0.318 & 0.072 \\ -0.318 & 0.107 & -0.027 \\ 0.072 & -0.027 & 0.007 \end{bmatrix} \begin{bmatrix} 951.500 \\ 6\,868.500 \\ 16\,033.645 \end{bmatrix}$$

$$= \begin{bmatrix} 104.91 \\ 0.5545 \\ -0.7602 \end{bmatrix}. \tag{3.63}$$

Units should always be included in the final answer. It is rare that a parameter in engineering does not have some physical units attached. The number of decimals to display is determined by the confidence interval, which will be determined later. Intermediate results should contain a reasonable amount of digits. All digits should be used to perform the calculations.

The standard deviation of the model can be obtained from

$$\hat{\sigma} = \sqrt{\frac{\vec{y}^T \vec{y} - \hat{\vec{\beta}}^T \mathcal{A}^T \vec{y}}{m - n}}$$

$$= \sqrt{\frac{\begin{bmatrix} 105.3 \\ 104.8 \\ 104.2 \\ 101.3 \\ 99.5 \\ 97.2 \\ 91.5 \\ 88.1 \\ 84.2 \\ 75.4 \end{bmatrix}^T \begin{bmatrix} 105.3 \\ 104.8 \\ 104.2 \\ 101.3 \\ 99.5 \\ 97.2 \\ 91.5 \\ 88.1 \\ 84.2 \\ 75.4 \end{bmatrix} - \begin{bmatrix} 104.91 \\ 0.5545 \\ -0.7602 \end{bmatrix}^T \begin{bmatrix} 1 & 1.00 & 0.25 \\ 1 & 3.00 & 2.25 \\ 1 & 4.00 & 4.00 \\ 1 & 6.00 & 9.00 \\ 1 & 7.00 & 12.20 \\ 1 & 8.00 & 16.00 \\ 1 & 10.00 & 25.00 \\ 1 & 11.00 & 30.20 \\ 1 & 12.00 & 36.00 \\ 1 & 14.00 & 49.00 \end{bmatrix}^T \begin{bmatrix} 105.3 \\ 104.8 \\ 104.2 \\ 101.3 \\ 99.5 \\ 97.2 \\ 91.5 \\ 88.1 \\ 84.2 \\ 75.4 \end{bmatrix}}{10 - 3}}$$

$$= \sqrt{\frac{91447.21 - 91447.18}{10 - 1}}$$

$$= 0.067\,668 \tag{3.64}$$

The standard deviation of the model, $\hat{\sigma}$, is 0.067 MPa. The 95% confidence intervals for the parameters can be can be obtained using

$$\hat{\beta}_i \pm t_{1-\frac{\alpha}{2},m-n}\,\hat{\sigma}\,\sqrt{(\mathcal{A}^T\mathcal{A})_{ii}^{-1}} \tag{3.65}$$

where $t_{0.975,\,10-3} = 2.365$. This gives the following confidence intervals:

$$\beta_0$$
$$104.91 \pm 2.365(0.067\,668)\sqrt{1.191}$$
$$104.91 \pm 0.175 \text{ MPa}$$
$$\beta_1$$
$$0.5545 \pm 2.365(0.067\,668)\sqrt{0.107}$$
$$0.5545 \pm 0.0523\frac{\text{MPa}}{\text{mg}}$$
$$\beta_2$$
$$-0.7602 \pm 2.365(0.067\,668)\sqrt{0.007}$$
$$-0.7602 \pm 0.0136\frac{\text{MPa}}{\text{mg}} \tag{3.66}$$

The standard deviation and confidence bounds should be reported to one or two digits. Based on the magnitude of the standard deviation/confidence bound, the mean value should be rounded to the same value. For β_0, the confidence bound can be written as 0.2, which implies that the mean is accurate to no more than three digits. Therefore, the value of β_0 should be reported as 104.9 ± 0.2 MPa. Similarly, the final values for the other parameters can be written as $\beta_1 = 0.55\pm0.05$ MPa·mg^{-1} and $\beta_2 = -0.76\pm0.01$ MPa·mg^{-1}.

For the point, $x_A = 5.00$ mg and $x_B = 10.00$ mg, the mean response confidence interval for the yield strength is given as

$$\vec{a}_{\bar{x}_d}\hat{\vec{\beta}} \pm t_{1-\frac{\alpha}{2},m-n}\hat{\sigma}\sqrt{\vec{a}_{\bar{x}_d}(A^T A)^{-1}\vec{a}_{\bar{x}_d}^T}$$

$$\begin{bmatrix} 1 \\ 5.00 \\ 10.00 \end{bmatrix}^T \begin{bmatrix} 104.91 \\ 0.5545 \\ -0.7602 \end{bmatrix} \pm 2.365(0.067\,668)\sqrt{\begin{bmatrix} 1 \\ 5.00 \\ 10.00 \end{bmatrix}^T \begin{bmatrix} 1.191 & -0.318 & 0.072 \\ -0.318 & 0.107 & -0.027 \\ 0.072 & -0.027 & 0.007 \end{bmatrix}\begin{bmatrix} 1 \\ 5.00 \\ 10.00 \end{bmatrix}} \tag{3.67}$$

$$100.085 \pm 2.365(0.067\,668)\sqrt{0.155\,616}$$

$$100.085 \pm 0.06312 \text{ MPa.}$$

Therefore, the mean response confidence interval is 100.09 ± 0.06 MPa. (Same rounding rules as for the parameter estimates apply here.) The predictive confidence interval is given as

$$\vec{a}_{\bar{x}_d}\hat{\vec{\beta}} \pm t_{1-\frac{\alpha}{2},m-n}\hat{\sigma}\sqrt{\vec{a}_{\bar{x}_d}(A^T A)^{-1}\vec{a}_{\bar{x}_d}^T}$$

$$\begin{bmatrix} 1 \\ 5.00 \\ 10.00 \end{bmatrix}^T \begin{bmatrix} 104.91 \\ 0.5545 \\ -0.7602 \end{bmatrix} \pm 2.365(0.067\,668)\sqrt{1+\begin{bmatrix} 1 \\ 5.00 \\ 10.00 \end{bmatrix}^T \begin{bmatrix} 1.191 & -0.318 & 0.072 \\ -0.318 & 0.107 & -0.027 \\ 0.072 & -0.027 & 0.007 \end{bmatrix}\begin{bmatrix} 1 \\ 5.00 \\ 10.00 \end{bmatrix}} \tag{3.68}$$

$$100.085 \pm 2.365(0.067\,668)\sqrt{1 + 0.155\,616}$$

$$100.085 \pm 0.172 \text{ MPa.}$$

Therefore, the predictive confidence interval is 100.1 ± 0.2 kg·min^{-1}. (Same rounding rules as for the parameter estimates apply here.) Also, note that due to the change in the size of the confidence interval, the rounding has changed the way the final result is reported.

Since $SSE = \hat{\sigma}^2(m-n)$, it is easy to find SSE from the data, namely,

$$SSE = \hat{\sigma}^2(m - n) = (0.067668)^2(10 - 3) = 0.0321. \tag{3.69}$$

TSS can be found as follows:

$$TSS = \sum (y_i - \bar{y})^2 = \vec{y}^T\vec{y} - \frac{1}{m}\left(\vec{y}^T\vec{1}\vec{1}^T\vec{y}\right)$$

$$
\begin{aligned}
&= \begin{bmatrix} 105.3 \\ 104.8 \\ 104.2 \\ 101.3 \\ 99.5 \\ 97.2 \\ 91.5 \\ 88.1 \\ 84.2 \\ 75.4 \end{bmatrix}^T \begin{bmatrix} 105.3 \\ 104.8 \\ 104.2 \\ 101.3 \\ 99.5 \\ 97.2 \\ 91.5 \\ 88.1 \\ 84.2 \\ 75.4 \end{bmatrix} - \frac{1}{10} \left(\begin{bmatrix} 105.3 \\ 104.8 \\ 104.2 \\ 101.3 \\ 99.5 \\ 97.2 \\ 91.5 \\ 88.1 \\ 84.2 \\ 75.4 \end{bmatrix}^T \begin{bmatrix} 1 \\ 1 \\ 1 \\ 1 \\ 1 \\ 1 \\ 1 \\ 1 \\ 1 \\ 1 \end{bmatrix} \begin{bmatrix} 1 \\ 1 \\ 1 \\ 1 \\ 1 \\ 1 \\ 1 \\ 1 \\ 1 \\ 1 \end{bmatrix}^T \begin{bmatrix} 105.3 \\ 104.8 \\ 104.2 \\ 101.3 \\ 99.5 \\ 97.2 \\ 91.5 \\ 88.1 \\ 84.2 \\ 75.4 \end{bmatrix} \right) \\[4pt]
&= 91447.2 - \frac{1}{10}(905352.3) \\[4pt]
&= 911.985.
\end{aligned}
\tag{3.70}
$$

Since $TSS = SSE + SSR$, SSR can be found by subtraction that is,

$$
\begin{aligned}
911.985 &= 0.0321 + SSR \\
SSR &= 911.953.
\end{aligned}
\tag{3.71}
$$

Since $R^2 = SSR\ /\ TSS$, for which we have both values, it gives that $R^2 = 911.953\ /\ 911.985 \approx 1$. Since the F-statistic is defined as

$$
F = \frac{SSR/k}{SSE/m - n}
\tag{3.72}
$$

where k is defined as the number of nonconstant functions (which in this case is 2), the value of the F-statistic is

$$
F = \frac{911.953/2}{0.0321/(10 - 3)} = 99\,580
\tag{3.73}
$$

The critical F-value has two degrees of freedom in the numerator and seven in the denominator, which gives $F_{0.95,\,2,\,7} = 4.74$. Since it is much smaller than the calculated F-value, it can be concluded that the model is significant.

3.3.5 Model Validation

Model validation is the process by which the least-squares model is examined to determine whether it is sufficiently good. Although many aspects can be formalised mathematically, some amount of intuition and experience is required in order to analyse the results. The intuition and experience can only be gained by examining and analysing many different regression cases, in order to understand the impact different factors have on the model.

Model validation consists of three parts: (1) testing the residuals, (2) testing the adequacy of the model, and (3) taking corrective action. Each of these three parts will now be examined in greater detail.

3.3.5.1 Residual Testing

It has been shown that for ordinary, least-squares analysis to hold, four assumptions are required about the distribution of the errors. These assumptions must be verified in order to determine the validity of the method. To refresh, the four key assumptions are:

1. The errors have a mean of zero.
2. The errors are independent.
3. The errors are homoscedastic, that is, they have the same variance.
4. The errors are normally distributed.

Since the true errors cannot be known, the residuals obtained from regression analysis are used instead. The following tests can be performed to determine the validity of the assumptions:

1. **Test for Zero Mean**: The easiest test is to compute the mean value of the residuals. If desired a formal hypothesis test can be performed.
2. **Tests for Normal Distribution**: The most common method to test normality is to plot a normal probability plot of the residuals. The points should lie along a straight line. Examples of good and bad normal probability plots are shown in Table 3.2. Alternatively, more advanced methods that consider the correlation properties of normally distributed errors can be used.
3. **Tests for Independence and Homoscedasticity**: These two aspects are most commonly tested together using various types of scatter plots. The most common scatter plots to examine are:

   a. Plotting a time-series plot of the residuals, that is, plotting the residuals as a function of time or sample.
   b. Plotting the residuals as a function of the variables, $\vec{x}$.
   c. Plotting the residuals against the fitted values, $\hat{y}$.
   d. Plotting the residuals against the measured values, y.
   e. Plotting the residuals against the previous residual.

   In cases (a) to (e), there should not be any discernible patterns in the plots. Common scatter plots are shown in Table 3.3. In this case, the bad scatter plots reveal the potential issues with the data and may suggest how to correct the situation. Graph (a) shows that there is a single outlier (point much further from the other points). Graph (b) shows that the residuals are increasing in value with

Table 3.2 Sample, normal probability plots

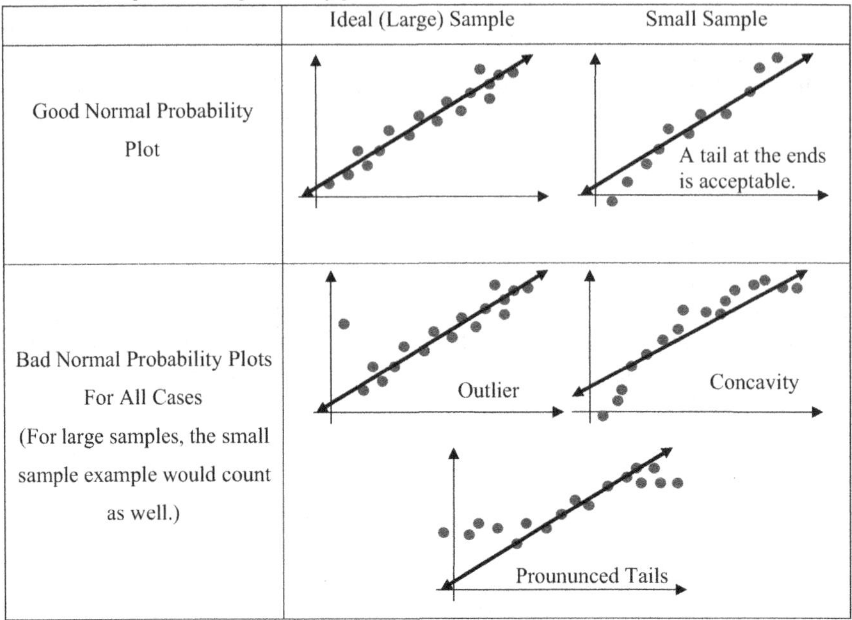

respect to the x-axis value. If the x-axis is a regressor, then it can be suggested that the error model depends on the given regressor and weighted, least-squares may need to be used. Similarly, if the x-axis represents time, then it can be concluded that the variance of the residuals depends on time. Graph (c) shows a sinusoidal pattern that suggests that a sinusoidal behaviour was missed when fitting the data. Graph (d) shows that the residuals are drifting as the x value increases. This suggests that either there were process issues or that the x value should be included in the analysis. Graph (e) shows a U-shaped (or quadratic) curve that potentially a quadratic term may be missing from the analysis. Finally, Graph (f) shows the results for a multivariate case, where multiple values were obtained at a single point. It can be seen here that the residuals at each of the points do not cluster about zero. Rather there seem to be points where they are smaller and points where they are larger.

3.3.5.2 Testing for Model Adequacy

Having determined that the residuals are well-behaved, it is now necessary to examine the quality of the model. The following methods can be used to achieve this.

1. **Using the confidence interval for each of the parameters, β_i:** If the confidence interval includes 0, then the parameter can be removed from the model. Ideally, a new regression analysis excluding that parameter would need to be performed and continued until there are no more parameters to remove.

Table 3.3 Sample scatter plots

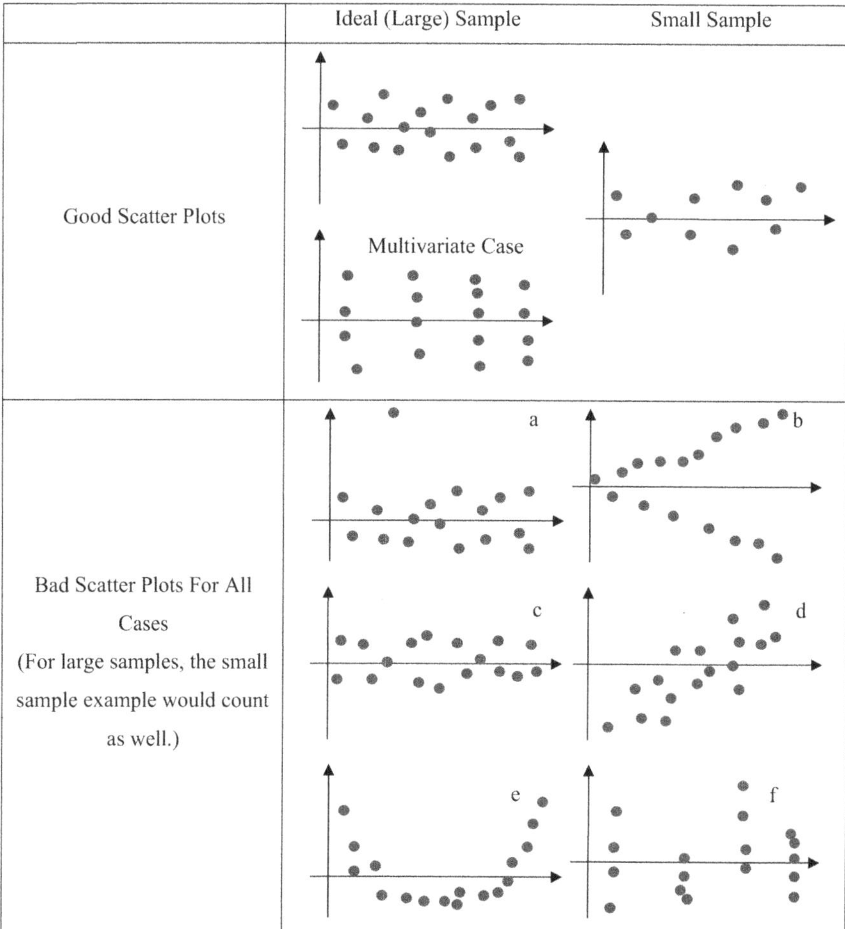

2. **Calculating Pearson's coefficient of regression**: The closer the value is to 1; the better the regression is. This coefficient gives what fraction of the observed behaviour can be explained by the given variables.

3. **Examining the observational and prediction confidence intervals**: The smaller they are the better. Note that if few samples or replicates are available, then the confidence intervals may be large purely due to the small sample size.

4. **Computing the F-statistic**: The F-test or analysis of variance (ANOVA) test considers which component, the variance in the error or the variance explained by the regression, is more prominent. If the error term is more significant, then the regression is likely to be spurious or unimportant.

5. **Examining outliers**: **Outliers** are defined as points that have residuals whose values differ greatly from surrounding values. Possible causes for outliers include typos when entering the data into the computer or errors in obtaining the data. Outliers tend to increase the confidence intervals producing "worse" results. Outliers can be spotted from any of the above plots that are used to check the model assumptions as points that are far from the expected behaviour. If a point is suspected to be an outlier, it should be removed and the regression redone.

6. **Examining influential points**: Some of the data points may strongly influence the regression model. Often this is a result of the fact that the given data points lie much further from the rest of the data. If the estimated model parameters change dramatically when such points are removed, then it can be stated that these points are influential. Formally, this can be determined using **Cook's Distance, D_i**, which is defined as

$$D_i = \frac{\left(\hat{\beta}_i - \vec{\beta}\right)^T A^T A \left(\hat{\beta}_i - \vec{\beta}\right)}{n\hat{\sigma}^2} \tag{3.74}$$

where $\hat{\beta}_i$ is the model parameters estimated by removing the ith data point. Practically, Cook's Distance can be calculated as follows

$$D_i = \frac{\varepsilon_i^2 h_{ii}}{n\hat{\sigma}^2(1 - h_{ii})^2} \tag{3.75}$$

where h_{ii} is the (i, i) entry in the $\mathcal{H}$-matrix, which is defined as

$$\mathcal{H} = A(A^T A)^{-1} A^T \tag{3.76}$$

If $D_i > 1$, then the given point can be considered to be influential, and the appropriateness of the model should be reconsidered.

7. **Examining the parity plot:** Plotting the predicted values as a function of y (the true values) can be very useful for dealing with large data sets. It can reveal which predicted values differ greatly from the measured values. Theoretically speaking, all the data should lie on the $y = x$ axis. Practically, due to noise, data points will lie in an ellipse or oval around the line. Table 3.4 shows some typical predicted as a function of measured value plots. Common problems include (the letters correspond to the graphs shown in Table 3.4 as problem graphs):

   a. **Outlier**: The presence of an outlier is easy to spot, as it will be located far from the $y = x$ line.

   b. **Cluster**: A cluster of points that lie far from the $y = x$ line is common when dealing with historical industrial data that does not contain any

Table 3.4 Sample, predicted as a function of true value plots

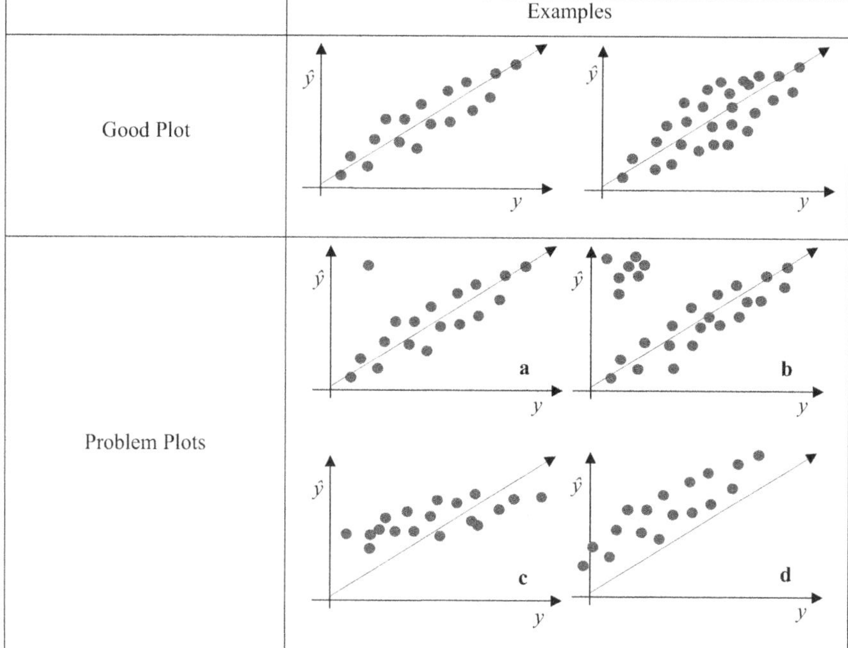

planned experiments. Such a cluster suggests that the operating conditions are different from the main set of data. This could be due to a plant shut-down, changes in process conditions, or other factors that should be investigated before removing the points from the data set.

c.   **Calibration**: In this case, the slope of the points is not 1. This suggests that either terms are missing from the model or that different models apply for different values of the measured values, since some of the values are overpredicted, while others are underpredicted.

d.   **Bias**: This suggests that the values obtained differ by some constant from the expected values. In linear regression, such a problem should not occur. However, it can happen with other more complex methods.

3.3.5.3   Taking Corrective Action

If the above tests fail or give inconclusive results or the assumptions in the model are not verified, then the analyst is left with the following options:

1.   **Changing the model** by adding additional parameters suggested by the analysis of the residual plots, using a completely different model, or changing the error structure.

a. **Regressor Selection**: Selecting appropriate regressors based on the data set provided is a complex subject that involves combining intuition or previous knowledge with various data mining methods. There are two main approaches to regressor selection:

   i. **Forward Selection**: In forward selection, the parameter estimates for the simplest model are first determined. If the resulting model analysis suggests that the model is sufficient for the required application, then it is adopted. Otherwise, additional regressors are added to the model until the model fit is adequate. The benefit of this approach is that a simple model is likely to be found. However, an important regressor could have been missed if it was not selected in the initial or subsequent steps.

   ii. **Backward Selection**: In backward selection, the full set of regressors is used to estimate the model. Regressors are then removed until the model is simplified to the smallest available model. The advantage of this approach is that all relevant regressors are likely to be in the model. However, the model could be larger than required for good fit.

b. **Error Structure**: In least-squares analysis, it is assumed that the error structure is additive and only applies to the measured output variable (y). In many engineering applications both the regressor and output variables are measured and can contain errors. Furthermore, the error may be proportional to some percentage of the measured value, which means that the error is not additive, but rather multiplicative. One way to correct for this error is to use weighted, least squares.

2. **Changing the regression method** by using either weighted, least-squares analysis, if the variance is not homoscedastic, or nonlinear, least-squares analysis to determine the parameter values.

3. **Obtaining more data** by running additional experiments, which can be useful if few experiments were performed and the results are ambiguous. Obtaining additional data helps resolve two potential issues:

   a. **Sample Size**: Small sample size can lead to large confidence intervals, especially if there is some noise in the original data. Increasing sample size can improve the accuracy of the parameter estimates. This helps to minimise **Type II** errors, which seeks to minimise the probability of rejecting the alternative hypothesis, even if it is true.

   b. **Replicates or Reproducibility**: Repeating an experiment at the same conditions allows a more accurate determination of the variance associated with the given point. As well, replicates can show how reproducible the results are at any one given point and if any additional factors could affect the results. Often, in practical cases, it may not be possible to obtain exactly the same conditions. However, if the conditions are similar then the results can be treated as replicates.

One area that needs to be considered when obtaining additional data is **experimental design**, that is, how is the procedure for obtaining the data are defined. Given the importance of this concept a separate chapter, Chap. 4, is devoted to this topic. One area of concern is **multicollinearity**. Multicollinearity is defined as a relationship between the columns of the A matrix. This multicollinearity can detract from finding the desired relationship between the regressors and the output and lead to a poor estimate of the individual parameters. It can be noted that the overall model may still be useful. Often multicollinearity may not be suspected by the experimenter or it can hold approximately. Multicollinearity is common in experiments were the variables cannot all be independently varied, for example mixture experiments where the total sum of component fractions must total 1.

A further area of concern is **data scaling**. This is an important issue, especially in nonlinear, least-squares analysis, where the orders of magnitude of the different regressors may be quite different. Since the optimisation routines will seek to minimise the absolute error, the parameter estimates corresponding to the larger order of magnitude regressors will be less accurate than those corresponding to smaller order of magnitude regressors, as their contribution to the overall error will be less. Consider the case where temperature ranges from 270 to 400 K and level ranges from 0.01 to 0.40 m are used. In this case, the levels are at least 10^3 times smaller than the temperatures. Furthermore, poorly scaled data may lead to issues with inverting the matrix in order to obtain the parameter estimates. There are two main methods by which scaling can be performed:

1. **Normalisation**: A common method is to introduce scaling with the following formula

$$\tilde{x} = \frac{x - \bar{x}}{\sigma} \tag{3.77}$$

    where $\tilde{x}$ is the scaled variable, $\bar{x}$ is the mean value of x, and σ is the standard deviation of x. The mean and standard deviation can be determined based on all the values of x available for the regression.

2. **Centring the data**: The following formula, commonly encountered in analysis of variance (ANOVA), centres the data so that it lies between -1 and 1:

$$\tilde{x} = \frac{x - 0.5(x_{\max} + x_{\min})}{0.5(x_{\max} - x_{\min})}. \tag{3.78}$$

3.3.6   Computational Example Part II: Model Validation

Continuing with the yield strength example from Sect. 3.3.4, model validation will now be performed by examining the residuals, the model quality, and taking appropriate corrective action.

Figure 3.2 shows the residuals as a function of amount of compound A, the yield strength, and the previous residual. In all cases, it seems that the residuals associated with a higher yield strength are larger than those associated with a smaller yield strength. This can clearly be seen in the large deviations for small values of x_A. This suggests that perhaps the error structure may need additional investigation.

The normal probability plot of the residuals is shown in Fig. 3.3. There does not seem to be any strong deviations from normality.

In order to determine model adequacy, all the previous computed data will be collected and then examined. The values are

$$F = 99,581 > 4.74$$
$$R^2 \approx 1.$$

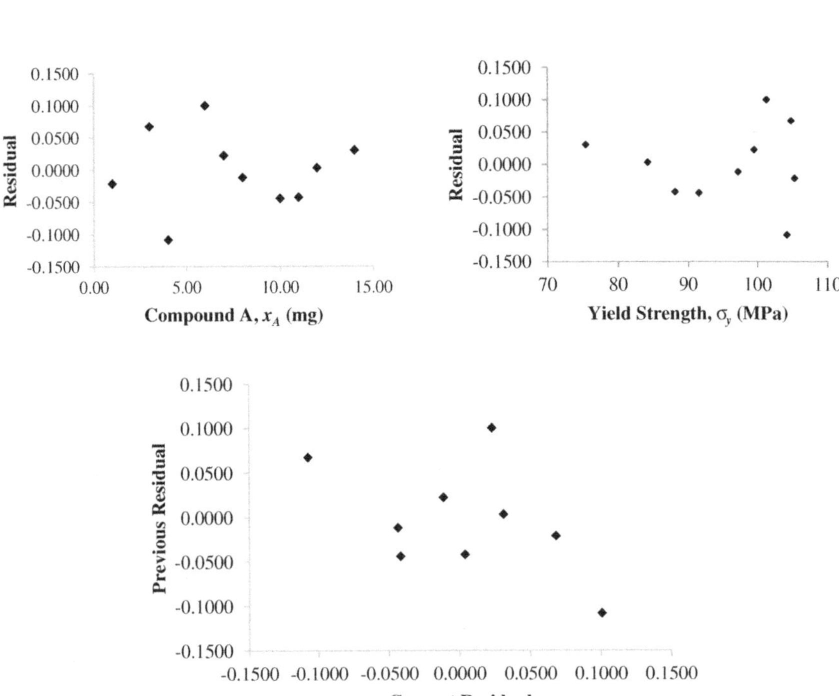

Fig. 3.2 Residuals as a function of the (top, left) Amount of compound A, (top, right) Yield strength, and (bottom) Previous residual

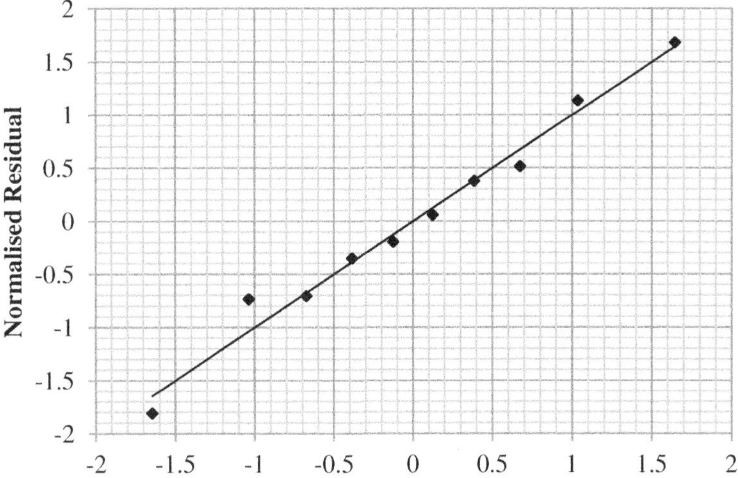

Fig. 3.3 Normal probability plot of the residuals

Influential analysis, using Cook's distance is given below. Cook's distance for each of the points is given in Table 3.5. It can be seen that none of the points is individually influential as all the values are less than 1.

Although there does seem to be some small issues with the residuals especially at high yield strengths, given the high value of R^2, the fact that the F-test is passed, and that the confidence intervals are small, it can be concluded that the model adequately describes the data.

Table 3.5 Calculating Cook's distance

σ_y (MPa)	x_A (mg)	x_B (mg)	Residual	Residual^2	Influence
105.3	1.00	0.25	−0.0208	4.34×10^{-4}	0.217
104.8	3.00	2.25	0.0680	46.18×10^{-4}	0.259
104.2	4.00	4.00	−0.1078	116.15×10^{-4}	1.97×10^{-4}
101.3	6.00	9.00	0.1006	101.11×10^{-4}	1.05×10^{-5}
99.5	7.00	12.20	0.0226	5.11×10^{-4}	6.55×10^{-6}
97.2	8.00	16.00	−0.0114	1.30×10^{-4}	3.64×10^{-6}
91.5	10.00	25.00	−0.0437	19.10×10^{-4}	2.00×10^{-10}
88.1	11.00	30.20	−0.0419	17.60×10^{-4}	0.152
84.2	12.00	36.00	0.0037	0.14×10^{-4}	2.17×10^{-12}
75.4	14.00	49.00	0.0308	9.51×10^{-4}	2.04×10^{-3}

3.3.7 *Weighted, Least-Squares Regression*

In weighted, least-squares analysis, it is assumed that the variance of the individual data points may be variable. In order to reduce this problem to the standard linear regression framework, a **weight, w_i,** is introduced for each observation that reflects how "good" the data point is. Thus, the regression model for weighted, least squares is

$$y = \sum_{i=1}^{n} \beta_i f_i(\vec{x}) + w^{-0.5}\varepsilon = \vec{a}\vec{\beta} + w^{-0.5}\varepsilon. \tag{3.79}$$

This implies that the objective function to be optimised in weighted, least squares is given as

$$\min_{\vec{\beta}} \left(\left(\mathcal{W}^{1/2}\vec{y} - \mathcal{W}^{1/2}\mathcal{A}\vec{\beta} \right)^T \left(\mathcal{W}^{1/2}\vec{y} - \mathcal{W}^{1/2}\mathcal{A}\vec{\beta} \right) \right). \tag{3.80}$$

Following a similar procedure as for the ordinary, least-squares case, Eq. (3.80) can be solved to obtain the unknown, estimated coefficients, $\hat{\vec{\beta}}$:

$$\mathcal{A}^T \mathcal{W} \mathcal{A} \hat{\vec{\beta}} = \mathcal{A}^T \mathcal{W} \vec{y} \tag{3.81}$$

where

$$\mathcal{A} = \begin{bmatrix} f_1(\vec{x}_1) & f_2(\vec{x}_1) & \cdots & f_n(\vec{x}_1) \\ f_1(\vec{x}_2) & f_2(\vec{x}_2) & \cdots & f_n(\vec{x}_2) \\ \vdots & & \ddots & \vdots \\ f_1(\vec{x}_m) & f_2(\vec{x}_m) & \cdots & f_n(\vec{x}_m) \end{bmatrix} \tag{3.82}$$

$$\hat{\vec{\beta}} = <\beta_1, \beta_2, \ldots, \beta_n>^T \tag{3.83}$$

$$\vec{y} = <y_1, y_2, \ldots, y_m>^T \tag{3.84}$$

$$\mathcal{W} = \begin{bmatrix} w_1 & 0 & \cdots & 0 \\ 0 & w_2 & 0 & 0 \\ 0 & 0 & \ddots & 0 \\ 0 & \cdots & 0 & w_m \end{bmatrix}. \tag{3.85}$$

The weighted **residual** is defined as the difference between the predicted and measured values, that is,

$$\vec{\varepsilon} = W^{1/2}\vec{y} - W^{1/2}\hat{\vec{y}} = W^{1/2}\vec{y} - W^{1/2}A\hat{\beta} = W^{1/2}\vec{y} - W^{1/2}A\left(A^T W A\right)^{-1}A^T W \vec{y}$$

$$= W^{1/2}\left(\mathcal{I} - A\left(A^T W A\right)^{-1}A^T W\right)\vec{y} \tag{3.86}$$

where $\mathcal{J}$ is the appropriately sized identity matrix.

The **standard deviation** for this model is given by

$$\hat{\sigma} = \sqrt{\frac{\vec{y}^T W \vec{y} - \hat{\beta}^T A^T W \vec{y}}{m - n}}. \tag{3.87}$$

The **$100(1-\alpha)\%$ Confidence Interval for β_i** is given by

$$\hat{\beta}_i \pm t_{1-\frac{\alpha}{2},m-n}\hat{\sigma}\sqrt{(A^T W A)_{ii}^{-1}} \tag{3.88}$$

where $(A^T W A)_{ii}^{-1}$ represents the value located at (i,i) in the matrix $(A^T W A)^{-1}$.

The **$100(1-\alpha)\%$ Mean Response Confidence Intervals**, that is, the region within which the mean value of repeated measurements will lie $100(1-\alpha)\%$ of the time, for the point $\vec{x}_d$ is given by

$$\vec{a}_{\vec{x}_d}\hat{\beta} \pm t_{1-\frac{\alpha}{2},m-n}\hat{\sigma}\sqrt{\vec{a}_{\vec{x}_d}(A^T W A)^{-1}\vec{a}_{\vec{x}_d}^T} \tag{3.89}$$

where $\vec{a}_{\vec{x}_d} = < f_1(\vec{x}_d),\, f_2(\vec{x}_d),\, \ldots,\, f_n(\vec{x}_d) >$.

The **$100(1-\alpha)\%$ Predictive Confidence Intervals**, that is, the region within which the actual value will lie $100(1-\alpha)\%$ of the time, for the point $\vec{x}_d$ is given by

$$\vec{a}_{\vec{x}_d}\hat{\beta} \pm t_{1-\frac{\alpha}{2},m-n-n_\sigma}\hat{\sigma}\sqrt{\frac{1}{w_0} + \vec{a}_{\vec{x}_d}(A^T W A)^{-1}\vec{a}_{\vec{x}_d}^T} \tag{3.90}$$

where w_0 is the predicted weight at $\vec{x}_d$ found based on the model with n_σ parameters (see Sect. 3.3.7.1 for additional information about creating models for the weights).

The sum of squares due to regression, SSR, can be calculated as follows:

$$SSR = \sum w_i(\hat{y}_i - \bar{y})^2 = \hat{\beta}^T A^T W \vec{y} - \frac{1}{m}\left(\vec{y}^T W^{1/2}\vec{1}\vec{1}^T W^{1/2}\vec{y}\right) \tag{3.91}$$

where $\vec{1}$ is the unit column vector with size $m \times 1$.

The error sum of squares, *SSE*, can be calculated as follows:

$$SSE = \sum w_i \left(y_i - \hat{y}_i\right)^2 = \vec{y}^T \mathcal{W}\vec{y} - \hat{\vec{\beta}}^T A^T \mathcal{W}\vec{y}.$$

(3.92)

The total sum of squares, *TSS*, can be calculated as follows:

$$TSS = \sum w_i (y_i - \bar{y})^2 = \vec{y}^T \mathcal{W}\vec{y} - \frac{1}{m}\left(\vec{y}^T \mathcal{W}^{1/2}\vec{1}\vec{1}^T \mathcal{W}^{1/2}\vec{y}\right).$$

(3.93)

Pearson's coefficient of regression, R^2, and the adjusted R^2 are calculated the same way as in ordinary, least-squares analysis. The same can be said of the F-statistic.

Model validation is performed the same way as for ordinary, least-squares analysis, replacing the residuals by their weighted residual values.

3.3.7.1 Determining the Weights

One of the hardest things in weighted, least-squares analysis is to determine the appropriate weights. There are two principle ways in which this can be achieved:

(1) **Replicates**: If for the same conditions, two or more values of y are obtained, then it is possible to calculate the variance of the values at the given point. The variance would be determined using Eq. (71). The weight would then be given as $w_i = 1 \,/\, \text{var}(y_i)$.

(2) **A-Priori Model**: If the model for the variance is known ahead of time, then it can be used to determine the appropriate weight. The weight would then be given as $w_i = 1 \,/\, \text{var}(x, y)$. If no suitable model can be determined, then an arbitrary model can be assumed, for example $w_i = 1 \,/\, x$ or $w_i = 1 \,/\, x^2$.

In predicting the actual response at x_i using Eq. (3.90), a model is required for the weights. If an *a-priori* model has been used, then it is relatively easy to calculate the "predicted" future weight value. If for this model, the value of the parameters is known ahead of time and is not estimated from the data obtained, then $n_\sigma = 0$. On the other hand, if the data or variances are used to estimate the *a-priori* model parameters using regression, then n_σ will equal the number of parameters in that model. If replicates are used, then a model needs to be fit to the variances or weights to obtain an appropriate model. The number of parameters in this model would equal n_σ.

Example 3.2 Determining the Weights for Weighted, Least-Squares Regression. Consider the following situations and determine appropriate weights for the given data sets.

Table 3.6 Replicated data for determining the weights

Run	h	Replicate 1	Replicate 2	Replicate 3
1	2	10.54	10.22	10.01
2	4	28.65	28.25	25.74
3	6	46.74	46.12	47.71

(1) The experiment with replicates shown in Table 3.6, and give appropriate weightings for each point.
(2) Assume that the standard deviation of the error is proportional to the square root of the height. What would be the appropriate weightings for each data point?

Solution

For the data shown in Table 3.6, calculate the mean of each run and then determine the variance of each point using the following formula

$$\text{var}(y_i) = (y_i - \overline{y})^2.$$

The weights for each point would then be $1 / \text{var}(y_i)$. The results are summarised in Table 3.7. It can be seen that those points which are located far from the mean have very small weights (consider replicates 2 and 3 of Run (3), while cases with generally large variability have small weights as well (consider Run 2, as an example).

Table 3.7 Weights for the example

Run	Mean	Variance of point			Weights		
1	10.25	0.082	9×10^{-4}	0.058	12.20	1,111	17.24
2	27.55	1.21	0.49	3.28	0.826	2.04	0.305
3	46.86	0.014	0.546	0.723	71.43	1.83	1.38

Therefore, the weighting matrix would be written as

$$\vec{y} = \langle y_{11}, y_{12}, y_{13}, y_{21}, y_{22}, y_{23}, y_{31}, y_{32}, y_{33}, \rangle^T$$

$$\mathcal{W} = \text{diag}(\langle 12.20, 1111, 17.24, 0.826, 2.04, 0.305, 71.43, 1.83, 1.38 \rangle).$$

For the second example, the weights would be $w = 1 / h$, since the variance would be proportional to h (as variance is the standard deviation squared). Therefore, the weighting matrix would be diagonal with the entry $1 / h$ evaluated for the corresponding value of the height.

3.4 Nonlinear Regression

In many cases, it may not be possible to obtain a valid linear regression model and it may be necessary to perform **nonlinear regression**. In general, since nonlinear regression can handle an arbitrarily complex function, there is really no need to make any simplifications about the form of the regression model. Therefore, the model to be identified can be written as

$$y = g\left(\vec{\beta}, \vec{x}, \varepsilon\right). \tag{3.94}$$

This ability to deal with general models means that much of the linear regression analysis cannot be performed exactly, since the underlying assumptions are not valid any more. Nevertheless, most of the linear regression results hold if the number of data points is much larger than the number of parameters to be estimated. The optimisation algorithm can be written as

$$\min_{\vec{\beta}} \sum_{i=1}^{m} w_i \left(y_i - g\left(\vec{\beta}, \vec{x}_i, \varepsilon_i\right)\right)^2 \tag{3.95}$$

where w_i is the weight. In most cases, $w_i = 1$ and the weights can be ignored.

All nonlinear regression approaches use numerical methods, such as the Gauss–Newton or Levenberg–Marquardt algorithms, to search for the optimal point.

The derivative matrix of this problem, called the grand Jacobian matrix, $\mathcal{J}$, plays a role similar to that of the $\mathcal{A}$ matrix in linear regression. The Jacobian, $\mathcal{J}'$, for the system can be calculated as

$$\mathcal{J}' = \left[\frac{\partial g\left(\vec{\beta}, \vec{x}, \varepsilon\right)}{\partial \beta_1} \quad \frac{\partial g\left(\vec{\beta}, \vec{x}, \varepsilon\right)}{\partial \beta_2} \quad \ldots \quad \frac{\partial g\left(\vec{\beta}, \vec{x}, \varepsilon\right)}{\partial \beta_n} \right]. \tag{3.96}$$

The value of $\mathcal{J}'$ is determined for each of the data points present to obtain the grand Jacobian matrix, $\mathcal{J}$. Thus, $\mathcal{J}$ can be written as

$$J = \begin{bmatrix} \mathcal{J}_1' \\ \mathcal{J}_2' \\ \vdots \\ \mathcal{J}_m' \end{bmatrix} = \begin{bmatrix} \dfrac{\partial g(\vec{\beta},\vec{x}_1,\varepsilon)}{\partial \beta_1} & \dfrac{\partial g(\vec{\beta},\vec{x}_1,\varepsilon)}{\partial \beta_2} & \cdots & \dfrac{\partial g(\vec{\beta},\vec{x}_1,\varepsilon)}{\partial \beta_n} \\ \dfrac{\partial g(\vec{\beta},\vec{x}_2,\varepsilon)}{\partial \beta_1} & \dfrac{\partial g(\vec{\beta},\vec{x}_2,\varepsilon)}{\partial \beta_2} & \cdots & \dfrac{\partial g(\vec{\beta},\vec{x}_2,\varepsilon)}{\partial \beta_n} \\ \vdots & \vdots & & \vdots \\ \dfrac{\partial g(\vec{\beta},\vec{x}_m,\varepsilon)}{\partial \beta_1} & \dfrac{\partial g(\vec{\beta},\vec{x}_m,\varepsilon)}{\partial \beta_2} & \cdots & \dfrac{\partial g(\vec{\beta},\vec{x}_m,\varepsilon)}{\partial \beta_n} \end{bmatrix}. \tag{3.97}$$

3.4.1 Gauss–Newton Solution for Nonlinear Regression

In order to show the numerical approach and understand some of the issues involved with solving nonlinear regression problems, the Gauss–Newton numerical solution will be examined. The Gauss–Newton method presented here is the generalisation to the multivariate case of the standard Newton method for finding roots of univariate functions. Using the grand Jacobian matrix, the values of the parameters are determined as follows:

(1) Using the initial guess for the parameters, $\vec{\beta}^{(0)}$, the grand Jacobian matrix is evaluated, and the predicted values are determined,

$$\hat{\vec{y}} = \begin{bmatrix} g\left(\vec{\beta}^{(0)}, \vec{x}_1\right) \\ g\left(\vec{\beta}^{(0)}, \vec{x}_2\right) \\ \vdots \\ g\left(\vec{\beta}^{(0)}, \vec{x}_m\right) \end{bmatrix}. \tag{3.98}$$

(2) Next, the difference between the predicted values, $\hat{\vec{y}}$, and the actual, measured values is determined, that is,

$$\Delta \vec{y} = \vec{y} - \hat{\vec{y}}. \tag{3.99}$$

(3) Using an appropriate numerical method, the following system of equation is solved for $\Delta\vec{\beta}$,

$$J^T J \Delta\vec{\beta} = J^T \Delta\vec{y}. \tag{3.100}$$

(4) Finally, the new estimated value for the parameters is determined as

$$\vec{\beta}^{(k+1)} = \vec{\beta}^{(k)} + \Delta\vec{\beta}. \tag{3.101}$$

(5)   The new value of the parameters, $\vec{\beta}^{(k+1)}$, then becomes the new guess, and the above procedure is repeated from Step 1. This procedure continues until the difference in values between the parameters from two consecutive steps is less than some predetermined accuracy or a certain number of iterations has been reached.

As with many numerical methods, the following are some common issues:

(1)   **Initial Guess**: Determining the initial guess can have a large impact on how quickly and accurately the values are obtained. If the function can be linearised, then a suitable initial guess can be the linearised parameter estimates. On the other hand, if the function cannot be linearised, then knowledge about the range of possible parameter values given the problem at hand can be used to provide reasonable initial estimates. The final option is to use multiple initial guesses and select the one providing the smallest objective value.

(2)   **Determining the Minimal Value**: As with any numerical method, the Gauss–Newton method only finds those points at which the derivative is zero. However, a derivative of zero does not necessarily imply that the point is the global minimum; instead it could be a local minimum. Therefore, one way around this problem is to use multiple initial guesses and then select the point that provides the lowest value.

3.4.2   *Useful Formulae for Nonlinear Regression*

Unlike in linear regression where exact results can be obtained under the stated assumptions, in nonlinear regression the results are only approximate. Furthermore, there do not exist nice matrix-based solutions for the various parameters. This section provides a convenient summary of the useful equations for nonlinear regression. In general, to compute the approximate confidence intervals for a nonlinear regression problem the final grand Jacobian matrix, $\mathcal{J}$, can be used in place of $\mathcal{A}$ and $\mathcal{J}'$ in place of $\vec{a}_{\bar{x}_d}$ in the linear regression formulae.

The model residuals can be computed as

$$\boxed{\varepsilon_i = y_i - \hat{y}_i = y_i - g\left(\hat{\vec{\beta}}, \vec{x}_i\right)}. \tag{3.102}$$

The standard deviation for this model, $\hat{\sigma}$, can be obtained as follows

$$\hat{\sigma} = \sqrt{\frac{\sum_{i=1}^{m} \left(y_i - \hat{y}_i\right)^2}{m - n}}. \tag{3.103}$$

The approximate **$100(1-\alpha)\%$ Confidence Interval for β_i** is given by

$$\boxed{\hat{\beta}_i \pm t_{1-\frac{\alpha}{2},m-n}\,\hat{\sigma}\,\sqrt{(\mathcal{J}^T\mathcal{J})_{ii}^{-1}}}. \tag{3.104}$$

where $(\mathcal{J}^T\mathcal{J})_{ii}^{-1}$ represents the value located at (i, i) in the grand Jacobian matrix $(\mathcal{J}^T\mathcal{J})^{-1}$.

The approximate **$100(1-\alpha)\%$ Mean ResponseConfidence Intervals**, that is, the region within which the mean value of repeated measurements will lie $100(1-\alpha)\%$ of the time, for the point $\vec{x}_d$ is given by

$$\boxed{f\left(\vec{x}_d,\hat{\vec{\beta}}\right) \pm t_{1-\frac{\alpha}{2},m-n}\,\hat{\sigma}\,\sqrt{\mathcal{J}'(\mathcal{J}^T\mathcal{J})^{-1}(\mathcal{J}')^T}}. \tag{3.105}$$

where $\mathcal{J}'$ is the value of the Jacobian evaluated at the point $\vec{x}_d$ and $\hat{\vec{\beta}}$.

The approximate **$100(1-\alpha)\%$ Predictive Confidence Intervals**, that is, the region within which the actual value will lie $100(1-\alpha)\%$ of the time, for the point $\vec{x}_d$ is given by

$$\boxed{f\left(\vec{x}_d,\hat{\vec{\beta}}\right) \pm t_{1-\frac{\alpha}{2},m-n}\,\hat{\sigma}\,\sqrt{1 + \mathcal{J}'(\mathcal{J}^T\mathcal{J})^{-1}(\mathcal{J}')^T}}. \tag{3.106}$$

The **sum of squares due to regression**, *SSR* and the **sum of squares due to the error**, *SSE*, are calculated using the definitions given by Eqs. (3.37) and (3.38). The **total sum of squares**, *TSS*, can be calculated as follows:

$$\boxed{TSS = \sum (y_i - \overline{y})^2 = \vec{y}^T\vec{y} - \frac{1}{m}\left(\vec{y}^T\vec{1}\vec{1}^T\vec{y}\right)}. \tag{3.107}$$

Model validation would be performed in the similar manner as that for ordinary, least-squares analysis bearing in mind that the computed confidence intervals are only approximate. R^2 cannot be used as a measure of performance in nonlinear regression, since the relationship between the sums of squares is no longer orthogonal.

3.4.3 Computational Example of Nonlinear Regression

In the development of a new drug, you are investigating the previously unknown reaction between two compounds X and Y to produce a valuable intermediary Z. One of the important tasks is to characterise the reaction properties. Luckily, X and Y are relatively easy to produce and so that multiple runs and trials could be performed. At each of the temperatures, three separate runs were performed. The results obtained are shown in Table 3.8. It is desired to fit Arrhenius's Equation to

Table 3.8 Reaction rate data

Temperature (K)	Reaction Rate (K, cm^3·s^{-1})		
	Run 1	Run 2	Run 3
200	122.9	122.5	121.9
250	145.5	146.6	146.0
300	163.9	165.5	164.5
350	180.4	179.4	179.3
400	191.9	191.5	191.4
450	201.4	199.8	201.0

this data set and determine the reaction constant and activation energy. Perform both linear and nonlinear regressions and compare the results using $\alpha = 0.05$.

Solution

The results will be presented without showing all the detailed calculations as they are relatively straightforward. From Example 3.1, we have that Arrhenius's reaction can be linearised as

$$\ln K = \ln A - \frac{E_a}{RT} \tag{3.108}$$

where $R = 8.314$ J·mol^{-1}·K^{-1}. Using Eq. (3.108), the linearised parameters are $\ln A = 5.700\,84 \pm 0.006\,53$ and $-E_a R^{-1} = -178.84 \pm 1.893\,51$. In order to obtain the original parameter estimates, there is a need to convert the values. Since both functions are monotonic, the conversion is relatively simple.

For A, the mean value can be computed as $\hat{A} = e^{5.70084} = 299.12$. The confidence interval can be computed by converting each of the bounds using the same formula. This gives $\hat{A}_{lower} = e^{5.700\,84 - 0.006\,53} = 297.2$ and $\hat{A}_{upper} = e^{5.700\,84 + 0.006\,53} = 301.08$. Therefore, the confidence for A can be written as $[297, 301]$ cm^3·s^{-1}. It should be mentioned that the confidence interval is not symmetric about the mean value. This is because the exponential function does not preserve distance.

For E_a, the mean value can be computed as $\hat{E}_a = -8.314 \times -178.84 = 1{,}486.91$. The confidence interval becomes $1.893\,51 \times -8.314 = (-)15.74$. Therefore, the confidence interval for E_a is $1{,}490 \pm 16$ J·mol^{-1}. Notice that, in this case, the confidence interval remains symmetric about the mean value. Model validation graphs will be shown combined with the nonlinear regression results.

For the nonlinear regression, the required derivatives are

$$\frac{\partial K}{\partial A} = e^{-\frac{E_a}{RT}}$$

$$\frac{\partial K}{\partial E_a} = -\frac{A e^{-\frac{E_a}{RT}}}{RT}. \tag{3.109}$$

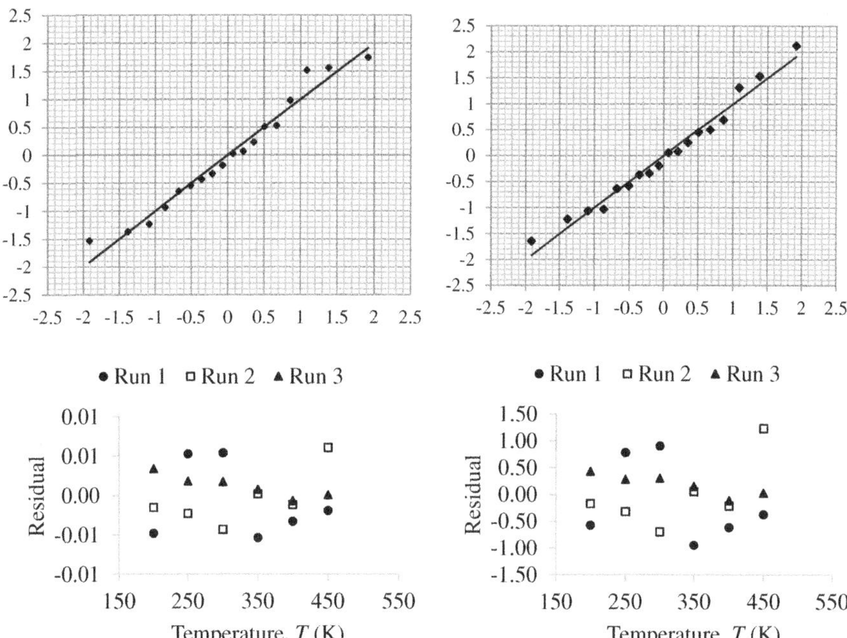

Fig. 3.4 (top) Normal probability plots of the residuals and (bottom) Residuals as a function of temperature for (left) Linearised and (right) Nonlinear models

After setting up the problem,[2] the following values are obtained: $\hat{A} = 299\pm2$ cm^3·s^{-1} and $\hat{E}_a = 1{,}490\pm18$ J·mol^{-1}. Approximate confidence intervals were computed based on the linearised formulae. The value of $\mathcal{J}^T \mathcal{J}$ is

$$(\mathcal{J}^T \mathcal{J}) = \begin{bmatrix} 2.483\,498 & 21.198\,8 \\ 21.198\,824 & 194.414 \end{bmatrix} \tag{3.110}$$

First, it can be seen that, in this case, the results are quite similar both in the estimated value and the confidence intervals. Secondly, Fig. 3.4 shows the normal probability plots and the residuals as a function of the temperature for both cases. From the normal probability plots, it would seem that the residuals for both models are quite similar. On the other hand, there do seem to be more abnormal points in the linearised model case, suggesting that the residuals may violate the assumption of normality. Examining the residual as a function of temperature plots shows some interesting results. Firstly, for the linearised model, Run 2 has a different behaviour from the other two runs. Secondly, there are few, if any, positive deviations compared with the large number of negative deviations. Based solely on the linearised model, one

---

[2] A detailed example on solving the nonlinear regression problem is given in Sect. 7.8.2 : Nonlinear Regression Example for MATLAB® and Sect. 8.7.2 : Nonlinear Regression Example for Excel®.

would have to conclude that Run 2 was abnormal, and the collection of the data would warrant additional scrutiny. On the other hand, when the nonlinear case is used, the results are quite different and a different pattern emerges. First, Run 2 is no longer usual, and there are now both positive and negative residuals in equal magnitude. Second, it would seem that the residuals depend on the temperature with a lower value around 350–400 K and higher values at the extremes. Since Arrhenius's equation is an accepted model for the observed behaviour, this feature could potentially be attributed to issues in experimental design, that is, the conditions and methods by which the data were obtained, for example, faulty measurements or an incomplete procedure.

It is interesting to note that, although both the linear and nonlinear methods provided similar parameter estimates and confidence intervals, the residual analysis is quite different. In the linear case, it would be concluded that Run 2 had some abnormal residuals and would require additional analysis. In the nonlinear case, it would be concluded that there seems to be some temperature dependency of the residuals. This shows the importance of selecting an appropriate method for the given problem.

3.5 Models and Their Use

Once regression analysis has been performed, it is often desired to use the model for predicting the behaviour of the system at other conditions. However, the validity of the results needs to be carefully examined. Firstly, the model should be used for **interpolation**, that is, to predict values that lie within the original region. The opposite term, **extrapolation**, denotes using the model outside the original region. However, determining whether or not extrapolation is an issue is not necessarily easy, especially in the multivariate case. Consider for example, the two variables shown in Fig. 3.5. The model was fit based on the data denoted as black circles in Fig. 3.5, which gives a range of $[10°C, 40°C]$ for the temperature (T) and $[10\,m, 20\,m]$ for the height (h). It can be seen that, for whatever reason, the temperatures and heights were not randomly selected across the range; rather they seem to fall in a certain region. If the resulting model was then used to predict the value for the green triangle, the results could be erroneous, as the point is located far outside the original region of the data. However, it can be noted that the point selected lies inside both of the ranges and would often be assumed to be okay. This means that it is necessary to check whether the selected data point lies within the original regression space.

3.6 Summative Regression Example

In order to apply all that has been considered in this chapter, a single summative example will be examined.

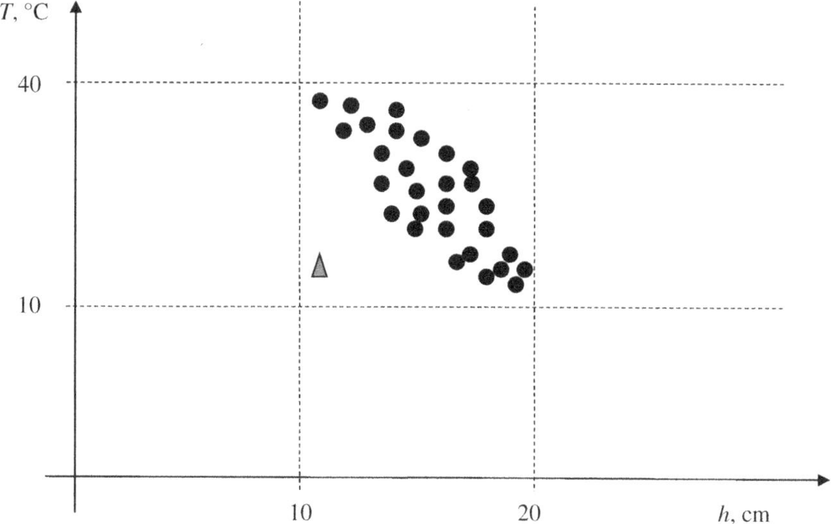

Fig. 3.5 Extrapolation in multivariate analysis

3.6.1 Data and Problem Statement

For the efficient and profitable operation, especially during the summer months, of electrical generating plants, there is a need to forecast the expected peak power load (P^*) as accurately as possible given the anticipated peak temperature. The data set in Table 3.9 is a random sample of thirty daily high temperatures (T, °F) and the peak power demand (P^*, MW). Perform the following analysis:

1. Fit a linear model ($P^* = aT + b$) to the data. Is the data set well-described by this model?
2. Fit a quadratic model ($P^* = aT^2 + bT + c$). Is the data set well-described by this model?
3. Using the best model, predict the peak power at $T = 50°F$ and $T = 105°F$? Compute the 95% mean response confidence intervals, which value you do trust more? Why?

3.6.2 Solution

3.6.2.1 Simple Linear Model

First, consider the simple linear model $P^* = aT + b$. The parameter estimates are obtained using ordinary, least-squares regression. The parameter estimates with 95% confidence intervals are

Table 3.9 Peak power and temperature

Temperature, T (°F)	Peak power, $P*$ (MW)	Temperature, T (°F)	Peak power, $P*$ (MW)
95	140.7	79	106.2
88	116.4	76	100.2
84	113.4	87	114.7
106	178.2	92	135.1
94	136.0	68	96.3
108	189.3	85	111.4
90	132.0	100	143.6
100	151.9	74	103.9
71	92.5	89	116.5
96	131.7	86	105.1
67	96.5	75	99.6
98	150.1	70	97.7
97	153.2	69	97.6
67	101.6	82	107.3
89	118.5	101	157.6

(*Data taken from Babutnde A. Ogunnaike (2010)*, Random Phenomena, Fundamentals of Probability and Statistics for Engineers)

$$a = 1.947 \pm 0.303$$
$$b = -44.54 \pm 26.3$$

R^2 is 0.8612 and the F-statistic is 173 ($F_{critical} = 4.195$). The residuals as a function of the temperature are shown in Fig. 3.6, while the normal probability plot of the residuals is shown in Fig. 3.7.

From Fig. 3.6, a quadratic pattern to the residuals is quite evident. This strongly suggests that a quadratic term is missing from the model and should be added. Notice how the confidence intervals are also quite large, especially for the intercept.

Fig. 3.6 Residuals as a function of temperature

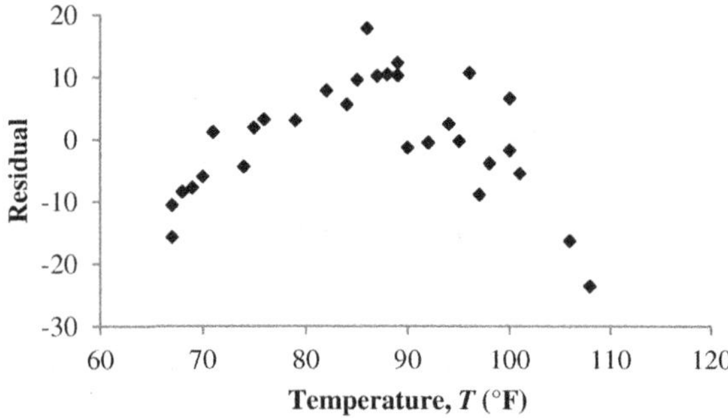

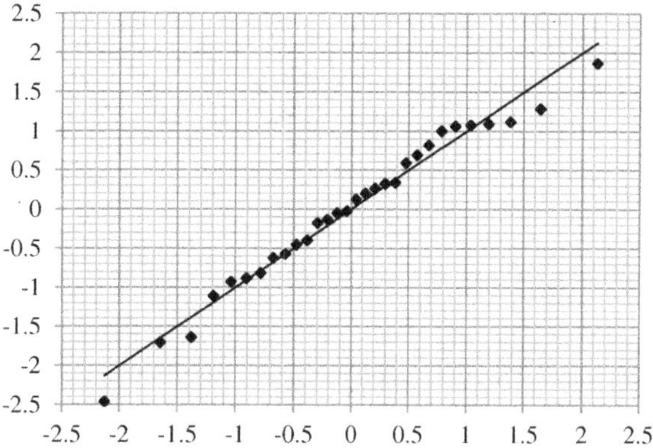

Fig. 3.7 Normal probability plot of the residuals

3.6.2.2 Quadratic Model

For the quadratic model $P^* = aT^2 + bT + c$, the parameter estimates with 95% confidence intervals are

$$a = 0.0598 \pm 0.0133$$
$$b = -8.295 \pm 2.27$$
$$c = 385.1 \pm 96.2$$

R^2 is 0.9667 and the F-statistic is 392.2 ($F_{critical} = 4.21$). The residuals as a function of the temperature are shown in Fig. 3.9, while the normal probability plot of the residuals is shown in Fig. 3.8.

It can be noted that adding a quadratic term has improved the size of the parameter estimate confidence intervals, as well as increasing R^2. Furthermore, the normal probability plot of the results seems to suggest that there could be some problems due to the clustering of values. The residuals as a function of temperature plot does

Fig. 3.8 Normal probability plot of the residuals for the quadratic case

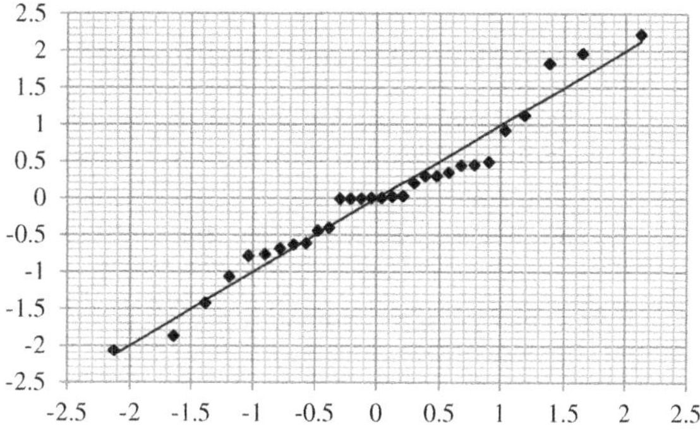

Fig. 3.9 Residuals as a function of the regressor for the quadratic case

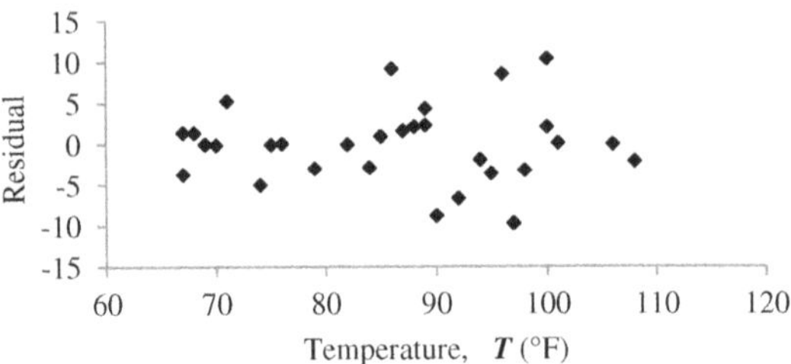

not show any real issues. There does seem to be a small increase in variability of the values as the temperature increases. There are no discernible parameters that would improve the fit.

3.6.2.3 Mean Response Intervals

For $T = 50°\mathrm{F}$, the predicted peak power and its 95% confidence interval is

$$120 \pm 15 \text{ MW},$$

while for $T = 105°\mathrm{F}$, the predicted peak power and its 95% confidence interval is

$$173.9 \pm 4.6 \text{ MW}.$$

The prediction for $T = 105°\mathrm{F}$ is more reliable, since it has a smaller confidence interval and there is no extrapolation. On the other hand, for $T = 50°\mathrm{F}$, there is extrapolation and the confidence interval is larger.

3.7 Further Reading

The following are references that provide additional information about the topic:

(1) **General Modelling and Linear Regression:**

    a. Montgomery, D. C., & Peck, E. A. (1982). *Introduction to Linear Regression Analysis* (1st ed.). New York, New York, United States of America: John Wiley & Sons.

    b. Ogunnaike, B. A. (2010). *Random Phenomena: Fundamentals of Probability and Statistics for Engineers.* Boca Raton, Florida, United States of America: CRC Press.

(2) **Weighted Regression**:

    a.   Zorn, M. E., Gibbons, R. D., & Sonzogni, W. C. (1997). Weighted Least-Squares Approach To Calculating Limits of Detection and Quantification by Modeling Variability as a Function of Concentration. *Analytical Chemistry, 69*(15), 3069–3075.

(3) **Nonlinear Regression**:

    a.   Seber, G. A., & Wild, C. J. (1989). *Nonlinear Regression.* New York, New York, United States of America: John Wiley & Sons.

3.8 Chapter Problems

Problems at the end of the chapter consist of three different types: (a) Basic Concepts (True/False), which seek to test the reader's comprehension of the key concepts in the chapter; (b) Short Exercises, which seek to test the reader's ability to compute the required parameters for a simple data set using simple or no technological aids. This section also includes proofs of theorems; and (c) Computational Exercises, which require not only a solid comprehension of the basic material, but also the use of appropriate software to easily manipulate the given data sets.

3.8.1 Basic Concepts

Determine if the following statements are true or false and state why this is the case.

(1) If the residuals are distributed so that they are increasing in magnitude as the x value increases, then it can be concluded that the model is adequate.

(2) Weighted, least-squares can correct for the error structure.

(3) If $R^2 \approx 0$, then it can be concluded that there is no relationship between the parameters of interest.

(4) If the residuals are normally distributed, then the calculated confidence intervals are valid.

(5) If the current residual depends on the past value of the residuals, then the regression analysis is valid.

(6) In order to analyse statistically the least-squares estimates, the residuals must be normally distributed.

(7) If the residuals as a function of the regressors have a quadratic pattern, then it can be concluded that a quadratic term should be added to the model.

(8) If the mean of the residuals is 7, then it can be concluded that the parameters are unbiased.

(9) If $\hat{b} = 1.25 \pm 10.5$ m, then it can be concluded that this parameter should be included in the model.

(10)   If there are more parameters than data points, then an estimate of the parameters cannot be obtained.

(11)   Linearising a nonlinear equation will always provide worse parameter estimates then performing nonlinear regression.

(12)   With nonlinear regression, exact confidence intervals can be found for the parameters.

(13)   When solving nonlinear regression using the Gauss–Newton method, the final parameter estimates are sensitive to the initial guesses.

(14)   In nonlinear regression, a good initial guess for the parameters is to use those obtained from a linear solution of the linearised model.

(15)   In nonlinear regression, all the regular tests apply, but they need not be satisfied for the model to be acceptable.

(16)   Especially in nonlinear regression, scaling can improve the results.

(17)   Decreasing the sample size will provide smaller confidence intervals.

(18)   Replicates allow for detecting outliers and potential issues with the error structure.

(19)   Multicollinearity implies that there are hidden relationships in the data that could impact on the invertibility of the information matrix $(A^\mathrm{T}A)-1$.

(20)   If the error structure is incorrect, then the least-squares parameters estimates are biased.

3.8.2 Short Exercises

These questions should be solved using only a simple, nonprogrammable, nongraphical calculator combined with pen and paper.

(21)   Derive the traditional, least-squares formulae shown in Section A3.1 for the two parameter case.

(22)   Derive the traditional, weighted, least-squares formulae shown in Section A3.2 for the two parameter case.

(23)   Consider the problem of trying to fit data to the following model

$$y_t + \alpha_1 y_{t-1} + \alpha_2 y_{t-2} = \beta_1 u_{t-1} + \beta_2 u_{t-2} + e_t \qquad (3.111)$$

where α and β are coefficients to be determined, t is a subscript representing the time of measurement, that is, t represents the current time and $t-1$ represents the time one sampling unit in the past. Show how this model can be set up in the standard least-squares system in order to estimate a sample of data (y_t, u_t) starting at $t = 1$ and going up until $t = 100$. (Note that not every entry need be listed).

(24)   Consider trying to verify the resistance of a resistor by measuring the voltage at different currents. The model of interest is based on Ohm's Law and can be written as

Table 3.10 Current and voltage for an unknown resistor (for Question 24)

Current (I) (Amp)	Voltage (V) (V)
1	1.95
2	3.78
3	6.21
7	13.98
8	16.61
10	22.83

$$V = IR + \varepsilon \tag{3.112}$$

where V is the voltage in Volts (V), I the current in Amperes (Amp), and R the resistance in Ohms (Ω). The data you obtained from the experiment is given in Table 3.10. Answer the following questions using this data set:

a. Set this problem up in the standard matrix-based, ordinary, linear regression format for an arbitrary number of data points.

b. Show that for the given model, the ordinary, least-squares estimate of the resistance ($\hat{R}$) is $\hat{R} = \dfrac{\sum\limits_{i=1}^{m} I_i V_i}{\sum\limits_{i=1}^{m} I_i^2}$ and $\hat{\sigma} = \sqrt{\dfrac{\sum\limits_{i=1}^{m} V_i^2 - \hat{R} \sum\limits_{i=1}^{m} I_i V_i}{m - n}}$.

c. Calculate $\hat{R}$ and $\hat{\sigma}$ using the derived equations.

d. Obtain the 95% confidence interval for the parameter estimate.

e. The stated resistance is 2 Ω. Does the experimental value obtained confirm the stated resistance?

f. You fit the model and obtained the residuals shown in Fig. 3.10. Is the fit good? If not, what could be the cause of the observed pattern?

(25) Continuing with the data in Table 3.10, but now consider the case where the variance of the errors is proportional to the current squared ($\text{var}(\varepsilon) \propto I^2$). Answer the following questions.

Fig. 3.10 Residuals as a function of current (for Question 24)

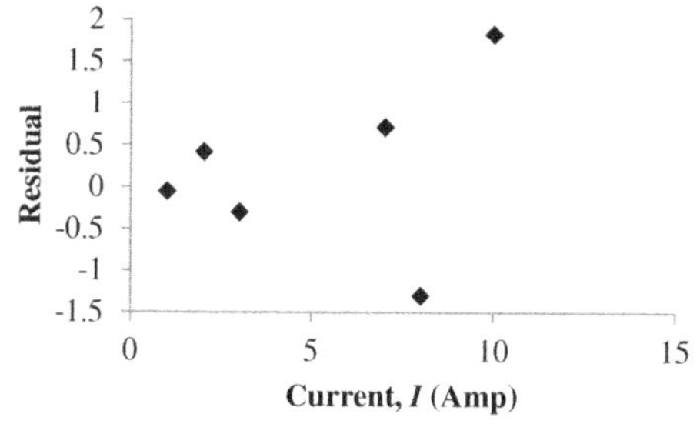

a. What is the weighting matrix? Set up this problem in the standard matrix-based weighted, least-squares, linear regression format for an arbitrary number of data points.

b. Show that, in this case, the estimate of the resistance can be calculated as $\hat{R} = \frac{1}{m} \sum_{i=1}^{m} \frac{V_i}{I_i}$. Calculate $\hat{R}$ using the derived equation.

c. How would you determine whether the provided data set is better described by the ordinary, least-squares estimate (Question 24) or the result obtained using weighted, least-squares?

d. When would this weighted, least-squares model be appropriate? Suggest why when computing the resistance, it is so common to simply take the average of all the available values.

(26) Using the data from Sect. 3.4.3, perform a more detailed analysis of the effect of linearisation on the model. Consider how linearisation changes the distribution of the original data points and how this could impact the parameter estimates obtained. What kind of transformations will cause this behaviour?

(27) Consider the problem of modelling the flow out of an orifice, where the relationship can be written as

$$\dot{m} = R\sqrt{h} \tag{3.113}$$

where $\dot{m}$ is the mass flow rate, R the resistance coefficient, and h the height in the tank. The data are provided in Table 3.11. Using these data, answer the following questions:

a. Determine the parameter estimates and their confidence intervals.
b. Validate the model.
c. Determine if there are any outliers and their impact on the regression.
d. Compute for the point, $x = 0.225$ m, the mean and predictive confidence intervals for the flow rate out of the tank.

Assume that $\alpha = 0.05$. Note that this problem can be solved without the use of complex matrix manipulations.

Table 3.11 Height and flow rate data

Height, h (m)	Flow Rate, $\dot{m}$ (kg/min)
0.35	16.4
0.30	15.4
0.25	14.3
0.20	12.7
0.15	11
0.10	9.6
0.05	6.4

Table 3.12 Freezing point of different ethylene glycol—water mixtures (for Question 28)

Mole percent ethylene glycol	Freezing point (K)
0.00	273.15
5.09	267.46
11.30	258.50
15.47	251.72
20.94	241.58
30.97	225.28
31.22	225.49
36.62	228.03
42.76	229.89
48.00	230.50
49.34	230.54
51.36	230.37
56.36	232.12
59.05	234.62

(Data taken from J. Bevan Ott, J. Rex Goates, and Hohn D. Lamb (1972), "Solid-liquid phase equilibria in water + ethylene glycol", Journal of Chemical Thermodynamics, 4, pp. 123–126)

3.8.3 *Computational Exercises*

The following problems should be solved with the help of a computer and appropriate software packages, such as MATLAB® *or* Excel®.

(28) Consider fitting a cubic (third-order) polynomial to determine the relationship between the freezing point of ethylene glycol and the weight percent of ethylene glycol in a water solution shown in Table 3.12.

   a. Determine the coefficients of the cubic polynomial using linear regression.
   b. Determine the 95% confidence intervals for the parameters.
   c. When we have 33.3 wt% of ethylene glycol in a solution, what is the freezing point? Calculate the 95% mean and predictive confidence intervals for the estimate. In general, why are the predictive confidence intervals larger?
   d. Plot the residuals as a function of weight percent and as a function of the freezing point. Include a normal probability plot of the residuals. Are there any issues with the model assumptions?
   e. Compute the R^2 and F-score for the model. Is the model good? Are there any physical explanations that could be provided for the observed behaviour?

(29) Consider trying to determine the calibration curve for gas chromatography based on some sample measurements with known ratios. Assume that the

Table 3.13 Gas chromatography calibration data (for Question 29)

Measured ratio (y)	Known ratio (x)
1.000	0.2
0.987	0.3
1.347	0.6
2.856	1.3
4.476	2.6
5.148	4.0

Table 3.14 Time constant (τ) as a function of the water level (h) (for Question 30)

Time constant, τ (s)	Level, h (m)
51.7	0.1
82.8	0.15
91	0.2
97.2	0.25
101.4	0.3
107.6	0.35

variance is proportional to x^2. Use the data provided in Table 3.13 to fit a linear model to the data. Analyse the residuals and model adequacy to determine how good the fit is

(30) Consider the data shown in Table 3.14, which seeks to determine the relationship between the time constant (residence time) in a tank and water level in the tank. Fit the following model to the data

$$\tau = K h^b \tag{3.114}$$

where τ is the time constant, K a parameter of interest, h the water level, and b the unknown power. Theoretically, the value of b should be 0.5 and $K = 2\rho A / R$, where ρ is the density, A the cross-sectional area of the tank, and R is the resistance coefficient.

a. Linearise the model given by Eq. (3.114) (ignore any error structure issues).
b. Fit the data to the linearised model to obtain the linearised parameter estimates.
c. Determine the confidence intervals of the linearised parameters.
d. Convert the linearised parameters into the true values. Can you obtain a confidence interval for the true value of K?
e. Compute R^2 and $\hat{\sigma}$.
f. Plot the time-series plot of the residuals and a normal probability plot of the residuals.
g. Is the value of $\hat{b}$ equal to 0.5?
h. If $\rho = 1{,}000$ kg/m^3 and $A = 0.0469$ m^2, compute R.

Table 3.15 Partial pressures of toluene at different temperatures (for Question 32)

Temperature, T (°C)	Vapour Pressure, P^{vap} (mm Hg)	
	Run 1	Run 2
−4.4	5.05	5.15
6.4	10.0	9.89
18.4	20.1	21.9
31.8	39.9	40.8
40.3	59.8	62.5
51.9	99.9	97.8
69.5	200	206
89.5	400	415
110.6	760	747
136.5	1502	1512

i. Repeat the above exercise for the original, nonlinear model. Obtain parameter estimates and confidence intervals for the nonlinear parameters. Compute R^2 and $\hat{\sigma}$. Plot the time-series plot of the residuals and a normal probability plot of the residuals.

j. Using the nonlinear model, determine if the value of $\hat{b}$ is equal to 0.5? Compute R using the nonlinear model if $\rho = 1{,}000$ kg/m^3 and $A = 0.0469$ m^2.

k. Compare the linearised results with the nonlinear ones? What are the main differences between the two fits. Plot both fits with the original data on the same plot and compare the results. Which fit is better?

(31) Using the friction factor data presented in Sect. 1.3, fit the nonlinear model and assess the quality of the result model. Pay special attention to the results of the different runs.

(32) Consider fitting the Antoine Equation to some vapour pressure as a function of temperature data that was obtained using toluene given in Table 3.15. The general form of the Antoine Equation can be written as

$$P^{vap} = 10^{A + \frac{B}{C+T}} \tag{3.115}$$

where A, B, and C are parameters, T is the temperature in °C, and P^{vap} is the vapour pressure of toluene in mm Hg. Two separate runs were performed using two different makes of measurement devices. By fitting a linearised model, a nonlinear model obtained by taking the $\log_{10}$ of Eq. (3.115), and a nonlinear model obtained using Eq. (3.115) to the data, and analysing the residuals, answer the following questions:

a. Are the errors for the two runs the same? How can this be determined?

b. Obtain separate parameter estimates for each of the runs and models.
    Which model best describes the data for the given run? What does this
    suggest about the appropriate error structure for each run?
c. Using the best parameter estimates for A, B, and C, compare them
    against the theoretical values of $A = 6.954\ 64$, $B = -1{,}344.8°C$, and
    $C = 219.482°C$ (Dean, 1999). Are the experimental values close to the
    accepted values?

*Hint: For the nonlinear models, it is suggested that the estimates obtained
using the linearised model be used as the initial guess for the nonlinear
method.*

Appendix A3: Nonmatrix Solutions to the Linear, Least-Squares Regression Problem

A3.1 Nonmatrix Solution for the Ordinary, Least-Squares Case

The nonmatrix solution only applies to the case of solving a simple model that can
be written as

$$y = a + bx. \tag{3.116}$$

Note that x can be replaced by $f(x)$ here and in all the following equations.

The ordinary, least-squares problem can be solved by first computing the following
two quantities:

$$s_x^2 = \frac{m \sum x^2 - \left(\sum x\right)^2}{m}$$
$$s_y^2 = \frac{m \sum y^2 - \left(\sum y\right)^2}{m}. \tag{3.117}$$

Then, the linear regression coefficients can be calculated as follows:

$$\hat{b} = \frac{m \sum xy - \sum x \sum y}{m \sum x^2 - \left(\sum x\right)^2}$$
$$\hat{a} = \frac{\sum y - \hat{b} \sum x}{m}. \tag{3.118}$$

The correlation coefficient is calculated using

$$R^2 = \frac{\left[m \sum xy - \left(\sum x\right)\left(\sum y\right)\right]^2}{\left[m \sum x^2 - \left(\sum x\right)^2\right]\left[m \sum y^2 - \left(\sum y\right)^2\right]}. \tag{3.119}$$

The standard deviation of the model is given as

$$\hat{\sigma} = \frac{1}{m-2}\left(s_y^2 - \hat{b}^2 s_x^2\right). \tag{3.120}$$

The standard deviation for coefficient b is given as

$$\hat{\sigma}\sqrt{\left(A^T A\right)_{22}^{-1}} = s_b = \frac{\hat{\sigma}}{s_x}. \tag{3.121}$$

The standard deviation of coefficient a is given as

$$\hat{\sigma}\sqrt{\left(A^T A\right)_{11}^{-1}} = s_a = \frac{\hat{\sigma}}{s_x}\sqrt{\frac{\sum x^2}{m}}. \tag{3.122}$$

The confidence interval for the mean response at a value of x_d is given by

$$\hat{y} \pm t_{1-\frac{\alpha}{2}, m-2}\,\hat{\sigma}\sqrt{\frac{1}{m} + \frac{\left(x_d - \frac{1}{m}\sum x\right)^2}{s_x^2}}. \tag{3.123}$$

The confidence interval for the prediction at a value of x_d is given by

$$\hat{y} \pm t_{1-\frac{\alpha}{2}, m-2}\,\hat{\sigma}\sqrt{1 + \frac{1}{m} + \frac{\left(x_d - \frac{1}{m}\sum x\right)^2}{s_x^2}}. \tag{3.124}$$

The total sum of squares would then be calculated using

$$TSS = \sum_{i=1}^{m}(y_i - \bar{y})^2 = \sum y^2 - \frac{1}{m}\left(\sum y\right)^2. \tag{3.125}$$

A3.2 Nonmatrix Solution for the Weighted, Least-Squares Case

The nonmatrix solution only applies to the case of solving a simple model that can be written as

$$y = a_w + b_w x. \tag{3.126}$$

Note that x can be replaced by $f(x)$ here and in all the following equations.

The weighted, least-squares problem can be solved by first computing the following two quantities:

$$s_{x_w}^2 = \frac{\sum w \sum w x^2 - \left(\sum wx\right)^2}{\left(\sum w\right)}$$

$$s_{y_w}^2 = \frac{\sum w \sum w y^2 - \left(\sum wy\right)^2}{\left(\sum w\right)}. \tag{3.127}$$

Then, the linear regression coefficients can be calculated as follows:

$$\hat{b}_w = \frac{\left(\sum w\right) \sum wxy - \sum wx \sum wy}{\left(\sum w\right) \sum wx^2 - \left(\sum wx\right)^2}$$

$$\hat{a}_w = \frac{\sum wy - \hat{b}_w \sum wx}{\left(\sum w\right)}. \tag{3.128}$$

The correlation coefficient is calculated using

$$R^2 = \frac{\left[\left(\sum w\right) \sum wxy - \left(\sum wx\right)\left(\sum wy\right)\right]^2}{\left[\left(\sum w\right) \sum wx^2 - \left(\sum wx\right)^2\right]\left[\left(\sum w\right) \sum wy^2 - \left(\sum wy\right)^2\right]}. \tag{3.129}$$

The standard deviation of the model is given as

$$\hat{\sigma}_w = \frac{1}{m-2}\left(s_{y_w}^2 - \hat{b}_w^2 s_{x_w}^2\right). \tag{3.130}$$

The standard deviation of coefficient b_w is given as

$$\hat{\sigma}\sqrt{\left(A^T W A\right)_{22}^{-1}} = s_{b_w} = \frac{\hat{\sigma}_w}{s_{x_w}} \tag{3.131}$$

The standard deviation of coefficient a is given as

$$\hat{\sigma}\sqrt{\left(A^T W A\right)_{11}^{-1}} = s_a = \frac{\hat{\sigma}_w}{s_{x_w}}\sqrt{\frac{\sum wx^2}{\sum w}}. \tag{3.132}$$

The confidence interval for the mean response at a value of x_d is given by

$$\hat{y} \pm t_{1-\frac{\alpha}{2},m-2}\hat{\sigma}_w \sqrt{\frac{1}{\left(\sum w\right)} + \frac{\left(x_d - \frac{1}{\left(\sum w\right)}\sum wx\right)^2}{s_{x_w}^2}}. \tag{3.133}$$

The confidence interval for the prediction at a value of x_d is given by

$$\hat{y} \pm t_{1-\frac{\alpha}{2},m-2-n_\sigma}\hat{\sigma}_w \sqrt{\frac{1}{w_d} + \frac{1}{\left(\sum w_i\right)} + \frac{\left(x_d - \frac{1}{\left(\sum w_i\right)}\sum w_i x_i\right)}{s_x^2}}. \tag{3.134}$$

It should be noted that the predicted weight at the given point, w_d, should be determined from a model with n_σ unknown parameters.

The total sum of squares would then be calculated using

$$TSS = \sum_{i=1}^{m} w_i(y_i - \bar{y})^2 = \sum wy^2 - \frac{1}{\left(\sum w_i\right)}\left(\sum wy\right)^2. \tag{3.135}$$

Chapter 4
Design of Experiments

Given the power of regression analysis, it would be great to design experiments or ways of obtaining the process information so that as much useful information is obtained with the smallest number of experiments. In any real system, running experiments costs money and wastes resources, since the process will not necessarily be producing at its optimal levels. The question is how to design experiments so that the maximal amount of information can be extracted and used in regression analysis. Basically, this problem reduces to developing various designs for the regression matrix. When the designed regression matrix contains certain desirable properties, then the computation and analysis of the parameters can be performed faster and better.

4.1 Fundamentals of Design of Experiments

In order to understand the methods by which optimal experiments may be designed, it is necessary to understand some of the factors that affect the results. The most important topics are **sensitivity**, **confounding and correlation between parameters**, **blocking**, and **randomization**. In general, the question in the design of experiments reduces to determining whether $A^T A$ is invertible and well conditioned. A well-conditioned matrix is required when performing computations on a computer, since there will always be inevitable round-off errors. In a well-conditions matrix, such round-off errors do not have a disproportionate impact on the final result.

4.1.1 Sensitivity

The **sensitivity**, S, is a measure of how easy it is to estimate a given parameter. The sensitivity of the process with respect to a given parameter β_i is

147

Y. A. W. Shardt, *Statistics for Chemical and Process Engineers*, https://doi.org/10.1007/978-3-030-83190-5_4

$$\vec{S}_{\beta_i} = \frac{\partial \vec{f}\left(\vec{x}, \vec{\beta}\right)}{\partial \beta_i}, \tag{4.1}$$

where $\vec{f}\left(\vec{x}, \vec{\beta}\right)$ is the regression function. The larger the value, the easier it is to estimate the parameters. Basically, if the sensitivity is small, then a large change is required in the regressors to obtain a statistically significant result. On the other hand, if the sensitivity is large, then a smaller change in the regressors is required to obtain a statistically significant result. This implies that the information matrix will be well conditioned.

Sensitivity is also important because there will always be noise (or unwanted changes) in the system that cannot be accounted for. In these cases, if the change in the system is not noticeable given the input then it will be difficult to identify the system effectively, that is, the **signal-to-noise** ratio needs to be larger for insensitive systems. Large (input or regressor) signals can have practical issues, in that they may not be feasible due to safety or process constraints, for example, a tank cannot be filled to more than its capacity or a valve cannot be opened to more than 100% (or fully open).

4.1.2 *Confounding and Correlation Between Parameters*

Correlation between parameters represents the degree to which two separate parameters can be identified independently of each other. Issues with correlation often arise when fitting theoretical models. For example, consider the following relationship

$$y = e^{-\frac{a}{bT}}, \tag{4.2}$$

where it is desired to estimate both a and b. In this example, one cannot estimate both parameters separately as they are completely correlated with one another. In such a situation, it is said that the two parameters are **confounded** with each other.

On the other hand, the situation where there is no correlation between the different parameters is referred to as **orthogonality**. The benefits of orthogonality are that it allows for easy analysis of the resulting models, including the effects of adding or removing parameters. In many experimental designs, orthogonality is a much sought-after quality of a model.

The correlation between parameters can be determined by an analysis of the $(\mathcal{A}^T \mathcal{A})$ matrix. If the off-diagonal entries of a given parameter are nonzero, then the variable is correlated with this other parameter. For example, for the matrix given as

$$(A^T A) = \begin{bmatrix} 2 & 0 & 1 \\ 0 & 3 & 0 \\ 1 & 0 & 1 \end{bmatrix} \tag{4.3}$$

the first parameter is correlated with the third parameter, while the second parameter is not correlated with any of the other two. If two or more columns are a linear combination of each other, then the parameters are said to be confounded with each other. In most cases, this will be seen as two columns that are multiples of each other.

4.1.3 Blocking

Blocking seeks to minimize the effect of known but uncontrollable variables that could obscure the desired relationships. The effect of these variables depends on the type of experiment being performed. In chemical engineering, uncontrollable variables can include changes in the daily feed composition, ambient conditions (changes in the seasons), and differences between different analysis methods. Depending on the type of experiment designed, the importance of and ways to implement blocking will be different.

A **block** is a single group of experiments that is run under nearly identical conditions, with separate blocks having potentially different conditions. The importance of blocking can be illustrated by the following examples.

4.1.3.1 Rabbit Weight Experiment

Consider an experiment designed to determine the effects of diets A and B on rabbits. There are eight cages arranged as shown in Fig. 4.1. There are a total of eight rabbits of the same species of which four are male and four female. Based on previous experiments, it is known that the sex of the rabbits and the location of the rabbit in the room will influence the weight of the rabbit. The question is how to design an experiment that will minimize the effect of the undesired variables (location and sex) and maximize the effect of the desired variables (diet) on the rabbits' weights.

If a diet regime is assigned randomly to each rabbit, which is then randomly assigned to the cages, it can easily be seen that it is possible that most of the females get Diet A and the males get Diet B. Since males gain more weight than females, the effect of the diets could be confounded with the effect of sex. A similar situation could apply to the partitioning of the rabbits in the different cages.

Another approach to take is to assign a pair of identical rabbits (same sex) to each station and then give one of the rabbits Diet A and the other Diet B. The analysis would then be performed by considering the difference between the two rabbits at each station. Since it can be assumed that each station has two, nearly identical rabbits

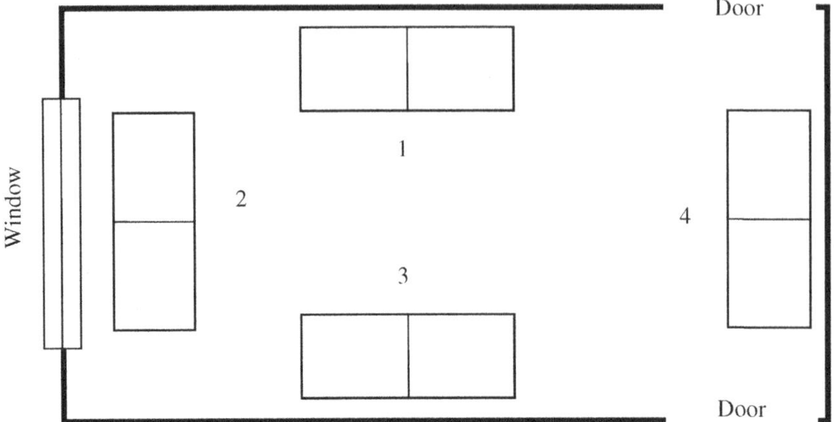

Fig. 4.1 Layout of the cages

with very similar environmental factors, the effect of these environmental variables on the two rabbits should be the same (or similar). Therefore, any difference between the pair can be attributed to the difference between the diets.

It should be noted that blocking has reduced the number of data points from 8 (in the naïve implementation) to 4 (in the blocked version). This reduction of available samples is often the result of implementing blocking. However, blocking can lead to an improvement in uncovering the relationship.

4.1.3.2   Shoe Wear Example

Consider the case of trying to determine whether a new material for the sole of a shoe is better than the previous material (Box, Hunter, & Hunter, 1978). In this case, it is obvious that the amount of wear a shoe experiences depends on the person. Therefore, giving one person either a shoe with the new material or the old material will not account for this problem. An ingenious solution to the problem is to give each person a pair of shoes where one sole is made of the new material and the other sole is of the old material, and then analyse the difference in wear between the two soles. Furthermore, randomizing on which foot (left or right) the new material will be used will allow for an even better analysis of the results.

4.1.4   Randomization

Randomization is the procedure by which the order of the runs in the experiment is determined randomly. This allows the effect of any *nuisance* or uncontrollable factors on the experiment to be minimized. This is especially true with replicates, as running a set of replicates one after another can often eliminate the benefit of having performed replication, since the conditions will often remain the same.

4.2 Types of Models

In the chemical and process industry, there exist three broad categories of models that can be used:

(1) **White-box or First-principle models**, which are developed based on a theoretical analysis of the system using mass and energy balances, as well as known physical constants. These models have the advantage that they are very general and can be applied to a wide range of similar system. Their main disadvantage is that often various limiting assumptions need to be taken in order to obtain a useful solution. As well, obtaining exact values for the parameters may be difficult.

(2) **Black-box models**, which are developed solely from experimental data. These models can accurately describe the given operating point, but cannot be applied to new conditions or operating points of the system.

(3) **Grey-box models**, which combine the first-principles models with experimental data. In these types of models, the general form of the model is obtained using a first-principles approach, and then experimental data are used to obtain the values of the different constants. The advantage of this approach is that it combines the advantages of the other approaches. It is a very common approach in chemical engineering.

4.2.1 Model Use

The models that are developed can be used for two different goals: **analysing (past) performance** or **forecasting future performance**. Depending on the goal, the methods used to validate the model, that is, show that it is sufficient, will be different. For the first goal of analysing performance, the method previously presented will be sufficient. However, when developing models for forecasting future performance, it is necessary to show that the model can accurately forecast future values using data that were not originally part of the model creation step. In many cases, a model that is sufficient for analysing performance may not give good forecasting performance. This topic is explored in greater detail in Chap. 6, where the development of black-box models for process control is considered.

4.3 Framework for the Analysis of Experiments

Irrespective of the type of experiment being analysed, the following procedure can be followed:

(I) **Pre-analysis**, which lays the groundwork for the following parts. In many ways, this is both the most difficult and the most important step in the whole framework. It consists of the following steps:

    a. Determine the type of experiment that was performed and relevant information about the model.

    b. Write down the model that will be estimated given the above information.

    c. If desired, determine an appropriate orthogonal or orthonormal basis for the model.

(II) **Calculating the values**, using appropriate formulae compute the parameter estimates, the normal probability plot of the parameters, and if appropriate, the SSR_i and F-value for each of the parameters.

(III) **Model reduction**, which seeks to remove all unnecessary parameters from the model. Three main methods can be used:

    a. **Normal probability plot of the parameters**, which seeks to determine which parameters are most significant based on their deviation from normality. This approach works as long as an orthonormal basis has been selected for the model.

    b. F_i**-test for the parameters**, which seeks to determine which parameters are most significant based on an analysis of variance method. This approach works as long as an orthogonal basis has been selected for the model and there are replicates with which to compute an estimate of the model variance.

    c. **Confidence intervals for the parameters**, which seeks to determine which confidence intervals of the parameters cover zero and hence should be rejected. This approach works as long as there are replicates with which to compute an estimate of the model variance.

(IV) **Residual analysis**, using the reduced model, the resulting residuals should be analysed to determine whether the assumptions underlying regression have been satisfied. If the residuals suggest problems, then further experiments or analysis may need to be performed to resolve the issue.

(V) **Conclusions**, which seek to answer based on the reduced model whatever questions the objectives of the exercise gave, for example, determining the optimal operating point.

4.4 Factorial Design

Factorial design seeks to determine a black-box model of the system that can accurately describe the behaviour of the system in the region studied that also includes interactions or combinations of the different variables (e.g. $x_1 x_2$). Factorial designs have the advantage that changes in the variables are not made sequentially, but following some type of pattern, so that interactions between the different variables can also be measured.

The basic factorial design consists of **k factors**, or independent variables, and l **different levels** or points at which the system will be tested. A factorial experiment with l levels and k factors is called an l^k factorial experiment. For the purposes of this discussion, it will be assumed that all factors have the same number of levels. The complete experimental design will be repeated n_R times, which is referred to as the **number of replicates**. A **treatment** refers to a single run of the factorial design with given values for each factor.

Traditionally, the factors are encoded using the centring formula presented as Eq. (175), so that the largest value of a factor has an encoded value of $+1$, and the smallest value of the factor has an encoded value of -1. Discrete or qualitative variables can be encoded by arbitrarily assigning a given value with a specific encoded value, for example, *cast iron* could be given a value of $+1$ and *stainless steel* a value of -1. For a two-level factorial design, it is traditional to denote the two levels as -1 and $+1$, which provides an orthonormal basis for regression analysis. For a three-level factorial design, it is traditional to denote the three levels as -1, 0, and $+1$. Unfortunately, such a coding is not orthogonal.

4.4.1 Factorial Design Models

The model that will be fit in factorial designs can be written abstractly as

$$y_i = \mu + \sum_{j=1}^{k} \tau_{ij} + \sum \gamma_{ig} + e_i, \tag{4.4}$$

where μ is the mean response, τ_{ij} is the main effect of the jth factor at the ith treatment level, and γ_{ig} is the gth interaction, y_i the observed output at the ith treatment level, and e is the error. Practically, the abstract model given by Eq. (4.4) can be rewritten as

$$y_i = \beta_0 + \sum_{j=1}^{k}\sum_{d=1}^{l-1}\beta_{j^d}x_j^d + \prod_{\text{in twos,}} \left(\beta_{..}\sum_{j=1}^{k}\sum_{d=1}^{l-1}x_j^d\right) + e_i, \qquad (4.5)$$

threes,

$\cdots,$

groups of k

where β are the parameters to be estimated, β_{j^d} is short-hand for writing $\beta_{\underbrace{j \ldots j}_{d-times}}$, and the product term represents the main effects taken in groups of two, three, and so on until a single group of all k parameters is taken. For an l^k experiment, there should be a total of l^k parameters to be estimated. For a two-level experiment, the model fit is given as

$$y = \beta_0 + \sum_{i=1}^{k}\beta_i x_i + \sum_{j=1}^{k}\sum_{p=j+1}^{k}\beta_{jp}x_j x_p + \ldots + \beta_{\underset{i=1}{\prod}k}\prod_{i=1}^{k}x_i. \qquad (4.6)$$

The **order of an interaction** is defined as the sum of the powers of the variables that multiply together to give the particular interaction. Thus, for example, the interaction given by x_1 has order 1, or is a first-order interaction, while the interaction given by $x_1 x_2 x_3$ has order 3, or is a third-order interaction. Likewise, $x_1^2 x_2$ has order 3, since x_1 is raised to the second power and x_2 is raised to the first power, which totals 3. First-order interactions are often referred to as the **main effects**. The zero-order interaction is commonly referred as the **mean response**. The coefficients for each interaction are denoted as follows: β_1 is the coefficient multiplying the interaction given by x_1, while β_{123} is the coefficient multiplying the interaction given by $x_1 x_2 x_3$. Note that the order of an interaction has nothing to do with the **order of a model**. The order of a model is defined as the highest power of a factor present in the model.

Therefore, based on the above discussion of the model to be determined, for a 2^3-factorial design, the model can be written as

$$y = \beta_0 + \beta_1 x_1 + \beta_2 x_2 + \beta_3 x_3 + \beta_{12}x_1 x_2 + \beta_{13}x_1 x_3 + \beta_{23}x_2 x_3 + \beta_{123}x_1 x_2 x_3 \qquad (4.7)$$

This implies that a single row of the $\mathcal{A}$-matrix can be written as

$$\mathcal{A} = \begin{bmatrix} 1 \; k \; factors \; groups \, of \, 2 \; groups \, of \, 3 \; \cdots \; all \, k \, factors \end{bmatrix}$$
$$= \begin{bmatrix} 1 \; x_1 \; x_2 \; x_3 \; x_1 x_2 \; x_1 x_3 \; x_2 x_3 \; x_1 x_2 x_3 \end{bmatrix} \qquad (4.8)$$

The values to the factors are assigned so that all possible combinations of levels and factors are obtained. For a 2^3-factorial experiment (i.e. there are two levels with three factors), the regression matrix ($\mathcal{A}$) would look like this:

$$
\mathcal{A} =
\begin{bmatrix}
\overbrace{1} & \overbrace{1} & \overbrace{1} & \overbrace{1} & \overbrace{1} & \overbrace{1} & \overbrace{1} & \overbrace{1} \\
1 & 1 & 1 & -1 & 1 & -1 & -1 & -1 \\
1 & 1 & -1 & 1 & -1 & 1 & -1 & -1 \\
1 & 1 & -1 & -1 & -1 & -1 & 1 & 1 \\
1 & -1 & 1 & 1 & -1 & -1 & 1 & -1 \\
1 & -1 & 1 & -1 & -1 & 1 & -1 & 1 \\
1 & -1 & -1 & 1 & 1 & -1 & -1 & 1 \\
1 & -1 & -1 & -1 & 1 & 1 & 1 & -1
\end{bmatrix},
$$

with column groups labelled 1, A, B, C | AB, AC, BC, ABC.

where $+1$ refers to one level (normally high) and -1 refers to the other level (normally low). Only the columns with the individual factors need to be entered separately. The rest can be calculated.

The output can be stored as follows. Define:

$$
\hat{\bar{y}} =
\begin{bmatrix}
y_{11} & \cdots & y_{1n_R} \\
\vdots & & \vdots \\
y_{l^k 1} & \cdots & y_{l^k n_R}
\end{bmatrix},
\tag{4.9}
$$

where y_{ij} are the individual replicate results corresponding to treatments given in $\mathcal{A}$. Furthermore, define

$$
\vec{\bar{y}} = \frac{1}{n_R}
\begin{bmatrix}
\sum\limits_{i=1}^{n_R} y_{1i} \\
\vdots \\
\sum\limits_{i=1}^{n_R} y_{l^k i}
\end{bmatrix}.
\tag{4.10}
$$

Example 4.1 *Full Factorial Design*

Consider a 2^2 full factorial experiment with two replicates and answer the following questions:

(1) How many levels and factors are there in this example?
(2) What model will be fit to the data?

(3) Which parameter(s) represent the mean response, the first-order interactions, and the second-order interactions?

(4) How many experiments will be run in this example?

(5) What is the regression matrix ($\mathcal{A}$) for this example?

Solution

(1) There are two levels and two factors. Traditionally, the factors are denoted using uppercase Latin letters in sequential order, that is A represents the first factor, B the second, C the third, and so on.

(2) The following model will be fit to the data:

$$y_t = \beta_0 + \beta_1 x_1 + \beta_2 x_2 + \beta_{12} x_1 x_2.$$

(3) The mean response is denoted by β_0, the first-order interactions by β_1 and β_2, and the second-order interaction by β_{12}.

(4) A total of $2^2 = 4$ experiments will be run in each replicate. Since there are two replicates, a grand total of eight experiments will be run.

(5) The regression matrix for this example is

$$\mathcal{A} = \begin{bmatrix} \overbrace{1}^{1} & \overbrace{1}^{A} & \overbrace{1}^{B} & \overbrace{1}^{AB} \\ 1 & 1 & -1 & -1 \\ 1 & -1 & 1 & -1 \\ 1 & -1 & -1 & 1 \end{bmatrix}$$

4.4.2 Factorial Analysis

It is possible to analyse the model based on an F-test analysis and matrices. A nonmatrix approach is presented in Appendix A4. Let:

$$\overline{\mathcal{A}} = \begin{bmatrix} \mathcal{A} \\ \vdots \\ n_R \; times \\ \mathcal{A} \end{bmatrix} \tag{4.11}$$

$$\vec{y} = \begin{bmatrix} \hat{\bar{y}}(1^{st}\ \text{column}) \\ \hat{\bar{y}}(2^{nd}\ \text{column}) \\ \vdots \\ \hat{\bar{y}}(n_R^{th}\ \text{column}) \end{bmatrix} \tag{4.12}$$

$$\hat{\vec{\beta}} = (\overline{A}^T \overline{A})^{-1} \overline{A}^T \vec{y}. \tag{4.13}$$

Then the standard deviation is defined as follows:

$$\hat{\sigma} = \sqrt{\frac{\vec{y}^T \vec{y} - \hat{\vec{\beta}}^T \overline{A}^T \vec{y}}{l^k (n_R - 1)}}. \tag{4.14}$$

The sum of squares due to errors, SSE, is given by

$$SSE = \vec{y}^T \vec{y} - \hat{\vec{\beta}}^T \overline{A}^T \vec{y}. \tag{4.15}$$

If an orthogonal basis is used for the levels, then the sum of squares due to each regressor, SSR_i, is given by

$$SSR_i = (\overline{A}^T \overline{A})_{ii} \hat{\beta}_i^2, \tag{4.16}$$

where $(\overline{A}^T \overline{A})_{ii}$ is the (i, i) (diagonal) entry in the $(\overline{A}^T \overline{A})$ matrix. If an orthonormal basis is used to encode the variables and their levels, then[1]

$$(\overline{A}^T \overline{A}) = l^k n_R \mathcal{I}. \tag{4.17}$$

When $l = 2$ and the proposed ± 1 encoding is used, then, by definition, the basis is orthonormal, and Eq. (4.17) reduces to

$$(\overline{A}^T \overline{A}) = 2^k n_R \mathcal{I}. \tag{4.18}$$

The total sum of squares, TSS, is given by

$$TSS = SSE + \sum_{i=1}^{l^k} (\overline{A}^T \overline{A})_{ii} \hat{\beta}_i^2. \tag{4.19}$$

To determine whether a given regressor should be in the model, calculate for each regressor the F-statistic as follows:

---

[1] Determining an orthogonal or orthonormal basis for an arbitrary level is explained fully in Sect. 4.7.

$$F_i = \frac{SSR_i}{\dfrac{SSE}{l^k(n_R-1)}}. \tag{4.20}$$

The value obtained from Eq. (4.20) is compared with the critical F-value denoted as $F(1-\alpha, 1, l^k(n_R-1))$. If F_i is greater than F-critical, then the regressor should be kept in the model. Otherwise, the given regressor can be eliminated from the model. The **effect** due to an interaction is equal to twice the corresponding calculated regression parameter.

4.4.3 Selecting Influential Parameters (Effects)

Since factorial designs are orthogonal designs, it is possible to analyse the significance of the individual parameters by examining their distribution. The easiest approach is to plot the parameters on a normal probability plot. Those parameters that are far from being normal (that is, they are "outliers") are most likely to be significant and should be retained in the final model, while those parameters that are close to being normal should not be retained in the final model. For example, consider the estimated parameters for a 2^4-factorial experiment shown in Fig. 4.2.

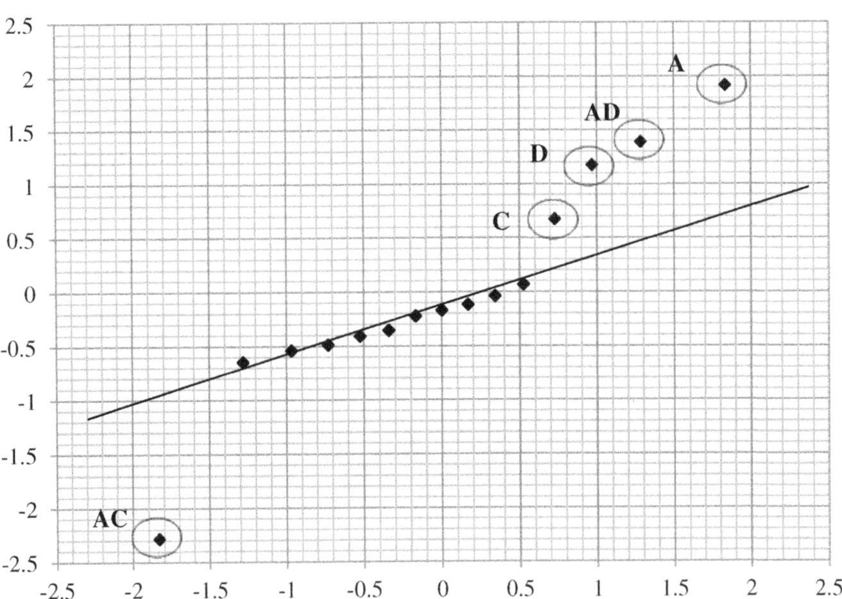

Fig. 4.2 Normal probability plot of parameters (effects) for a 2^4 experiment with significant points highlighted and labelled

Those points that have been circled and labelled are significant, in that they lie far from the main central cluster of points that define a line. Therefore, it can be seen that there are only five significant parameters (effects) that should be considered in the final model.

4.4.4 Projection

The property of **projection** states that if one of the factors is removed from a factorial experiment (due to whatever reasons), then the remaining experiment is still a factorial experiment. Formally, this can be stated that, if the original experiment is l^k and $p < k$ factors are removed, then the design becomes an l^{k-p} experiment with $p + 1 \, (= n_R)$ replicates. Consider the following 2^3 experiment where the first factor (x_1) is removed. Initially, the regression matrix can be written as

$$
\mathcal{A} =
\begin{bmatrix}
\overset{1}{1} & \overset{A}{1} & \overset{B}{1} & \overset{C}{1} & \overset{AB}{1} & \overset{AC}{1} & \overset{BC}{1} & \overset{ABC}{1} \\
1 & 1 & 1 & -1 & 1 & -1 & -1 & -1 \\
1 & 1 & -1 & 1 & -1 & 1 & -1 & -1 \\
1 & 1 & -1 & -1 & -1 & -1 & 1 & 1 \\
1 & -1 & 1 & 1 & -1 & -1 & 1 & -1 \\
1 & -1 & 1 & -1 & -1 & 1 & -1 & 1 \\
1 & -1 & -1 & 1 & 1 & -1 & -1 & 1 \\
1 & -1 & -1 & -1 & 1 & 1 & 1 & -1
\end{bmatrix}
$$

Removing the columns that contain the A factor gives

$$
\mathcal{A} =
\begin{bmatrix}
\overset{1}{1} & \overset{A}{1} & \overset{B}{1} & \overset{C}{1} & \overset{AB}{1} & \overset{AC}{1} & \overset{BC}{1} & \overset{ABC}{1} \\
1 & 1 & 1 & -1 & 1 & -1 & -1 & -1 \\
1 & 1 & -1 & 1 & -1 & 1 & -1 & -1 \\
1 & 1 & -1 & -1 & -1 & -1 & 1 & 1 \\
1 & -1 & 1 & 1 & -1 & -1 & 1 & -1 \\
1 & -1 & 1 & -1 & -1 & 1 & -1 & 1 \\
1 & -1 & -1 & 1 & 1 & -1 & -1 & 1 \\
1 & -1 & -1 & -1 & 1 & 1 & 1 & -1
\end{bmatrix}
\Rightarrow
\begin{bmatrix}
\overset{1}{1} & \overset{B}{1} & \overset{C}{1} & \overset{BC}{1} \\
1 & 1 & -1 & -1 \\
1 & -1 & 1 & -1 \\
1 & -1 & -1 & 1 \\
1 & 1 & 1 & 1 \\
1 & 1 & -1 & -1 \\
1 & -1 & 1 & -1 \\
1 & -1 & -1 & 1
\end{bmatrix}
$$

where there are now two replicates of the new experiment. This feature is extremely useful because it means that removing an unnecessary factor does not mean that the experiment was wasted. In fact, the remaining factors can be analysed as if they had been taken from an original full factorial designed experiment with replicates.

Example 4.2 Analysis of a Full Factorial Experiment

A series of experiments have been performed on a plant distillation column to determine the effects of different parameters on the overall purity of the overhead product. The variables of interest are reboiler duty (A), feed temperature (B), reflux ratio (C), and feed location (D). The purity of the product is expressed in a proprietary scale where 150 is absolutely pure and 50 is 70% pure. The data obtained from this 2^4-factorial experiment with no replicates are shown in Table 4.1. Perform the following analysis of the data set:

Table 4.1 Factorial design data for a plant distillation column

y	A (x_1)	B (x_2)	C (x_3)	D (x_4)
45	−1	−1	−1	−1
71	1	−1	−1	−1
48	−1	1	−1	−1
65	1	1	−1	−1
68	−1	−1	1	−1
60	1	−1	1	−1
80	−1	1	1	−1
65	1	1	1	−1
43	−1	−1	−1	1
100	1	−1	−1	1
45	−1	1	−1	1
104	1	1	−1	1
75	−1	−1	1	1
86	1	−1	1	1
70	−1	1	1	1
96	1	1	1	1

(1)   What is the full model being fit?
(2)   Plot a normal probability plot of the effects. Which effects are significant? Which factor does not seem to influence the results at all?

(3)   What simplified model could be fit?

(4)   Analyse the simplified model and determine if it is sufficient?

Solution

(1)   The model for the 2^4-factorial experiment can be written as

$$y = \beta_0 + \beta_1 x_1 + \beta_2 x_2 + \beta_3 x_3 + \beta_4 x_4 + \beta_{12} x_1 x_2 + \beta_{13} x_1 x_2 + \beta_{14} x_1 x_4 + \beta_{23} x_2 x_3$$
$$+ \beta_{24} x_2 x_4 + \beta_{34} x_3 x_4 + \beta_{123} x_1 x_2 x_3 + \beta_{124} x_1 x_2 x_4 + \beta_{134} x_1 x_3 x_4 + \beta_{234} x_2 x_3 x_4$$
$$+ \beta_{1234} x_1 x_2 x_3 x_4$$

Since $2^4 = 16$, there should be 16 parameters in the model. The regression matrix can be written as

(1)	A	B	C	D	AB	AC	AD	BC	BD	CD	ABC	ABD	ACD	BCD	ABCD
1	-1	-1	-1	-1	1	1	1	1	1	1	-1	-1	-1	-1	1
1	1	-1	-1	-1	-1	-1	-1	1	1	1	1	1	1	-1	-1
1	-1	1	-1	-1	-1	1	1	-1	-1	1	1	1	-1	1	-1
1	1	1	-1	-1	1	-1	-1	-1	-1	1	-1	-1	1	1	1
1	-1	-1	1	-1	1	-1	1	-1	1	-1	1	-1	1	1	-1
1	1	-1	1	-1	-1	1	-1	-1	1	-1	-1	1	-1	1	1
1	-1	1	1	-1	-1	-1	1	1	-1	-1	-1	1	1	-1	1
1	1	1	1	-1	1	1	-1	1	-1	-1	1	-1	-1	-1	-1
1	-1	-1	-1	1	1	1	-1	1	-1	-1	-1	1	1	1	-1
1	1	-1	-1	1	-1	-1	1	1	-1	-1	1	-1	-1	1	1
1	-1	1	-1	1	-1	1	-1	-1	1	-1	1	-1	1	-1	1
1	1	1	-1	1	1	-1	1	-1	1	-1	-1	1	-1	-1	-1
1	-1	-1	1	1	1	-1	-1	-1	-1	1	1	1	-1	-1	1
1	1	-1	1	1	-1	1	1	-1	-1	1	-1	-1	1	-1	-1
1	-1	1	1	1	-1	-1	-1	1	1	1	-1	-1	-1	1	-1
1	1	1	1	1	1	1	1	1	1	1	1	1	1	1	1

The parameter estimates can be obtained by noting that $\mathcal{A}^T \mathcal{A} = 2^4 \mathcal{J}$ and that $\mathcal{A}^T y$ can be computed by taking the sign from the appropriate column in the above table and placing it in front of the y matrix. This gives

$$\hat{\vec{\beta}} = 2^{-4} \mathcal{A}^T \vec{y}$$

$$= \begin{bmatrix} 70.06\ 10.81\ 1.56\ 4.94\ 7.31\ 0.0625\ -9.06\ 8.31\quad 1.19\quad -0.188\quad \dots \\ \dots \qquad\qquad -0.563\ 0.938\ 2.063\ -0.813\ -1.32\ 0.688 \end{bmatrix}^T$$

(b)	A normal probability plot of the effects is shown in Fig. 4.3. The effects that lie far from the expected normal distribution values are those that are significant because they are not chance values. The most significant effects have been circled and labelled. Therefore, the significant effects are those denoted as A, C, D, AC, and AD. The effect due to B is negligible.

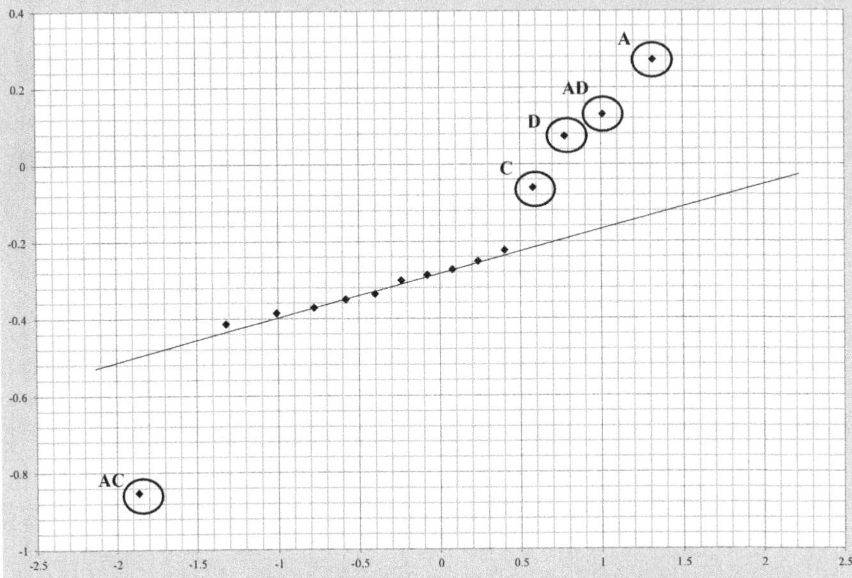

Fig. 4.3 Normal probability plot of the effects

(iii)	Dropping the B factor will produce a 2^3-factorial experiment with two replicates. In addition to dropping the terms associated with the B factor, all other terms will also be dropped. Since the design is orthogonal, we can drop the terms, without needing to recalculate anything. Therefore, the simplified model is given as

$$y = 70.06 + 10.8x_1 + 4.94x_3 + 7.31x_4 - 9.06x_1x_3 + 8.31x_1x_4$$

(iv)	The residuals for this case are shown in Fig. 4.4. It can be seen that they are more or less normally distributed. Furthermore, since the reduced model has an $R^2 = 0.966$ with all significant parameter values, it can be concluded that the results are probably good.

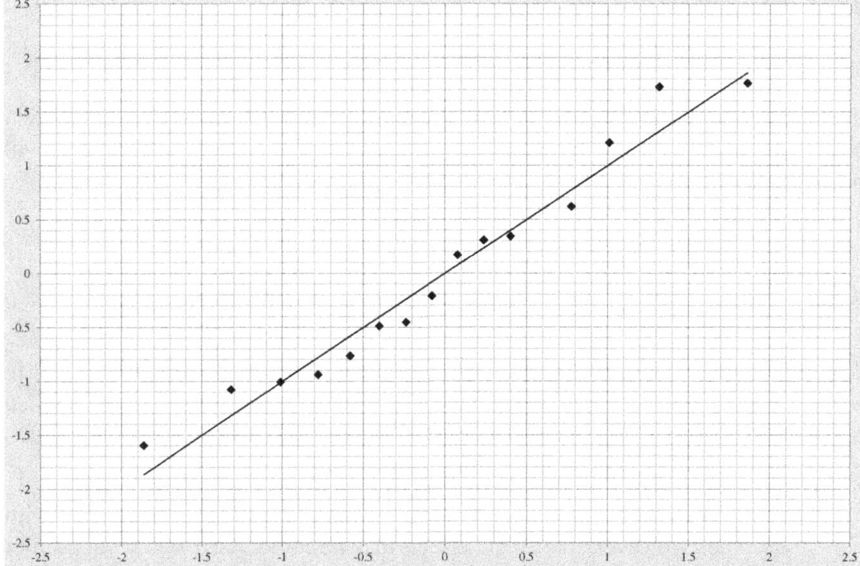

Fig. 4.4 Normal probability plot of the residuals for the reduced model

The solution to this problem using Excel® is presented in Sect. 8.7.3: Factorial Design Examples.

4.5   Fractional Factorial Design

For a large number of factors, it may be inconvenient to perform all the necessary experiments to determine a full fractional experiment, for example, for a 10-factor experiment, a total of $2^{10} = 1{,}024$ experiments need to be run. Since in many cases, a single experiment can take a few hours to run, then it could easily take more than two months to complete the experiment. At the same time, it may be known that many of the higher-order interactions may be negligible and can, thus, be ignored. This assumption is necessary, since by performing fewer experiments, some of the effects will be **confounded**, or **aliased**, with other measured effects. If the confounded interactions are small, then they will not affect the values obtained for the main effects. Therefore, the goal of this section is to determine the confounding pattern so that only the most important parameters are included in the model and the confounded parameters are kept to a minimum.

4.5.1   Notation for Fractional Factorial Experiments

Given l levels, k factors, and p fractions, then l^{k-p} interactions of the original l^k interactions can be estimated. For example, an experiment where $l = 2$, $k = 5$, and $p = 1$ is often called a half-fraction experiment, since half as many experiments will be performed.

4.5.2   Resolution of Fractional Factorial Experiments

The resolution of an experiment shows the manner in which the confounding in a given experiment occurs. The resolution is denoted using a subscripted Roman numeral, for example, 2_{IV}^{6-2} is a quarter fraction of a 2^6-factorial experiment with a resolution of IV. There are three common **resolutions**:

(1)   **Resolution III:** In these experiments, no main effects are confounded with each other. However, some main effects may be confounded with second-order interactions. Finally, some second-order interactions may be confounded with each other.

(2)   **Resolution IV:** In these experiments, no main effects are confounded with each other or second-order interactions. However, some second-order interactions are confounded with each other.

(3)   **Resolution V:** In these experiments, no main effects or second-order interactions are confounded with each other. However, second-order interactions are confounded with third-order interactions.

Determining the resolution requires looking at the complete confounding pattern for the given fractional factorial experiment and determining the term with the smallest number of variables multiplied together.

4.5.3   Confounding in Fractional Factorial Experiments

One of the most important concepts in fractional factorial design is **confounding** or **aliasing**. Confounding occurs when two or more interactions share the same column space, that is, the column entries for the interactions are the same. Only a full factorial experiment does not have confounding. By reducing the number of experiments performed, not all of the parameters can be estimated. In a p-fractional experiment, $l^{k-p}(l^p-1)$ of the interactions will be confounded. Confounding implies that the estimate of a given parameter is actually the estimate of two or more unbiased (unaliased) parameters. However, if it is assumed that higher-order interactions are negligible, then the effect of the higher interactions on the lower interactions' estimates will be small, and the parameter that will be estimated will be close to the true lower-interaction value. This assumption can be made due to Occam's razor or the sparsity-of-effects principle, which states that the smaller the number of interactions, the more likely the model is to be a good description of the system.

4.5.3.1 · Background Information

In order to determine the manner in which the different variables are confounded, it is first necessary to consider two mathematical concepts: identity vector and modular arithmetic.

For an orthogonal basis, let I be defined as a vector of 1's. The vector I forms the basis for the constant term, β_0, in the factorial experiment. Irrespective of the factorial design, the vector I can be treated as representing the identity vector for the system under pointwise multiplication denoted by $\odot$.[2]

Modular arithmetic denoted as $x \bmod y$, where x is the divisor and y is the dividend (or base) seeks to determine the remainder when x is divided by y, for example, 7 mod 2 will be equal to 1, since the remainder when 7 is divided by 2 is 1 ($7 = 3 \times 2 + 1$). When seeking to determine the confounding pattern in fractional factorial experiments and higher-order terms are encountered, then reduction of these terms is performed using l-base modular arithmetic, where l, as before, is the number of levels in the design.

Example 4.3 Modular Arithmetic

Consider a three-level design with the term $x_1^3 x_2 x_3^4$. Determine the reduced form.

Solution

Since this is a three-level experiment, all reductions will be performed using modular arithmetic with a base of 3. Therefore, the following reductions will be performed.

x_1: $3 \bmod 3 = 0 \Rightarrow x_1 = I$.

x_2: $1 \bmod 3 = 1 \Rightarrow x_2$ remains as is.

x_3: $4 \bmod 3 = 1 \Rightarrow x_3^4$ becomes x_3.

Therefore, the reduced form for $x_1^3 x_2 x_3^4$ is $x_2 x_3$.

---

[2] Pointwise multiplication of two vectors, also called the Schur or Hadamard product, and denoted in this work by $\odot$ (U + 2299), is defined as the multiplication of two vectors by taking each entry of the two vectors and multiplying them together, that is, $z_k = x_k y_k$, where k are the index locations.

4.5.3.2 Generators for Fractional Factorial Experiments

A **generator** is defined as a set of variables which when multiplied together will yield I. Ideally, it is desired that only the variable multiplied by itself will yield I. However, this can only occur in the case where a full factorial design is used. If a fractional factorial design is used, then some subset of the variables will yield I. The relationship showing this subset is called the **defining relationship**. The defining relationship can be determined from the method presented to create a fractional factorial experiment, by multiplying each generator by the new variable to yield I, for example, if the generator is $x_4 = x_1 x_2 x_3$, then, for $l = 2$, the defining relationship would be obtained as follows:

$$x_4 = x_1 x_2 x_3 \Rightarrow x_4 x_4 = x_1 x_2 x_3 x_4 \Rightarrow I = x_1 x_2 x_3 x_4.$$

The term "$x_1 x_2 x_3 x_4$" is called a **word**. The resolution of the design is equal to the number of terms (letters) in the smallest word that is used as a generator. Further note that each of the generators used must be independent generators. In any fractional factorial experiment, p **independent** generators will be required.

Although most experiments assume that the defining relationship is positive (as above), it is also possible to define it as a negative value, that is, $I = -x_1 x_2 x_3 x_4$. Such a design is less commonly seen.

> **Example 4.4** Generators and Defining Relationships for a Fractional Factorial Design
>
> Determine a suitable defining relationship and generator for an experiment where $l = 2$, $p = 1$, and $k = 5$.
>
> **Solution**
>
> For this experiment, since $p = 1$, there will be a need to determine a single generator. Although there are many different options, the best generator would be $x_5 = x_1 x_2 x_3 x_4$, since it combines the largest number of variables together. This generator implies that the signs/levels for x_5 will be determined as the product of the signs of the other four variables.
>
> The defining relationship would then be obtained by multiplying the generator by x_5 and reducing all powers by modulo 2 arithmetic. Therefore,
>
> $$x_5 x_5 = x_1 x_2 x_3 x_4 x_5 \Rightarrow I = x_1 x_2 x_3 x_4 x_5.$$
>
> Note that $x_5 x_5$ has a power of two, which becomes zero and hence drops out.
>
> Since the length of the word is five, the resolution is V. It is always true that the resolution of a fractional factorial design is equal to the length of the smallest word in the defining relationship.

4.5.3.3 Complete Defining Relationship for Fractional Factorial Experiments

The **complete defining relationship** presents the **confounding pattern** for (or gives those variables that are confounded with) the mean response. It can be determined as follows. Since each defining relationship is equal to I, all the defining relationships can be equated (as they are confounded with the same parameter—the mean response). However, this does not determine all the possible defining relationships that are equal to I, since multiplying any two defining relationships will produce another defining relationship that equals I. Thus, if there are n defining relationships, then the **complete defining relationship** can be determined by taking all possible products of two defining relationships taken together, three defining relationships taken together, and continuing until all n defining relationships have been taken together. The equality as it stands will give one of the confounding patterns, stating that the mean response will be confounded with the given interactions.

4.5.3.4 Complete Confounding Pattern for Fractional Factorial Experiments

To determine the **complete confounding pattern**, the complete defining relationship is then multiplied by each of the variables singly and is reduced according to the rules. This will give the confounding between the first-order interactions and higher-order interactions. Next multiply the complete defining relationship by each of the variables taken in groups of two and simplify. If any group of m variables occurs in a word of length $m + 1$, then it can be ignored. This is repeated until, if k is even, $(k/2)$ variables are being multiplied or, if k is odd, $[(k-1)/2]$ variables are being multiplied.

It should be strongly emphasized that the parameters that are equated in the complete confounding pattern cannot be estimated independently. In practice, the lowest-order interaction is assumed to be the most significant and fit, while the higher-order interactions are assumed to be negligible (or zero). However, this lower-order interaction will be influenced by the value of the other interactions with which it is confounded, especially if they are not zero.

Example 4.5 Complete Defining Relationship and Confounding Pattern for a Half-Fractional Factorial Example

Continuing with the same fractional factorial experiment as in Example 4.4, determine the complete defining relationship and the complete confounding pattern for the experiment. The defining relationship is given as

$$I = x_1 x_2 x_3 x_4 x_5,$$

which happens to also be equal to the complete defining relationship. This states that the zero-order interaction, constant term, β_0, is confounded with the highest order interaction, β_{12345}. Multiplying by each of the variables singly gives

$$x_1 I = x_1(x_1 x_2 x_3 x_4 x_5) \Rightarrow x_1 = x_2 x_3 x_4 x_5$$
$$x_2 I = x_2(x_1 x_2 x_3 x_4 x_5) \Rightarrow x_2 = x_1 x_3 x_4 x_5$$
$$x_3 I = x_3(x_1 x_2 x_3 x_4 x_5) \Rightarrow x_3 = x_1 x_2 x_4 x_5$$
$$x_4 I = x_4(x_1 x_2 x_3 x_4 x_5) \Rightarrow x_4 = x_1 x_2 x_3 x_5$$
$$x_5 I = x_5(x_1 x_2 x_3 x_4 x_5) \Rightarrow x_5 = x_1 x_2 x_3 x_4.$$

This states that β_1, a first-order interaction, is confounded with the fourth-order interaction, β_{2345}. Furthermore, β_2, which is also a first-order interaction, is confounded with the (different) fourth-order interaction, β_{1345}. A similar analysis can be performed with the remaining cases.

Multiplying the identity by each of the variables taken in pairs will give

$$x_1 x_2 I = x_1 x_2(x_1 x_2 x_3 x_4 x_5) \Rightarrow x_1 x_2 = x_3 x_4 x_5$$
$$x_1 x_3 I = x_1 x_3(x_1 x_2 x_3 x_4 x_5) \Rightarrow x_1 x_3 = x_2 x_4 x_5$$
$$x_1 x_4 I = x_1 x_4(x_1 x_2 x_3 x_4 x_5) \Rightarrow x_1 x_4 = x_2 x_3 x_5$$
$$x_1 x_5 I = x_1 x_5(x_1 x_2 x_3 x_4 x_5) \Rightarrow x_1 x_5 = x_2 x_3 x_4$$
$$x_2 x_3 I = x_2 x_3(x_1 x_2 x_3 x_4 x_5) \Rightarrow x_2 x_3 = x_1 x_4 x_5$$
$$x_2 x_4 I = x_2 x_4(x_1 x_2 x_3 x_4 x_5) \Rightarrow x_2 x_4 = x_1 x_3 x_5$$
$$x_2 x_5 I = x_2 x_5(x_1 x_2 x_3 x_4 x_5) \Rightarrow x_2 x_5 = x_1 x_3 x_4$$
$$x_3 x_4 I = x_3 x_4(x_1 x_2 x_3 x_4 x_5) \Rightarrow x_3 x_4 = x_1 x_2 x_5$$
$$x_3 x_5 I = x_3 x_5(x_1 x_2 x_3 x_4 x_5) \Rightarrow x_3 x_5 = x_1 x_2 x_4$$
$$x_4 x_5 I = x_4 x_5(x_1 x_2 x_3 x_4 x_5) \Rightarrow x_4 x_5 = x_1 x_2 x_3.$$

This states that the β_{12}, a second-order interaction, is confounded with the third-order interaction, β_{345}. Likewise, β_{13} is confounded with β_{245}.

Since $k = 5$ and is odd, all the possible confounded variables have been found. Thus, the resolution of this method is V, since no second-order terms are confounded with each other. As well, note that the length of the smallest word in the complete defining relationship is 5, which is equal to the resolution. Such a relationship between the resolution and the smallest word in the complete defining relationship always holds.

Example 4.6 Confounding Pattern for a Quarter-Fractional Factorial Example

Determine the confounding pattern for a quarter-fractional factorial experiment, where $l = 2$, $p = 2$, and $k = 5$. Select a suitable set of generators.

Solution

Since $p = 2$, more than one generator is required. A suitable choice would be the following set of generators:

$$x_4 = x_1 x_3 \text{ and } x_5 = x_1 x_2$$

which give the defining relationship to be

$$I = x_1 x_3 x_4 \text{ and } I = x_1 x_2 x_5.$$

Setting the two defining relationships equal to each other will give the following equality:

$$I = x_1 x_3 x_4 = x_1 x_2 x_5.$$

In order to obtain the complete defining relation, the two generators can be multiplied together to give the complete defining relationship as

$$I = x_1 x_3 x_4 = x_1 x_2 x_5 = x_2 x_3 x_4 x_5.$$

This implies that the zero-order interaction is confounded with two third-order interactions, β_{134} and β_{125} and one fourth-order interaction, β_{2345}. It also implies that the resolution of this experiment is III, since this is the length of the shortest word.

Multiplying the complete defining relationship by each of the variables singly gives

$$x_1 I = x_1(x_1 x_3 x_4) = x_1(x_1 x_2 x_5) = x_1(x_2 x_3 x_4 x_5) \Rightarrow x_1 = x_3 x_4 = x_2 x_5 = x_1 x_2 x_3 x_4 x_5$$
$$x_2 I = x_2(x_1 x_3 x_4) = x_2(x_1 x_2 x_5) = x_2(x_2 x_3 x_4 x_5) \Rightarrow x_2 = x_1 x_2 x_3 x_4 = x_1 x_5 = x_3 x_4 x_5$$
$$x_3 I = x_3(x_1 x_3 x_4) = x_3(x_1 x_2 x_5) = x_3(x_2 x_3 x_4 x_5) \Rightarrow x_3 = x_1 x_4 = x_1 x_2 x_3 x_5 = x_2 x_4 x_5$$
$$x_4 I = x_4(x_1 x_3 x_4) = x_4(x_1 x_2 x_5) = x_4(x_2 x_3 x_4 x_5) \Rightarrow x_4 = x_1 x_3 = x_1 x_2 x_4 x_5 = x_2 x_3 x_5$$
$$x_5 I = x_5(x_1 x_3 x_4) = x_5(x_1 x_2 x_5) = x_5(x_2 x_3 x_4 x_5) \Rightarrow x_5 = x_1 x_3 x_4 x_5 = x_1 x_2 = x_2 x_3 x_4.$$

This implies that the first-order interactions are confounded with both second-, fourth-, or fifth-order interactions. For example, β_3 is confounded with β_{14}, β_{245}, and β_{1235}.

Multiplying by pairs of the variables will give, after ignoring terms $x_1 x_2$, $x_1 x_3$, $x_1 x_4$, $x_1 x_5$, $x_2 x_5$, and $x_3 x_4$, which all can be found in the smallest words:

$$\left.\begin{array}{l} x_2x_3I = x_2x_3\left(x_1x_3x_4\right) = x_2x_3\left(x_1x_2x_5\right) = x_2x_3\left(x_2x_3x_4x_5\right) \Rightarrow x_2x_3 = x_1x_2x_4 = x_1x_3x_5 = x_4x_5 \\ x_2x_4I = x_2x_4\left(x_1x_3x_4\right) = x_2x_4\left(x_1x_2x_5\right) = x_2x_4\left(x_2x_3x_4x_5\right) \Rightarrow x_2x_4 = x_1x_2x_3 = x_1x_4x_5 = x_3x_5 \\ x_3x_5I = x_3x_5\left(x_1x_3x_4\right) = x_3x_5\left(x_1x_2x_5\right) = x_2x_4\left(x_2x_3x_4x_5\right) \Rightarrow x_3x_5 = x_1x_4x_5 = x_1x_2x_3 = x_3x_5 \\ x_4x_5I = x_4x_5\left(x_1x_3x_4\right) = x_4x_5\left(x_1x_2x_5\right) = x_4x_5\left(x_2x_3x_4x_5\right) \Rightarrow x_4x_5 = x_1x_3x_5 = x_1x_2x_4 = x_2x_3 \end{array}\right\} \boxed{1} \left.\right\} \boxed{2}$$

It should be noted that the first and last lines are the same, as are the two middle lines. Thus, the complete confounding pattern can be given as

Pattern	Number of Terms
$I = x_1x_3x_4 = x_1x_2x_5 = x_2x_3x_4x_5$	3
$x_1 = x_3x_4 = x_2x_5 = x_1x_2x_3x_4x_5$	3
$x_2 = x_1x_2x_3x_4 = x_1x_5 = x_3x_4x_5$	3
$x_3 = x_1x_4 = x_1x_2x_3x_5 = x_2x_4x_5$	3
$x_4 = x_1x_3 = x_1x_2x_4x_5 = x_2x_3x_5$	3
$x_5 = x_1x_3x_4x_5 = x_1x_2 = x_2x_3x_4$	3
$x_2x_3 = x_1x_2x_4 = x_1x_3x_5 = x_4x_5$	3
$x_2x_4 = x_1x_2x_3 = x_1x_4x_5 = x_3x_5$	3
Total	24

It should be noted that the number of terms confounded will always be equal to $l^{k-p}(l^p-1)$, which in this case is, $2^{5-2}(2^2-1) = 8 \times 3 = 24$. This is an easy way to check that all the confounded terms have been considered.

4.5.3.5 Higher-Level Designs

When dealing with higher-level designs, some of the rules and observations need to be changed slightly. The two biggest changes are

(1) Modulo l arithmetic must be used.
(2) The complete defining relationship is obtained by following the above procedure and adding the following step: once the defining relationship has been created as above, it must be augmented by all the powers of itself up to $l-1$. Thus, the complete defining relationship would be $I = I = I^2 = I^3 \ldots = I^{l-1}$.

Example 4.7 Complete Confounding Pattern for a Three-Level Experiment

Consider the following 3^{3-1} factorial experiment with a generator given as

$$x_3^2 = x_1x_2^2.$$

Determine the complete defining relationship, the complete confounding pattern, and the model that can be fit given the confounding pattern.

Solution

The complete defining relationship for this experiment can be found as follows. First, the given generator has to be converted to give

$$x_3\left(x_3^2 = x_1 x_2^2\right) \Rightarrow I = x_1 x_2^2 x_3.$$

If this were a two-level experiment, then one would stop at this point. However, for higher-level experiments, powers of the above equation must be taken and reduced with modulo l arithmetic. Therefore, squaring this defining relationship and reducing everything modulo 3 gives

$$I^2 = x_1 x_2^2 x_3 x_1 x_2^2 x_3 = x_1^2 x_2 x_3^2.$$

Therefore, the complete defining relationship is

$$I = x_1 x_2^2 x_3 = x_1^2 x_2 x_3^2.$$

The complete confounding pattern can be found as follows (Note that all reductions are modulo 3)

$$x_1\left(I = x_1 x_2^2 x_3 = x_1^2 x_2 x_3^2\right) \quad x_2\left(I = x_1 x_2^2 x_3 = x_1^2 x_2 x_3^2\right)$$
$$x_1 = x_1^2 x_2^2 x_3 = x_2 x_3^2 \ (2) \qquad x_2 = x_1 x_3 = x_1^2 x_2^2 x_3^2 \ (2)$$

$$x_3\left(I = x_1 x_2^2 x_3 = x_1^2 x_2 x_3^2\right) \quad x_1^2\left(I = x_1 x_2^2 x_3 = x_1^2 x_2 x_3^2\right)$$
$$x_3 = x_1 x_2^2 x_3^2 = x_1^2 x_2 \ (2) \qquad x_1^2 = x_2^2 x_3 = x_1 x_2 x_3^2 \ (2)$$

$$x_2^2\left(I = x_1 x_2^2 x_3 = x_1^2 x_2 x_3^2\right) \quad x_3^2\left(I = x_1 x_2^2 x_3 = x_1^2 x_2 x_3^2\right)$$
$$x_2^2 = x_1 x_2 x_3 = x_1^2 x_3^2 \ (2) \qquad x_3^2 = x_1 x_2^2 = x_1^2 x_2 x_3 \ (2)$$

$$x_1 x_2\left(I = x_1 x_2^2 x_3 = x_1^2 x_2 x_3^2\right) \quad x_2 x_3\left(I = x_1 x_2^2 x_3 = x_1^2 x_2 x_3^2\right)$$
$$x_1 x_2 = x_1^2 x_3 = x_2^2 x_3^2 \ (2) \qquad x_2 x_3 = x_1 x_3^2 = x_1^2 x_2^2 \ (2)$$

$$x_1 x_2^2 x_3\left(I = x_1 x_2^2 x_3 = x_1^2 x_2 x_3^2\right)$$
$$x_1 x_2^2 x_3 = x_1^2 x_2 x_3^2 = I \ (2).$$

Therefore, a total of eighteen terms are confounded with the last entry being equal to the original confounding pattern. It can be noted that $l^{k-p}(l^p-1) = 3^{3-1}(3^1-1) = 9 \times 2 = 18$, which is as expected.

A second-order model can be fitted with the given experiment. Ignoring any higher-order terms that are confounded, the fitted model will have the form:

$$y = \beta_0 + \beta_1 x_1 + \beta_2 x_2 + \beta_3 x_3 + \beta_{11} x_1^2 + \beta_{22} x_2^2 + \beta_{33} x_3^2 + \beta_{12} x_1 x_2 + \beta_{23} x_2 x_3.$$

It can be noted that not all of the second-order interaction terms can be estimated (namely, $x_1 x_3$ is confounded with x_2). This shows that the design is of Resolution III. This can be confirmed by noting that the smallest word has length 3.

4.5.4 Design Procedure for Fractional Factorial Experiments

The following steps will allow the design of a fractional factorial experiment:

(1) Design a full l^{k-p} factorial experiment for the $k-p$ variables. These $k-p$ variables are called the **independent** or **basic** variables.

(2) Create p new variables, which are independent combinations of some or all of the $k - p$ factors. The combinations should be determined using the above information on confounding, resolution, confounding pattern, and any additional information known about the process.

(3) Add p columns to the table created in Step (1), each containing one of the combinations determined above. Set each of the p variables equal to one of the new columns. Determine the appropriate level for each new column by considering the combination chosen for that row. It should be noted that there are many different methods to accomplish this creation of the table.

(4) Thus, the actual experiment will consist of l^{k-p} runs (or treatments), where each row corresponds to a single run. For each run, the values of the factors are determined based on the corresponding column values in the table obtained in Step 3).

Example 4.8 Analysing the Structure of a Fractional Factorial Experiment

Consider the factorial design shown in Table 4.2. What are the independent factors, the dependent factors, the generators, the complete defining relationship, the resolution, and the aliases for A and for AB? What type of factorial design is it?

Table 4.2 Design for the fractional factorial experiment

Run	A	B	C	D	Value
1	−	−	−	−	219
2	+	−	−	+	214
3	−	+	−	+	154
4	+	+	−	−	150
5	−	−	+	+	124
6	+	−	+	−	132
7	−	+	+	−	114
8	+	+	+	+	134

Solution

Independentand Dependent Factors: Looking at the way the data is presented, it can be seen that one of the factors depends on the others. Since the first three factors have the form of the classical full factorial experiment, it will be assumed that factor D depends on the other factors. Therefore, the basic factors are A, B, and C, while the dependent factor is D.

Generator: The generator is $D = ABC$. Once it has been determined that D is the dependent factor, it is all that is required to try different combinations of the other factors to determine the appropriate generator.

Complete Defining Relationship: The complete defining relationship is obtained as follows:
$$D \times D = D \times ABC \Rightarrow I = ABCD.$$

Resolution: Since the smallest word in the complete defining relationship is four letters (factors) long, the resolution is IV, that is, the second-order interactions are confounded with each other.

Aliases for A: They can be obtained by multiplying the defining relationship by A and reducing everything modulus 2, that is,
$$A \times I = A \times ABCD \Rightarrow A = A^2 BCD \Rightarrow A = BCD.$$

Aliases for AB: Similar to the aliases for A, the results are.
$$AB \times I = AB \times ABCD \Rightarrow AB = A^2 B^2 CD \Rightarrow AB = CD.$$

This is expected because it was determined that second-order interactions can be confounded with each other. This implies that both AB and CD cannot be estimated simultaneously.

Complete Description: This is a 2^{4-1}_{IV} factorial experiment with the complete defining relationship $I = ABCD$. Note that a complete description requires that all the necessary parameters (l, p, resolution, and complete defining relationship) be provided.

4.5.5 *Analysis of Fractional Factorial Experiments*

Define, as before,

$$\overline{\mathcal{A}} = \begin{bmatrix} \mathcal{A} \\ \vdots \\ n_R \, times \\ \mathcal{A} \end{bmatrix} \tag{4.21}$$

$$\vec{y} = \begin{bmatrix} \hat{\vec{y}}(1^{st}\ \text{column}) \\ \hat{\vec{y}}(2^{nd}\ \text{column}) \\ \vdots \\ \hat{\vec{y}}(n_R^{th}\ \text{column}) \end{bmatrix} \tag{4.22}$$

$$\hat{\vec{\beta}} = (\overline{\mathcal{A}}^T \overline{\mathcal{A}})^{-1} \overline{\mathcal{A}}^T \vec{y}. \tag{4.23}$$

Then the standard deviation is defined as follows:

$$\boxed{\hat{\sigma} = \sqrt{\frac{\vec{y}^T \vec{y} - \hat{\vec{\beta}}^T \overline{\mathcal{A}}^T \vec{y}}{l^{k-p}(n_R - 1)}}.} \tag{4.24}$$

The sum of squares due to errors, *SSE*, is given by

$$\boxed{SSE = \vec{y}^T \vec{y} - \hat{\vec{\beta}}^T \overline{\mathcal{A}}^T \vec{y}.} \tag{4.25}$$

If an orthogonal basis is used for the levels, then the sum of squares due to each regressor, SSR_i, is given by[3]

$$SSR_i = \left(\overline{\mathcal{A}}^T \overline{\mathcal{A}}\right)_{ii} \hat{\beta}_i^2. \tag{4.26}$$

where $\left(\overline{\mathcal{A}}^T \overline{\mathcal{A}}\right)_{ii}$ is the (i, i) (diagonal) entry in the $\left(\overline{\mathcal{A}}^T \overline{\mathcal{A}}\right)$ matrix. If an orthonormal basis is used to encode the variables and their levels, then

$$\left(\overline{\mathcal{A}}^T \overline{\mathcal{A}}\right)_{ii} = l^{k-p} n_R. \tag{4.27}$$

---

[3] Determining an orthogonal basis for an arbitrary level is explained fully in Sect. 4.7.

When $l = 2$ and the proposed ± 1 encoding is used, then, by definition, the basis is orthonormal, and Eq. (4.27) reduces to

$$\left(\overline{A^T A}\right)_{ii} = 2^{k-p} n_R.$$
(4.28)

The total sum of squares, TSS, is given by

$$\boxed{TSS = SSE + \sum_{i=1}^{l^{k-p}} \left(\overline{A^T A}\right)_{ii} \hat{\beta}_i^2.}$$
(4.29)

To determine whether a given regressor should be in the model, calculate for each regressor the F-statistic as follows:

$$\boxed{F_i = \frac{SSR_i}{\dfrac{SSE}{l^{k-p}(n_R-1)}}.}$$
(4.30)

The value obtained from Eq. (4.30) is compared with the critical F-value denoted as $F(1-\alpha, 1, l^{k-p}(n_R-1))$. If F_i is greater than F-critical, then the regressor should be kept in the model. Otherwise, the given regressor can be eliminated from the model. The **effect** due to an interaction is equal to twice the corresponding calculated regression parameter.

4.5.6 Framework for the Analysis of Factorial Designs

The analysis of a factorial experiment can be summarized as follows:

(I) **Pre-analysis**, which characterizes the experiment and determines the appropriate method to be followed.

    a. Determine the number of factors, k, the number of levels, l, and the number of replicates, n_R.

    b. Determine the number of dependent variables, p.

    c. If $p \neq 0$, determine the generators, the complete defining relationship, and the complete confounding pattern.

    d. Write down the model that will be estimated given the above information.

    e. Determine an appropriate orthogonal or orthonormal basis for the model.

(II) **Calculating the values**, using appropriate formulae compute the parameter estimates, the normal probability plot of the parameters, and if appropriate, the SSR_i and F-value for each of the parameters.

(III) **Model reduction**, which seeks to remove all unnecessary parameters from the model. Three main methods can be used:

    a. **Normal probability plot of the parameters**, which seeks to determine which parameters are most significant based on their deviation from normality. This approach works as long as an orthonormal basis has been selected for the model.

    b. **F_i-test for the parameters**, which seeks to determine which parameters are most significant based on an analysis of variance method. This approach works as long as an orthogonal basis has been selected for the model and there are replicates with which to compute an estimate of the model variance.

    c. **Confidence intervals for the parameters**, which seeks to determine which confidence intervals of the parameters cover zero and hence should be rejected. This approach works as long as there are replicates with which to compute an estimate of the model variance.

(IV) **Residual analysis**, using the reduced model, the resulting residuals should be analysed to determine whether the assumptions underlying regression have been satisfied. If the residuals suggest problems, then further experiments may need to be performed to resolve the issue.

(V) **Conclusions**, which seek to answer based on the reduced model whatever questions the objectives of the exercise had, for example, determining the optimal operating point.

Example 4.9 Detailed Analysis of a Fractional Factorial Experiment

Consider the following experiment to obtain the best beef stew rations. Note that the company does not care about the taste, but cares solely on how well the product can be cooked and hence the focus will be on the Heating Index. The following five factors will be considered: A = sauce viscosity; B = residual gas; C = solid/liquid ratio; D = net weight; and E = rotation speed. Given the large number of factors and limited number of available samples, a fractional factorial experiment has been conducted with a single replicate. The data are shown in Table 4.3. Answer the following questions:

Table 4.3 Preparing beef stew ration data

Heating Index		A	B	C	D	E
Replicate I	Replicate II					
8.46	9.61	−1	−1	−1	−1	1
15.68	14.68	1	−1	−1	−1	−1
14.94	13.09	−1	1	−1	−1	−1
12.52	12.71	1	1	−1	−1	1
17.00	16.36	−1	−1	1	−1	−1
11.44	11.83	1	−1	1	−1	1
10.45	9.22	−1	1	1	−1	1
19.73	16.94	1	1	1	−1	−1
17.7	16.36	−1	−1	−1	1	−1
14.98	11.93	1	−1	−1	1	1
8.40	8.16	−1	1	−1	1	1
19.08	15.40	1	1	−1	1	−1
13.07	10.55	−1	−1	1	1	1
18.57	20.53	1	−1	1	1	−1
20.59	21.19	−1	1	1	1	−1
14.03	11.31	1	1	1	1	1

(1)   Using this data set, determine the generators, the complete defining relationship, and the type of experiment.

(2)   Using a normal probability plot of the estimated model parameters, determine the significant parameters. Analyse this reduced model.

(3)   Using the F-test approach to determining the significant parameters, find a reduced model and analyse its residuals ($\alpha = 0.05$). Compare with the results from part (2). Which method is better and when can each be used?

(*Data taken from R. C. Wurl and S. L. Albin (1999), "A Comparison of Multiresponse Optimization Sensitivity to Parameter Selection"*, Quality Engineering.)

Solutions

Question 1

After some experimentation, it can be determined that the basic factors are A, B, C, and D, while the dependent factor is E. The generator for this experiment can be written as E = ABCD. The complete defining relationship is then I = ABCDE. Based on the above analysis, it can be concluded that this is a ½-fractional factorial experiment with a resolution of V, that is, 2_V^{5-1}. It should be noted here that not all of the parameters can be estimated since they will be confounded with others. Without going into the details here, all of the zero-, first-, and second-order interactions are estimable. They will be confounded with various higher-order interactions.

Question 2

Without going into the details of the calculation of all of the values, as an Excel® spreadsheet was used to obtain the values, the summary results are presented here. The normal probability plot of the parameters is shown in Fig. 4.5. From this figure, it can be seen that there are two really significant values (which could well not be white noise), β_0 and β_5, which implies that the only significant factor is E. This suggests that the model can be written as

$$y = 14.7 - 3.1x_5.$$

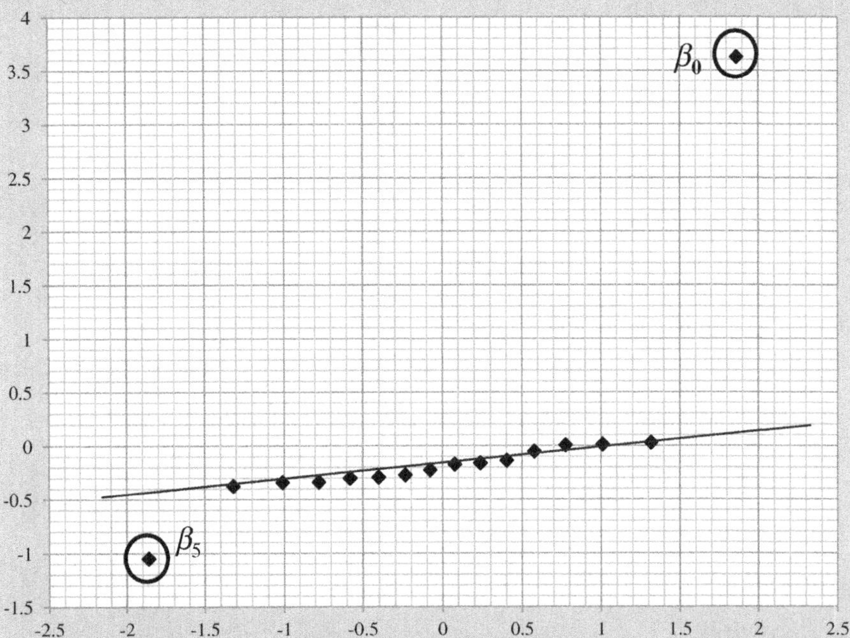

Fig. 4.5 Normal probability plot of the parameters

Table 4.4 Reduced model statistics for beef stew ration example

Statistic	Value	Statistic	Value
$\Sigma \varepsilon^2$	148	$\hat{\sigma}_{model}$	3.05
SSE	148	TSS	455
R^2	0.67	SSR	307
F-critical, model $F_{0.95,\,1,\,30}$	4.17	F-test, model	62

The analysis of the reduced model is shown in Fig. 4.6 and Table 4.4.

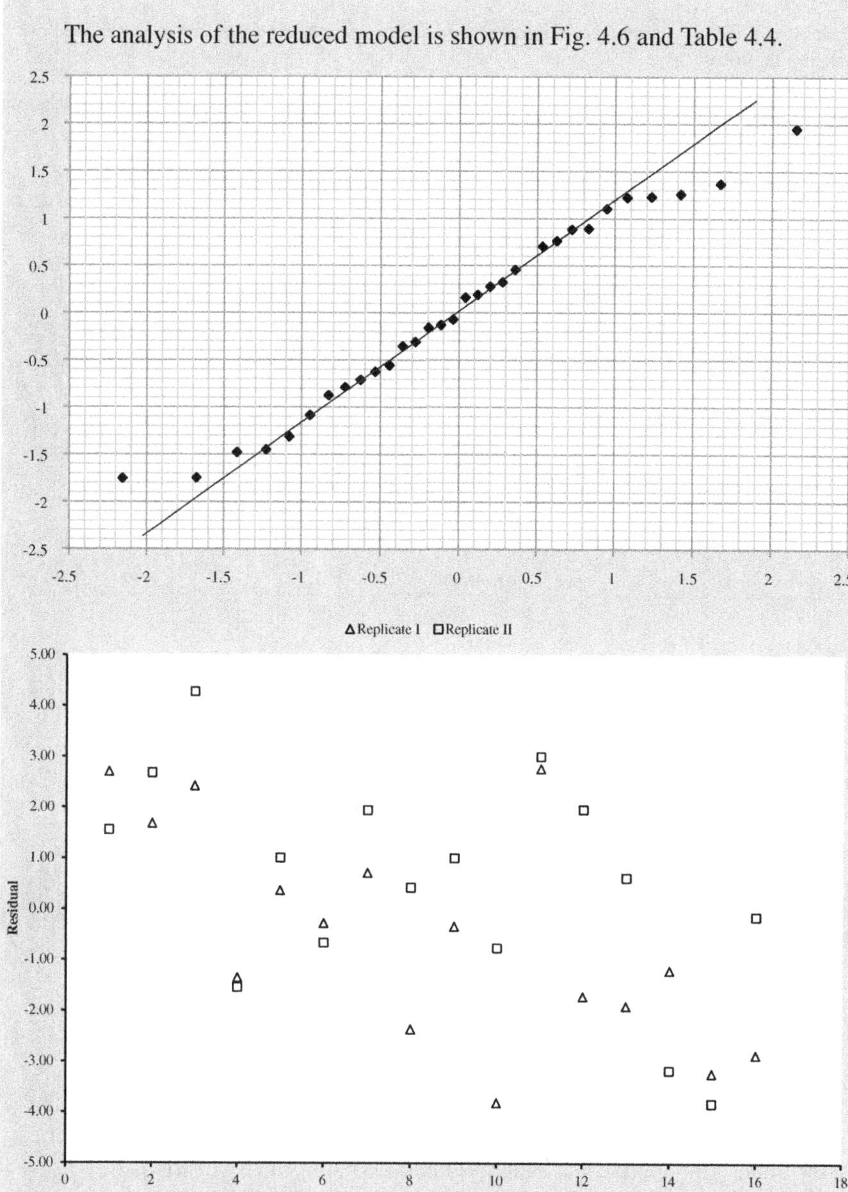

Fig. 4.6 (Top) Normal probability plot of the residuals and (bottom) Time-series plot of the residuals with the different replicates clearly shown

Table 4.5 Model parameters and statistical scores for the beef stew ration model reduced using the F-test

Parameters	Value	SSR_i	F_i
β_0	14.3	6610	**3580**
β_1	0.819	21.5	**11.8**
β_2	−0.031	0.03	0.02
β_3	0.910	26.5	**14.5**
β_4	0.850	23.1	**12.7**
β_5	−3.10	307	**168**
β_{12}	0.161	0.83	0.46
β_{13}	−0.447	6.40	3.51
β_{14}	−0.206	1.36	0.75
β_{15}	0.608	11.8	**6.49**
β_{23}	0.288	2.65	1.46
β_{24}	−0.315	3.17	1.74
β_{25}	−0.286	2.62	1.44
β_{34}	0.205	1.34	0.74
β_{35}	−0.589	11.1	**6.10**
β_{45}	−0.463	6.85	3.76
Statistic	Value	Statistic	Value
$\Sigma\varepsilon^2$	39.8	$\hat{\sigma}_{model}$	1.34
SSE	39.7	TSS	455
R^2	0.913	SSR	416
F-critical, model, $F_{0.95,\,15,\,16}$	2.34	F-test, model	25.5
F-critical, parameter, $F_{0.95,\,1,\,16}$	4.49		

Based on these figures, it can be seen that it does not look like the model captures all of the variation present in the data. Specifically, the residuals do not seem to be normally distributed because there is a significant tail especially at the top end. Furthermore, the residuals seem to be decreasing in value as the runs increase (that is, run 16 has consistently smaller variance than run 1). This could suggest that additional variables may need to be considered. On the other hand, the coefficient of regression is 0.67, and the F-test for the model suggests that the model is significant (as F-test $> F_{critical}$). If we desire a very simple model, this is it (one parameter to explain 67% of the variability in the data is quite good). On the other hand, there do seem to be some issues with the underlying model, since the residuals are not normally distributed.

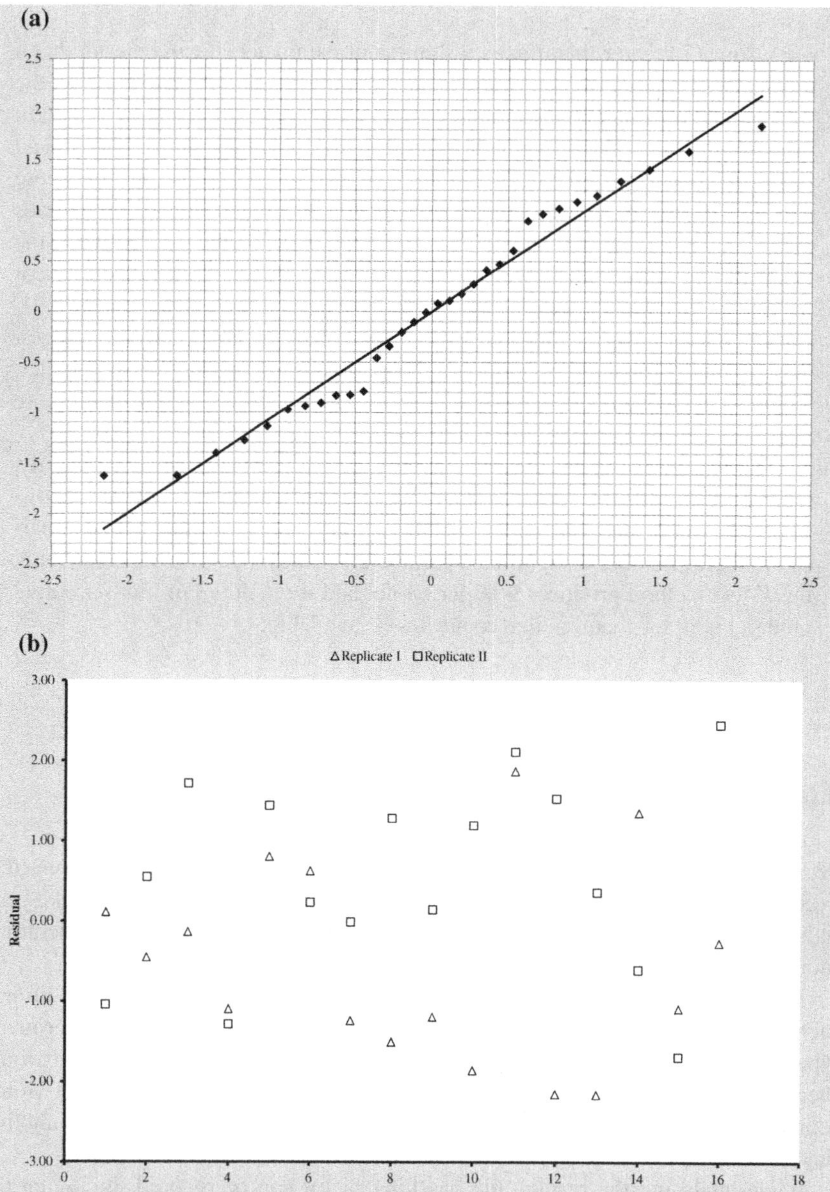

Fig. 4.7 (Top) Normal probability plot of the residuals and (bottom) Time-series plot of the residuals with the different replicates clearly shown for the model reduced using the F-test

Question 3

Repeating the analysis but using the F-test method gives the following results. The required values are given in Table 4.5. The critical value of the F-test is

4.49. Any F_i greater than this is a significant parameter. It can be seen that a much larger number of parameters is now significant. However, examining the "discarded" parameters, it can be seen that factor B is completely irrelevant to the experiment, as all its interactions are zero. Model analysis for this reduced model is shown in Fig. 4.7. It can be seen that the residuals are now more normally distributed and that there is no tail. On the other hand, the time-series plot of the residuals still seems to show some weird behaviour between the two different replicates. Now it seems to suggest that one replicate is different from the other. On the other hand, the coefficient of regression is now 0.91, and the model is significant given the F-test. On the whole, it would suggest that the model obtained by reducing the parameters using the F-test is better. However, it can be noted that the number of parameters selected is much more than in the normal probability plot method.

Finally, some brief comments regarding the different methods used. Firstly, it can be noted that the normal probability plot method can be used irrespective of whether replicates were performed. In this particular case, the method underestimates the number of significant parameters in the model. On the other hand, the F-test method produces a larger model and it would seem more accurate model, but it does require that replicates be available.

4.6 Blocking and Factorial Design

When blocking has to be implemented due to either a very large number of runs that need to be run over multiple days or equipment limitations that could make each individual run not be identical, it is necessary to develop an appropriate structure to minimize these effects on the overall design.

The easiest approach is to treat the blocking variable (day or run) as an additional factor in the experiment and then design a $l^{(k+1)-p-1}$ fractional factorial experiment, where the blocking factor will be treated as the dependent factor. The runs would then be segregated based on the values of the parameters, for example, $+1$ would represent all runs done on one day, and -1 would represent all runs done on another day.

When analysing the results, the blocking factor can be ignored. Including the blocking factor in the model will allow a determination of the effect of blocking on the overall system. The principles from fractional factorial design can be used to design the optimal blocking pattern.

Table 4.6 Design for a blocked, full factorial experiment. All experiments with $(-)$ in the final column would be run on one day and those with a $(+)$ in the final column would be run on another day

Run	A	B	C	D	E = ABC (blocking factor)
1	−	−	−	−	−
2	+	−	−	−	+
3	−	+	−	−	+
4	+	+	−	−	−
5	−	−	+	−	+
6	+	−	+	−	−
7	−	+	+	−	−
8	+	+	+	−	+
9	−	−	−	+	−
10	+	−	−	+	+
11	−	+	−	+	+
12	+	+	−	+	−
13	−	−	+	+	+
14	+	−	+	+	−
15	−	+	+	+	−
16	+	+	+	+	+

Example 4.10 Blocking and Full Factorial Design

Consider the case where we have a full 2^4-factorial experiment that must be run in two blocks of eight runs (two days). Assume that it is known that the ABC interaction is zero. Design an appropriate experiment that maximizes the information that can be extracted.

Solution

In general, whenever one is faced with a blocking issue with known variables, then the problem can be reduced and analysed as if it were a fractional factorial design with additional dummy variables. In this particular example, let the original factors be A, B, C, and D and let the blocking variable be an additional fifth factor, E. Since we have been told that the ABC interaction is zero, in order to minimize the confounding let E = ABC. All runs where E is positive (+) would be run on one day and all runs where E is negative (−) would run on the other day. The analysis would be performed using the original factors. The factor ABC that would be fit would basically represent the contribution of blocking to the design (and should hopefully be close to zero). Practically, this can be written as shown in Table 4.6.

Example 4.11 Blocking and Fractional Factorial Design

Consider a 2^{4-1} factorial experiment with the complete defining relationship I = ABCD. Determine an appropriate blocking pattern for this experiment given that AB is known to be zero and determine the new complete defining relationship.

Solution

Given the information, let the blocking variable be E. The generator will then be E = AB with the defining relationship being I = ABE. The new complete defining relationship can be obtained using the same procedure as before. First, setting the given defining relationships equal to each other,

$$I = ABCD = ABE.$$

Second, multiplying together the last two members gives

$$ABCD \times ABE = A^2B^2CDE = CDE.$$

Therefore, the complete defining relationship is

$$I = ABCD = ABE = CDE.$$

It can be noted that the resolution of the experiment has decreased from IV to III by the introduction of a blocking variable. This is expected given the general nature of blocking.

4.7   Generalized Factorial Design

In order to effectively apply the proposed methods to solving higher-level experiments, there is a need to develop the tools for the creation of orthogonal bases for arbitrary polynomial functions. Such an approach has the benefit that the previously obtained results from linear regression can be applied to solving such problems. Therefore, this section will provide the methods for obtaining such as solution.

Consider an l-level experiment with a single factor x, which has been scaled so that its values lie between $[-1, 1]$. Furthermore, assume that the spacing between each of the treatment points is equal. The spacing between successive treatment points, δ_l, can be determined as follows:

$$\delta_l = \frac{2}{l-1}. \tag{4.31}$$

Finally, if there are replicates, assume that all treatments/runs have been replicated, so that $n_R l$ experiments have been performed, where n_R is the number of replicates. The model for such an experiment can be written as

$$y = \beta_0 + \beta_1 x + \beta_{11} x^2 + \beta_{111} x^3 + \cdots + \beta_{1^{l-1}} x^{l-1}. \tag{4.32}$$

If the $\mathcal{A}$-matrix is created for the problem as it is currently written, it is easy to see that the resulting $\mathcal{A}^T \mathcal{A}$ matrix will not be orthogonal. In order to obtain an orthogonal matrix, it is necessary to obtain an appropriate orthogonal basis for the set of polynomials given as $\{1, x, x^2, x^3, \ldots, x^{l-1}\}$. For example, for a three-level, single factor experiment, equally spaced treatment points would be $-1, 0$, and 1, and the model fit would be

$$y = \beta_0 + \beta_1 x + \beta_{11} x^2. \tag{4.33}$$

Therefore, there is a need to re-arrange Eq. (4.32) so that the individual terms are orthogonal to each other, that is

$$y = \beta_0 L_0(x) + \beta_1 L_1(x) + \beta_{11} L_2(x), \tag{4.34}$$

where L_0, L_1, and L_2 are some orthogonal functions on $[-1, 1]$ that depend on the original parameter x. In general, L_0 is always assumed to be the unit function, that is, $L_0 = 1$. The exact form of the remaining functions will now be determined.

4.7.1 Obtaining an Orthogonal Basis

In order for the basis to be orthogonal, it is necessary that the following two conditions hold:

$$\sum_{i=1}^{l} L_j(x_i) = 0, \forall j > 0 \tag{4.35}$$

$$\sum_{i=1}^{l} L_j(x_i) L_k(x_i) = \begin{cases} 0 & j \neq k \\ d & j = k \end{cases}, \tag{4.36}$$

where x_i is the coded value of the treatment point and d is any arbitrary value greater than zero. The first condition expresses the orthogonality of the zero-order, constant polynomial from all the other terms. Although it can be obtained from the second condition, it is useful to keep it separate. A polynomial basis where $d = l$ has computational and theoretical advantages. Such a basis, where $d = l$, is called an **orthonormal** basis. The values of the basis at the given treatment point will be referred to as the experimental coefficient, γ, that is,

$$\gamma_{ji} = L_j(x_i) \tag{4.37}$$

and the subscript on the polynomial basis refers to which power the polynomial is a basis. The general form of the polynomial can be written as[4]

$$L_j(x) = \begin{cases} \displaystyle\sum_{i=1}^{\frac{(j-1)}{2}+1} \beta_{ji} x^{2i-1} & j \text{ odd} \\[2ex] \displaystyle\sum_{i=0}^{\frac{j}{2}} \beta_{ji} x^{2i} & j \text{ even} \end{cases}. \tag{4.38}$$

Combining Eqs. (4.37) and (4.38) gives that for the case of an even power the individual experimental coefficients can be written as

$$\gamma_{ji} = \sum_{k=0}^{\frac{j}{2}} \beta_{jk} x_i^k. \tag{4.39}$$

The case for an odd power is similar, *mutatis mutandis*.

4.7.2 Orthogonal Bases for Different Levels

In general, if an l-level factorial design is of interest, then the orthogonal basis for this factorial design will be formed by the set of polynomials from 0 to $l-1$, evaluated at each of the treatment levels. The zero-order polynomial, L_0, will be used for the zero-order interaction (β_0). The first-order polynomial, L_1, will be used for the factors whose powers are 1. Similarly, the second-order polynomials, L_2, will be used for the factors whose powers are 2. The results can be obtained by solving Eqs. (4.35) and (4.36). In the following sections, the results for $l = 2$, 3, and 4 will be provided, as well as generalized orthonormal polynomials for $l = 2$ and 3. Examples will be provided as appropriate.

---

[4] The form of the polynomials is similar to the standard, discrete Gram polynomials.

4.7.2.1 Case $l = 2$

Consider first the situation of determining the orthogonal basis for the simplest factorial design with two levels. Since the two treatment points are -1 and 1, and the basis function has the form

$$L_1(x) = \beta_{11}x \tag{4.40}$$

the treatment coefficients can be written as

$$
\begin{aligned}
\gamma_{11} &= \beta_{11}x_1 = \beta_{11}(-1) \\
\gamma_{12} &= \beta_{11}x_2 = \beta_{11}(1).
\end{aligned}
\tag{4.41}
$$

Note that, by condition (4.35), $\gamma_{11} + \gamma_{12} = 0$. This gives that Eq. (4.41) can be written as

$$
\begin{aligned}
\gamma_{11} &= \beta_{11}(-1) \\
-\gamma_{11} &= \beta_{11}(1)
\end{aligned}
\tag{4.42}
$$

for which there is an infinite number of valid solutions. Taking the simplest solution $\gamma_{11} = -1$ and $\gamma_{12} = 1$, gives that $\beta_{11} = 1$. This is clearly the same result as before. Condition (4.36), which can be written as

$$\sum_{i=1}^{l} \gamma_{ji}^2 = d \tag{4.43}$$

gives that with this basis, $d = 2 = l$, which is good. Thus, for $l = 2$, the basis polynomial is orthonormal and can be written as

$$L_1(x) = x. \tag{4.44}$$

4.7.2.2 Case $l = 3$

Now, consider the case of determining the orthogonal basis for the factorial design with three levels. The three equispaced coded treatment points are $-1, 0$, and 1 ($\delta_l = 1$). Ignoring the zero-order basis function, there will be two additional basis functions of the general form:

$$
\begin{aligned}
L_1(x) &= \beta_{11}x \\
L_2(x) &= \beta_{21}x^2 + \beta_{20}.
\end{aligned}
\tag{4.45}
$$

For the factors raised to the first power, the treatment coefficients can be written as

$$
\begin{aligned}
\gamma_{11} &= \beta_{11} x_1 = \beta_{11}(-1) \\
\gamma_{12} &= \beta_{11} x_2 = \beta_{11}(0) \\
\gamma_{13} &= \beta_{11} x_3 = \beta_{11}(1).
\end{aligned}
\tag{4.46}
$$

Since the situation is similar to that previously obtained for the two-level case, the following values can be obtained[5]:

$$
\gamma_{11} = -1, \gamma_{12} = 0, \gamma_{13} = 1, \beta_{11} = 1.
\tag{4.47}
$$

However, note that $d = 2$, which means that this basis is not orthonormal. An orthonormal basis can be determined by solving the following system of equations

$$
\begin{cases}
\gamma_{11}^2 + \gamma_{13}^2 = 3 \\
\gamma_{11} + \gamma_{13} = 0
\end{cases}
\tag{4.48}
$$

which gives after substitution of the second equation into the first

$$
\gamma_{11} = -\sqrt{\frac{3}{2}}, \gamma_{13} = \sqrt{\frac{3}{2}}.
\tag{4.49}
$$

This implies that $\beta_{11} = 1.5^{0.5}$.

L_2, which forms the basis for the second-power terms, can be obtained by first noting that the treatment coefficients can be written as

$$
\begin{aligned}
\gamma_{21} &= \beta_{21} x_1^2 + \beta_{20} = \beta_{21}(-1)^2 + \beta_{20} \\
\gamma_{22} &= \beta_{21} x_2^2 + \beta_{20} = \beta_{21}(0)^2 + \beta_{20} \\
\gamma_{23} &= \beta_{21} x_3^2 + \beta_{20} = \beta_{21}(1)^2 + \beta_{20}
\end{aligned}
\tag{4.50}
$$

which can be rewritten as the following system of equations

$$
\begin{aligned}
\gamma_{21} &= \beta_{21} + \beta_{20} \\
\gamma_{22} &= \beta_{20} \\
\gamma_{23} &= \beta_{21} + \beta_{20}
\end{aligned}
\tag{4.51}
$$

subject to the usual constraint of $\gamma_{11} + \gamma_{12} + \gamma_{13} = 0$. It can be noted that this is an indeterminate system with one degree of freedom (six unknowns, but five equations).

---

[5] Note that γ_{12} must always equal zero given the setup of the problem.

Arbitrarily setting $\gamma_{13} = \gamma_{11} = 1$ gives the following solution,

$$\gamma_{21} = \gamma_{23} = 1, \gamma_{22} = -2,$$
$$\beta_{21} = 3, \beta_{20} = -2. \tag{4.52}$$

Therefore, the factors raised to the first power will be encoded by L_1, while the factors raised to the second power will be encoded by L_2. Thus, for $l = 3$, the basis polynomial is orthogonal and can be written as

$$L_1(x) = x$$
$$L_2(x) = 3x^2 - 2. \tag{4.53}$$

Example 4.12 *Orthogonal Basis for a Mixed-Level Factorial Experiment*

Consider the following two-factor experiment where the first factor (x_1) has three levels and the second factor (x_2) has two levels. Assume that a total of 6 ($= 3 \times 2$) experiments have been performed and the following model will be fit:

$$y_t = \beta_0 + \beta_1 x_1 + \beta_2 x_2 + \beta_{12} x_1 x_2 + \beta_{11} x_1^2 + \beta_{112} x_1^2 x_2.$$

Assume that the treatment points for x_1 are $-1, 0$, and 1, while for x_2 they are -1 and 1. It should be noted that for x_1 the encoding provided by L_1 will be used, while for x_1^2 the encoding provided by L_2 will be used. As well, unless an orthonormal basis is desired, then the same L_1 can be used to encode those factors containing the factors x_1 and x_2 at the first power. Otherwise, different first-order polynomials will be used. This implies that the actual model being fit is

$$y_t = \beta_0 + \beta_1 L_1(x_1) + \beta_2 L_1(x_2) + \beta_{12} L_1(x_1)L_1(x_2) + \beta_{11} L_2(x_1) + \beta_{112} L_2(x_1)L_1(x_2).$$

Based on this discussion, the regression matrix can be given as

$$
\mathcal{A} =
\begin{array}{c}
\begin{array}{cccccc}
\beta_0 & \beta_1 & \beta_2 & \beta_{12} & \beta_{11} & \beta_{112}
\end{array} \\
\left[
\begin{array}{cccccc}
1 & -1 & -1 & 1 & 1 & -1 \\
1 & 0 & 1 & 0 & -2 & -2 \\
1 & 1 & -1 & -1 & 1 & -1 \\
1 & -1 & 1 & -1 & 1 & 1 \\
1 & 0 & -1 & 0 & -2 & 2 \\
1 & 1 & 1 & 1 & 1 & 1
\end{array}
\right]
\end{array}.
$$

The column headed by β_0 is encoded using L_0. The column headed by β_1 is encoded using L_1 evaluated at the three treatment points $(-1, 0, 1)$, while the column headed by β_2 is encoded using L_1 evaluated at the two treatment points $(-1, 1)$. The column headed by β_{12} is encoded by multiplying the corresponding rows in β_1 and β_2, for example, the first row would be $-1 \times -1 = 1$. The column headed by β_{11} is encoded by evaluating L_2 at the three treatment points $(-1, 0, 1)$. Finally, the column headed by β_{112} is encoded by multiplying the corresponding rows in β_{11} and β_2, for example, the first row would be $1 \times -1 = -1$. The inverse of the information matrix can be written as

$$
A^T A = \begin{bmatrix} 6 & & & & & \\ & 4 & & & & \\ & & 6 & & & \\ & & & 4 & & \\ & & & & 12 & \\ & & & & & 12 \end{bmatrix}.
$$

Notice how the diagonal values are all different. This is a result of the fact that a nonnormal basis was used to encode the values.

4.7.2.3   Case $l = 4$

Now, consider the case of determining the orthogonal basis for the factorial design with four levels. The four equispaced treatment points are -1, $-\frac{1}{3}$, $\frac{1}{3}$, and 1 $(\delta_l = \frac{2}{3})$. Ignoring the zero-order basis function, there will be three additional basis functions of general form:

$$
\begin{aligned}
L_1(x) &= \beta_{11}x \\
L_2(x) &= \beta_{21}x^2 + \beta_{20} \\
L_3(x) &= \beta_{32}x^3 + \beta_{31}x.
\end{aligned}
\tag{4.54}
$$

For the factors raised to the first power, the treatment coefficients can be written as

$$
\begin{aligned}
\gamma_{11} &= \beta_{11}x_1 = \beta_{11}(-1) \\
\gamma_{12} &= \beta_{11}x_2 = \beta_{11}\left(-\frac{1}{3}\right) \\
\gamma_{13} &= \beta_{11}x_3 = \beta_{11}\left(\frac{1}{3}\right)
\end{aligned}
$$

$$\gamma_{13} = \beta_{11}x_3 = \beta_{11}(1). \tag{4.55}$$

Since the solution the situation is similar to that previously obtained for the two- and three-level cases, the following values can be obtained:

$$\gamma_{11} = -1, \gamma_{12} = -\frac{1}{3}, \gamma_{12} = \frac{1}{3}, \gamma_{13} = 1$$
$$\beta_{11} = 1. \tag{4.56}$$

However, note that $d = 7/3$, which means that this basis is not orthonormal. In general, for factors to the first power, it is possible to use the same basis irrespective of the number of levels. However, this is not generally true, especially for higher-order basis functions.

An orthonormal basis can be determined by solving the following system of equations

$$\begin{cases} \gamma_{11}^2 + \gamma_{12}^2 + \gamma_{13}^2 + \gamma_{14}^2 = 4 \\ \gamma_{11} + \gamma_{12} + \gamma_{13} + \gamma_{14} = 0 \end{cases} \tag{4.57}$$

with the relationships obtained from Eq. (4.55), that is,

$$\gamma_{11} = -\beta_{11}, \gamma_{12} = -\frac{\beta_{11}}{3}, \gamma_{13} = \frac{\beta_{11}}{3}, \gamma_{13} = \beta_{11}. \tag{4.58}$$

Since the second part of Eq. (4.57) is always satisfied with the given combination of parameters, the first part of Eq. (4.57) gives

$$\beta_{11} = \sqrt{\frac{36}{20}} = \frac{3}{\sqrt{5}}. \tag{4.59}$$

This gives that

$$\gamma_{11} = -\frac{3}{\sqrt{5}}, \gamma_{12} = -\frac{1}{\sqrt{5}}, \gamma_{13} = \frac{1}{\sqrt{5}}, \gamma_{13} = \frac{3}{\sqrt{5}} \tag{4.60}$$

which is indeed an orthonormal basis.

L_2, which forms the basis for the second-power terms, can be obtained by first noting that the treatment coefficients can be written as

$$\gamma_{21} = \beta_{21}x_1^2 + \beta_{20} = \beta_{21}(-1)^2 + \beta_{20} = \beta_{21} + \beta_{20}$$
$$\gamma_{22} = \beta_{21}x_2^2 + \beta_{20} = \beta_{21}\left(-\frac{1}{3}\right)^2 + \beta_{20} = \frac{1}{9}\beta_{21} + \beta_{20}$$

$$\gamma_{23} = \beta_{21}x_3^2 + \beta_{20} = \beta_{21}\left(\frac{1}{3}\right)^2 + \beta_{20} = \frac{1}{9}\beta_{21} + \beta_{20}$$

$$\gamma_{24} = \beta_{21}x_4^2 + \beta_{20} = \beta_{21}(1)^2 + \beta_{20} = \beta_{21} + \beta_{20} \tag{4.61}$$

subject to the usual constraint of $\gamma_{11} + \gamma_{12} + \gamma_{13} + \gamma_{14} = 0$. It can be noted that this is an indeterminate system with one degree of freedom (six unknowns, but five equations). Firstly, it can be noted that $\gamma_{13} = \gamma_{12}$ and $\gamma_{14} = \gamma_{11}$. Arbitrarily setting $\gamma_{13} = \gamma_{12} = -1$ gives the following solution:

$$\gamma_{22} = \gamma_{23} = -1, \gamma_{21} = \gamma_{24} = 1$$
$$\beta_{21} = 2.25, \beta_{20} = -1.25. \tag{4.62}$$

L_3, which forms the basis for the factors raised to the third power, can be obtained by first noting that the treatment coefficients can be written as

$$\gamma_{31} = \beta_{32}x_1^3 + \beta_{31}x_1 = \beta_{32}(-1)^3 + \beta_{31}(-1) = -\beta_{32} - \beta_{31}$$

$$\gamma_{32} = \beta_{32}x_2^3 + \beta_{31}x_2 = \beta_{32}\left(-\frac{1}{3}\right)^3 + \beta_{31}\left(-\frac{1}{3}\right) = -\frac{1}{27}\beta_{32} - \frac{1}{3}\beta_{31}$$

$$\gamma_{33} = \beta_{32}x_3^3 + \beta_{31}x_3 = \beta_{32}\left(\frac{1}{3}\right)^3 + \beta_{31}\left(\frac{1}{3}\right) = \frac{1}{27}\beta_{32} + \frac{1}{3}\beta_{31}$$

$$\gamma_{34} = \beta_{32}x_4^3 + \beta_{31}x_3 = \beta_{32}(1)^3 + \beta_{31}(1) = \beta_{32} + \beta_{31} \tag{4.63}$$

subject to the usual constraint of $\gamma_{11} + \gamma_{12} + \gamma_{13} + \gamma_{14} = 0$. It can be noted that this is an indeterminate system with two degrees of freedom (six unknowns, but only 4 independent equations). First, it can be noted that $\gamma_{13} = -\gamma_{12}$ and $\gamma_{14} = -\gamma_{11}$, which implies that the constraint is immediately satisfied given the values. Arbitrarily setting $\gamma_{11} = -1$, which implies that $\gamma_{14} = 1$, and $\gamma_{12} = 1$, which implies that $\gamma_{13} = -1$, gives the following solution,

$$\gamma_{31} = -1, \gamma_{34} = 1, \gamma_{32} = 1, \gamma_{33} = -1$$
$$\beta_{32} = 4.5, \beta_{31} = -3.5. \tag{4.64}$$

Therefore, the factors raised to first power will be encoded by L_1, the factors raised to the second power by L_2, and the factors raised to the third power by L_3.

Thus, for $l = 4$, the basis polynomial is orthogonal and can be written as

$$L_1(x) = x$$
$$L_2(x) = 2.25x^2 - 1.25$$
$$L_3(x) = 4.5x^3 - 3.5x. \tag{4.65}$$

4.7.2.4 Generalized Orthonormal Basis Functions for First- and Second-Order Terms

Deriving the required forms for the orthonormal basis functions can be quite challenging in the general case. This section will present generalized orthonormal basis functions for first- and second-order terms for an arbitrary number of levels. The derivation of these results follows the same pattern as for the specific examples.

The generalized orthonormal $L_1(x)$ function can be written as

$$L_1(x) = \sqrt{\frac{3l-3}{l+1}}x. \tag{4.66}$$

The generalized orthonormal $L_2(x)$ function can be written as

$$L_2(x) = \frac{3(l-1)}{2}\sqrt{\frac{5(l-1)}{(l+1)(l-2)(l+2)}}x^2 - \frac{1}{2}\sqrt{\frac{5(l-1)(l+1)}{(l-2)(l+2)}}. \tag{4.67}$$

The generalized orthonormal $L_3(x)$ function can be written as

$$L_3(x) = \frac{(l-1)^{2.5}}{4}\sqrt{\frac{201}{(2l-1)(l+1)(l-2)\left(11l^2-11l-10\right)}}x^3$$
$$- \frac{l(l-1)^{1.5}}{2}\sqrt{\frac{201}{(2l-1)(l+1)(l-2)\left(11l^2-11l-10\right)}}x. \tag{4.68}$$

Higher-order generalized functions can be derived using a similar pattern.

4.7.3 Sum of Squares in Generalized Factorial Designs

When fitting a model that contains factors raised to powers higher than 1, it is common practice to combine the terms into a single term. The basic rule for combining the terms is to ignore any powers on the factors and see what the corresponding interaction is, for example, if we have the following set of interactions {A, B, AB, A^2, A^{2}B}, then both A and A^2 would be represented by the sum of squares due to A (SS_A), while AB and A^{2}B would be represented by the sum of squares due to AB (SS_{AB}). These terms would be found by combining the appropriate individual terms, that is,

$$SS_A = SS_{A_L} + SS_{A_Q}, \tag{4.69}$$

where SS_{A_L} is obtained using the standard formula for the SSR of β_1, while SS_{A_Q} is obtained using the standard formula for the SSR of β_{11}. The number of terms that are combined would equal the degrees of freedom for the given component, for example, SS_A in the above example would have two degrees of freedom.

4.7.4 Detailed Mixed-Level Example

A soft drink bottler is interested in obtaining more uniform fill heights in the bottles produced by the manufacturing process. The process engineer can control three variables: the percent carbonation (x_1), the operating pressure in the filler (x_2), and the bottles produced per minute or the line speed (x_3). Due to various control issues, the percent carbonation was selected at three different levels (10, 12, and 14%), while the pressure (25 and 30 psi) and line speed (200 and 250 rpm) were selected at two different levels. Two replicates will be performed and the deviation from the correct fill height noted. The results are shown in Table 4.7. Based on the provided results, analyse the model using the methods provided in the above discussion to determine the best model for the process. Be certain to analyse the residuals to determine the adequacy of the model. (*Data taken from D. Montgomery (2007)*, Design and Analysis of Experiments, 6th *Edition (2007).*)

Table 4.7 Optimizing the performance of a bottling process

Run	Percent Carbonation (x_1)	Line Speed (x_3)	Operating Pressure (x_2)	Deviation from Expected Height (inches)	
				I	II
1	10	200	25	−3	−1
2	10	250	25	−1	0
3	10	200	30	−1	0
4	10	250	30	1	1
5	12	200	25	0	1
6	12	250	25	2	1
7	12	200	30	2	3
8	12	250	30	6	5
9	14	200	25	5	4
10	14	250	25	7	6
11	14	200	30	7	9
12	14	250	30	10	11

4.7.4.1 Analysis

In order to solve this problem, the procedure will be split into the following steps:

(1) Preprocessing the given data, that is converting it into the desired format.
(2) Determining the general form of the model.
(3) Obtaining an appropriate basis for each of the factors.
(4) Performing linear regression to obtain the parameter estimates.
(5) Analysing the results using the F-test to determine an appropriate model.
(6) Analysing the residuals and if necessary revising the model.

An explanation of how to implement this problem in Excel® is presented in Sect. 8.7.3: Factorial Design Examples.

4.7.4.2 Preprocessing the Data

Before the data can be analysed, it must be converted into the format required for analysis, that is, it must vary between -1 and 1 (inclusive). Using the formula for centring the data (Eq. (4.70)), gives the following formulae:

$$\tilde{x}_1 = \frac{x_1 - 0.5(14 + 10)}{0.5(14 - 10)} = 0.5x_1 - 6$$

$$\tilde{x}_2 = \frac{x_3 - 0.5(25 + 30)}{0.5(30 - 25)} = 0.4x_2 - 11$$

$$\tilde{x}_3 = \frac{x_2 - 0.5(200 + 250)}{0.5(250 - 200)} = 0.04x_3 - 9. \tag{4.70}$$

Using this encoding will place all of the variables in the $[-1, 1]$ range.

4.7.4.3 Determining the General Model Form

The general model can be written as

$$y = \beta_0 + \beta_1\tilde{x}_1 + \beta_2\tilde{x}_2 + \beta_3\tilde{x}_3 + \beta_{12}\tilde{x}_1\tilde{x}_2 + \beta_{13}\tilde{x}_1\tilde{x}_3 + \beta_{23}\tilde{x}_2\tilde{x}_3 + \beta_{123}\tilde{x}_1\tilde{x}_2\tilde{x}_3$$
$$+ \beta_{11}\tilde{x}_1^2 + \beta_{112}\tilde{x}_1^2\tilde{x}_2 + \beta_{113}\tilde{x}_1^2\tilde{x}_3 + \beta_{1123}\tilde{x}_1^2\tilde{x}_2\tilde{x}_3. \tag{4.71}$$

It can be noted that since x_1 has three levels, terms raised to the second power can be included in the model. Note that there are $3 \times 2 \times 2 = 12$ terms, which is equal to the product of the individual number of levels for each factor.

4.7.4.4   Obtaining a Basis

Using the previously developed methods, the first factor will be transformed using the $l = 3$ results, while the other two factors will be transformed using the standard $l = 2$ results (as has been done previously). Therefore, for $l = 3$, the basis functions will be $L_1(\tilde{x}_1) = \tilde{x}_1$ and $L_2(\tilde{x}_1) = 3\,\tilde{x}_1^2 - 2$, while for $l = 2$, the basis function will be $L_1(\tilde{x}_2) = \tilde{x}_2$ (or $\tilde{x}_3$).

4.7.4.5   Defining the Linear Regression Problem

Based on the above results, the following matrices will be defined

$$\mathcal{A}_{part} = \begin{bmatrix}
1 & -1 & -1 & -1 & 1 & 1 & 1 & -1 & 1 & -1 & -1 & 1 \\
1 & -1 & -1 & 1 & 1 & -1 & -1 & 1 & 1 & -1 & 1 & -1 \\
1 & -1 & 1 & -1 & -1 & 1 & -1 & 1 & 1 & 1 & -1 & -1 \\
1 & -1 & 1 & 1 & -1 & -1 & 1 & -1 & 1 & 1 & 1 & -2 \\
1 & 0 & -1 & -1 & 0 & 0 & 1 & 0 & -2 & 2 & 2 & 2 \\
1 & 0 & -1 & 1 & 0 & 0 & -1 & 0 & -2 & 2 & -2 & 2 \\
1 & 0 & 1 & -1 & 0 & 0 & -1 & 0 & -2 & -2 & 2 & -2 \\
1 & 0 & 1 & 1 & 0 & 0 & 1 & 0 & -2 & -2 & -2 & 1 \\
1 & 1 & -1 & -1 & -1 & -1 & 1 & 1 & 1 & -1 & -1 & -1 \\
1 & 1 & -1 & 1 & -1 & 1 & -1 & -1 & 1 & -1 & 1 & -1 \\
1 & 1 & 1 & -1 & 1 & -1 & -1 & -1 & 1 & 1 & -1 & 1 \\
1 & 1 & 1 & 1 & 1 & 1 & 1 & 1 & 1 & 1 & 1 & 1
\end{bmatrix} \tag{4.72}$$

$$\mathcal{A} = \begin{bmatrix} \mathcal{A}_{part} \\ \mathcal{A}_{part} \end{bmatrix} \tag{4.73}$$

$$\vec{y} = [-3 \ -1 \ -1 \ 1 \ 0 \ 2 \ 2 \ 6 \ 5 \ 7 \ 7 \ 10 \ -1 \ 0 \ 0 \ 1 \ 1 \ 3 \ 1 \ 5 \ 4 \ 9 \ 6 \ 1 \ 1]^T. \tag{4.74}$$

Solving the linear regression problem will give

$$\hat{\vec{\beta}} = \begin{bmatrix} 3.125 & 3.938 & 1.375 & 0.958 & 0.563 & 0.188 & 0.208 & 0.063 & 0.313 & -0.063 & -0.208 & -0.146 \end{bmatrix}^T. \tag{4.75}$$

The diagonal entries of the $(\mathcal{A}^T \mathcal{A})$ matrix can be written as

$$\mathrm{diag}(\mathcal{A}^T \mathcal{A}) = \begin{bmatrix} 24 & 16 & 24 & 24 & 16 & 16 & 24 & 16 & 48 & 48 & 48 & 48 \end{bmatrix}. \tag{4.76}$$

It can be noted that the values are not the same. This is expected since an orthonormal basis was not used.

4.7.4.6 Determining the Model

From Eq. (4.16), the sum of squares due to regression can be written as

$$\boxed{SSR_i = \left(A^T A\right)_{ii}\hat{\beta}_i^2} \tag{4.77}$$

and the F-test will have the following form

$$F_i = \frac{SSR_i}{\frac{SSE}{3(2)^2(n_R-1)}} = \frac{SSR_i}{\frac{SSE}{24-12}}, \tag{4.78}$$

where SSE is equal to 8.50 (from sum of residuals). The results are summarized in Table 4.8. The critical value of F at 95% with one and twelve degrees of freedom is Sect. 4.7.

Based on the values in Table 4.8, and comparing the final column with the critical value gives that the values in bold are significant ($F_i > F_{critical}$). Therefore, the reduced model can be written as

$$y = 3.13 + 3.94\tilde{x}_1 + 1.38\tilde{x}_2 + 0.958\tilde{x}_3 + 0.563\tilde{x}_1\tilde{x}_2 + 0.313L_2\left(\tilde{x}_1^2\right). \tag{4.79}$$

Replacing the basis function by its actual value gives the following model

$$y = 3.13 + 3.94\tilde{x}_1 + 1.38\tilde{x}_2 + 0.958\tilde{x}_3 + 0.563\tilde{x}_1\tilde{x}_2 + 0.313\left(3\tilde{x}_1^2 - 2\right)$$
$$y = 2.50 + 3.94\tilde{x}_1 + 1.38\tilde{x}_2 + 0.958\tilde{x}_3 + 0.563\tilde{x}_1\tilde{x}_2 + 0.939\tilde{x}_1^2. \tag{4.80}$$

Table 4.8 F-test values—values in bold are significant at the 95% level

Parameters	Value	SSR_i	F_i
β_0	3.125	234	**331**
β_1	3.938	248	**350**
β_2	1.375	45.4	**64.1**
β_3	0.958	22.0	**31.1**
β_{12}	0.563	5.063	**7.147**
β_{13}	0.188	0.563	0.794
β_{23}	0.208	1.042	1.471
β_{123}	0.063	0.063	0.088
β_{11}	0.313	4.688	**6.618**
β_{112}	−0.063	0.188	0.265
β_{113}	−0.021	0.021	0.029
β_{1123}	−0.146	1.021	1.441

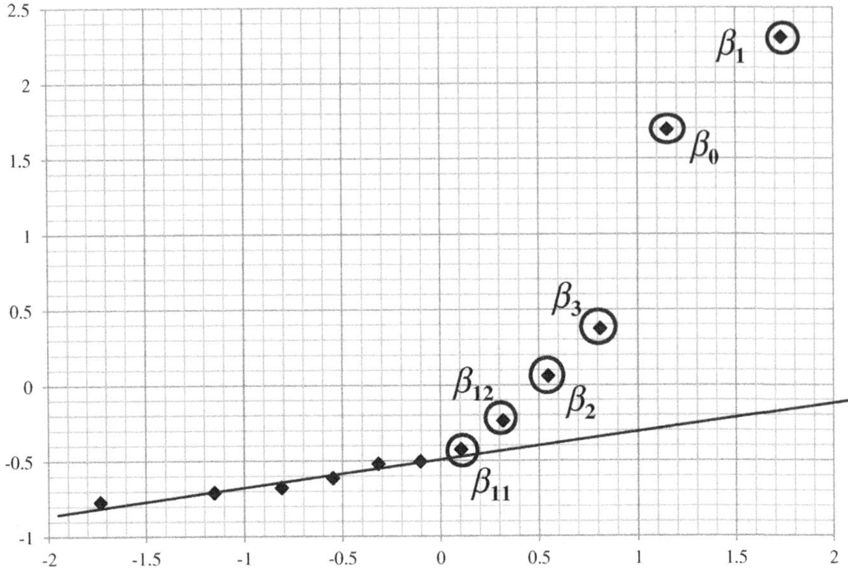

Fig. 4.8 Normal probability plot of the parameters for the mixed factorial example

If model reduction was to be performed using a normal probability plot of the parameter values, the results would be a bit more ambiguous. Figure 4.8 shows the normal probability plot of the parameters. In such cases, an orthonormal basis is preferred as it would weigh each of the points equally.

4.7.4.7   Analysing the Residuals

For the reduced model, $R^2 = 0.966$ and the F-statistic for the reduced model is 103 (>2.77 at the 95% confidence level). The residual plots are shown in Figs. 4.9,

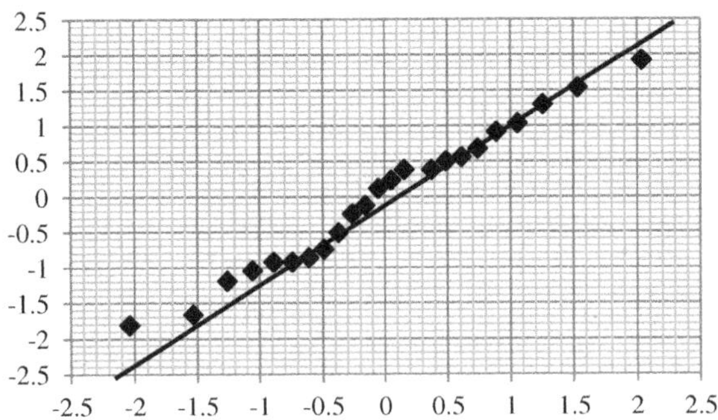

Fig. 4.9 Normal probability plot of the residuals

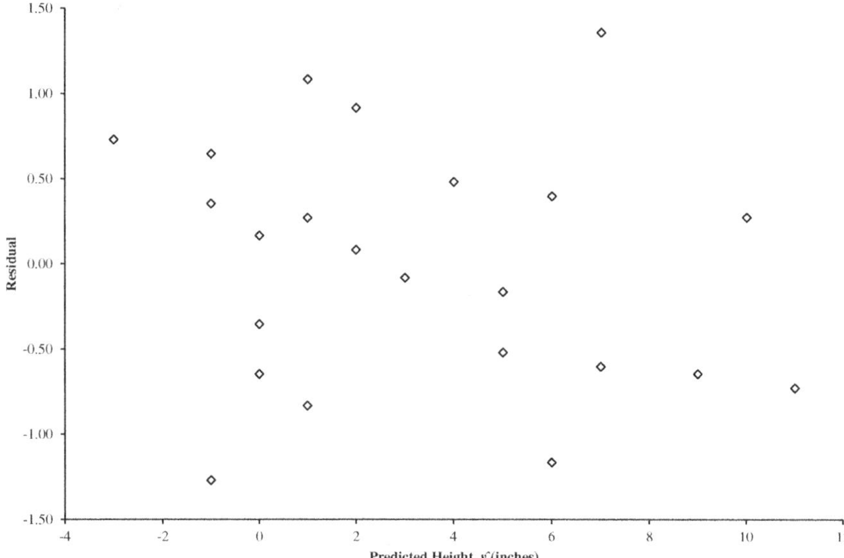

Fig. 4.10   Residuals as a function of $\hat{y}$

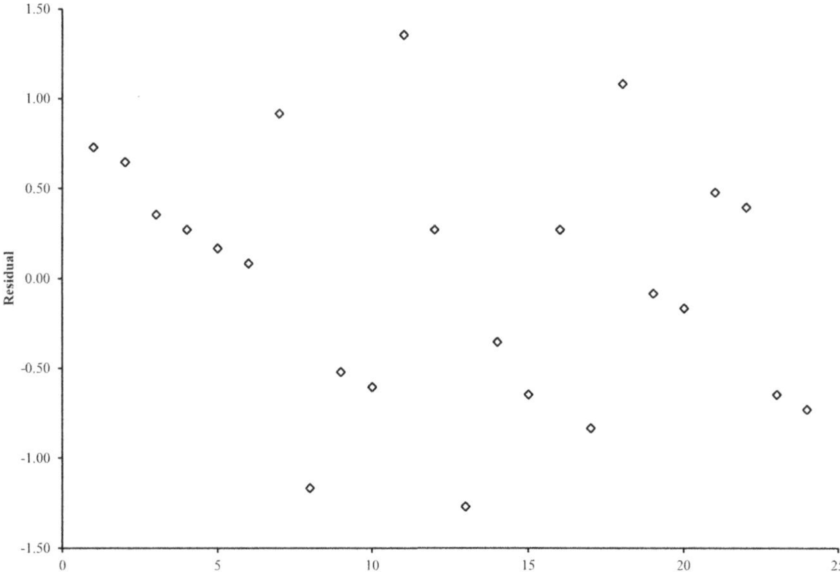

Fig. 4.11   Time-series plot of the residuals

4.10, and 4.11. It would seem that the residuals are normally distributed without any significant trends. Therefore, it can be concluded that the reduced-order model accurately describes the given data.

4.8 2^k-Factorial Designs with Centre Point Replicates

Although higher-order factorial designs are useful for fitting higher-order models, it is easy to see that the number of experiments required can quickly become too large to handle effectively given various constraints, such as available time and cost. Therefore, methods have been developed that can fit higher-order models without necessarily dealing with a full factorial experiment. One simple approach is to add centre points, that is, runs where all the coded variables are equal to zero, to the design of the experiment. This approach has two advantages. First, it allows for the computation of a variance for the model, which implies that the F-test can be used even with a full factorial experiment with no replicates. Second, it can provide a measure of the amount of curvature in the model. The model for this design consists of two parts: the standard $l = 2$ factorial design model combined with a curvature term denoted as $\beta_c x^2$, that is,

$$y = \beta_0 + \sum_{i=1}^{k} \beta_i x_i \underbrace{+ \ldots +}_{\substack{\text{remaining} \\ \text{factorial} \\ \text{terms}}} \beta_c x^2. \tag{4.81}$$

It can be noted that it is impossible to determine the individual contributions to the curvature as all the second-order terms are confounded with each other, for example, both x_1^2 and x_2^2 will have a column of 2^k ones followed by as many zeroes as there are centre point replicates.

As well, multiple replicates at the centre point are performed, so that an estimate of the variance can be obtained. Clearly, if there is only a single centre point replicate, then no estimate of the model variance can be obtained.

4.8.1   Orthogonal Basis for 2^k-Factorial Designs with Centre Point Replicates

Assuming that the model is coded as usual for an $l = 2$ design, it is easy to show that the curvature term is not orthogonal with the rest of the factorial design, since the curvature term is correlated with the first-order responses $(\beta_1, \beta_2, \ldots)$. Furthermore, there is an uneven number of experiments performed. Therefore, let $L_1(x)$ represent an orthogonal basis for the curvature term, that is, Eq. (4.81) can be rewritten as

$$y = \beta_0 + \sum_{i=1}^{2^k} \beta_i x_i \underbrace{+ \ldots +}_{\substack{\text{remaining} \\ \text{factorial} \\ \text{terms}}} \beta_c L_1(x). \tag{4.82}$$

In order to obtain an orthogonal basis in this case, it is necessary to note that the results from Sect. 4.7 need to be modified to take into consideration the fact that an arbitrary number of experiments will be performed. In this case, the key constraint can be written as

$$\sum_{i=1}^{2^k n_R + n_C} L_1(x_i) = 0, \tag{4.83}$$

where n_C is the number of centre point replicates. It is easy to show that the factorial component will satisfy this constraint as the factorial component must sum to zero and the centre points are all encoded as 0, so that the sum remains 0. On the other hand, the curvature term will not be orthogonal as it will contain $2^k n_R$ ones and n_C zeroes, which will not equal zero. Note that for the curvature basis, there are only two independent values γ_1 and γ_2, and Eq. (4.83) can be rewritten as

$$2^k n_R \gamma_1 + n_C \gamma_2 = 0. \tag{4.84}$$

Since this equation has two unknowns and only one equation, we are free to arbitrary assign any value to one of the two parameters. For the sake of consistency,[6] let $\gamma_1 = 1$, which implies that

$$\boxed{\gamma_2 = -\frac{2^k n_R}{n_C}}. \tag{4.85}$$

The basis function can be written as

$$L_1(x) = \beta_1 x^2 + \beta_2 \tag{4.86}$$

which given the two treatment values gives

$$L_1(0) = \beta_2 = \gamma_2 = -\frac{2^k n_R}{n_C}$$

$$L_1(\pm 1) = \beta_1 + \beta_2 = 1 \Rightarrow \beta_1 = 1 + \frac{2^k n_R}{n_C} = \frac{n_C + 2^k n_R}{n_C}. \tag{4.87}$$

Therefore, the basis function for the curvature can be written as

---

[6] This will leave the factorial component unchanged.

$$L_1(x) = \left(\frac{n_C + 2^k n_R}{n_C} \right) x^2 - \frac{2^k n_R}{n_C}. \tag{4.88}$$

4.8.2   Factorial Design with Centre Point Example

Consider a chemical engineer who is studying the yield of a chemical process. There are two variables of interest: reaction time and reaction temperature. Since there is some uncertainty regarding the appropriateness of a linear model, a single unreplicated 2^2-factorial experiment was performed with five centre point replicates. The results of the experiment are shown in Table 4.9. Based on the provided results, analyse the model using the methods provided in the above discussion to determine the best model for the process. Be certain to analyse the residuals to determine the adequacy of the model. (*Data taken from D. Montgomery (2007), Applied Statistics and Probability for Engineers, 4th Edition.*)

An Excel®-based solution to this problem is presented in Sect. 8.7.3.

A similar procedure to that used to analyse the results in Sect. 4.7.4.1 will be used in this example.

4.8.2.1   Determining the General Model

The model that will be analysed can be written as

$$y = \beta_0 + \beta_1 x_1 + \beta_2 x_2 + \beta_{12} x_{12} + \beta_c x^2. \tag{4.89}$$

Table 4.9 Improving chemical plant yield data set

Run	Reaction Time (min)	Temperature (°C)	Yield (%)
1	-1	-1	39.3
2	-1	1	40.0
3	1	-1	40.9
4	1	1	41.5
5	0	0	40.3
6	0	0	40.5
7	0	0	40.7
8	0	0	40.2
9	0	0	40.6

4.8.2.2 Selecting the Orthogonal Basis

Since $n_C = 5$ and $k = 2$, an appropriate basis function for the curvature tern can be written as

$$L_1(x) = \left(\frac{5 + 2^2(1)}{5}\right)x^2 - \frac{2^2(1)}{5}$$
$$= 1.8x^2 - 0.8 \tag{4.90}$$

which gives that the model in terms of the orthogonal basis functions can be written as

$$y = \beta_0 + \beta_1 x_1 + \beta_2 x_2 + \beta_{12} x_{12} + \beta_c\left(1.8x^2 - 0.8\right). \tag{4.91}$$

4.8.2.3 Defining the Linear Regression Problem

Based on the above results, the following matrices will be defined

$$A = \begin{bmatrix} 1 & -1 & -1 & 1 & 1 \\ 1 & -1 & 1 & -1 & 1 \\ 1 & 1 & -1 & -1 & 1 \\ 1 & 1 & 1 & 1 & 1 \\ 1 & 0 & 0 & 0 & -0.8 \\ 1 & 0 & 0 & 0 & -0.8 \\ 1 & 0 & 0 & 0 & -0.8 \\ 1 & 0 & 0 & 0 & -0.8 \\ 1 & 0 & 0 & 0 & -0.8 \end{bmatrix} \tag{4.92}$$

$$\vec{y} = \begin{bmatrix} 39.3 & 40.0 & 40.9 & 41.5 & 40.3 & 40.5 & 40.7 & 40.2 & 40.6 \end{bmatrix}^T. \tag{4.93}$$

Solving the linear regression problem will give

$$\hat{\vec{\beta}} = \begin{bmatrix} 40.44 & 0.775 & 0.325 & -0.025 & -0.019 \end{bmatrix}^T. \tag{4.94}$$

The diagonal entries of the $(A^T A)$ matrix can be written as

$$\text{diag}(A^T A) = \begin{bmatrix} 9 & 4 & 4 & 4 & 7.2 \end{bmatrix}. \tag{4.95}$$

It can be noted that the values are not the same. This is expected since an orthonormal basis was not used.

Table 4.10 F-test values—values in bold are significant at the 95% level

Parameters	Value	SSR_i	F_i
β_0	40.44	14,700	**342,000**
β_1	0.775	2.40	**55.9**
β_2	0.325	0.423	**9.83**
β_{12}	−0.025	0.0025	0.058
β_c	−0.019	0.002 72	0.063

4.8.2.4 Determining the Model

From Eq. (4.16), the sum of squares due to regression can be written as

$$SSR_i = \left(A^T A\right)_{ii} \hat{\beta}_i^2 \tag{4.96}$$

and the F-test will have the following form

$$F_i = \frac{SSR_i}{\frac{SSE}{m-n}} = \frac{SSR_i}{\frac{SSE}{2^k+n_C-2^k-1}} = \frac{SSR_i}{\frac{SSE}{n_C-1}} = \frac{SSR_i}{\frac{SSE}{4}}, \tag{4.97}$$

where SSE is equal to 0.172 (from sum of residuals). The results are summarized in Table 4.10. The critical value of F at 95% with one and four degrees of freedom is 7.71.

Based on the values in Table 4.10, and comparing the final column with the critical value gives that the values in bold are significant ($F_i > F_{critical}$). Therefore, the reduced model can be written as

$$y = 40.4 + 0.775x_1 + 0.325x_2. \tag{4.98}$$

Furthermore, it can be concluded that the effect of curvature is minimal in this system as β_c is not significant.

With so few parameters, using a normal probability plot of the parameters will not give meaningful results.

4.8.2.5 Analysing the Residuals

For the reduced model, $R^2 = 0.941$ and the F-statistic is 47.8 (> 5.14 at the 95% confidence level). The residual plots are shown in Figs. 4.12, 4.13, 4.14, and 4.15. It would seem that the residuals are normally distributed without any significant trends. Therefore, it can be concluded that the reduced-order model accurately describes the given data.

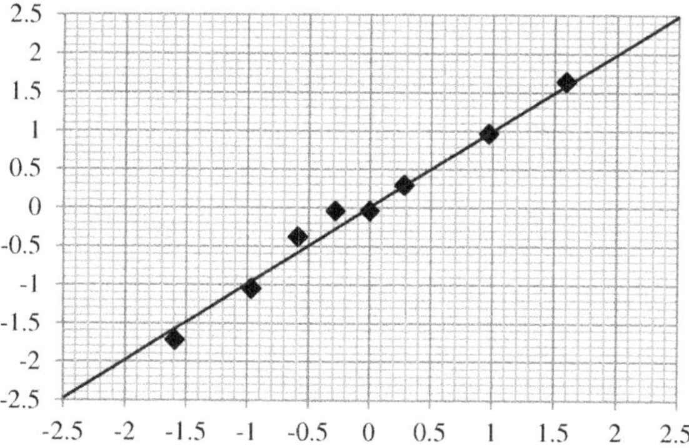

Fig. 4.12 Normal probability plot of the residuals for the reduced model

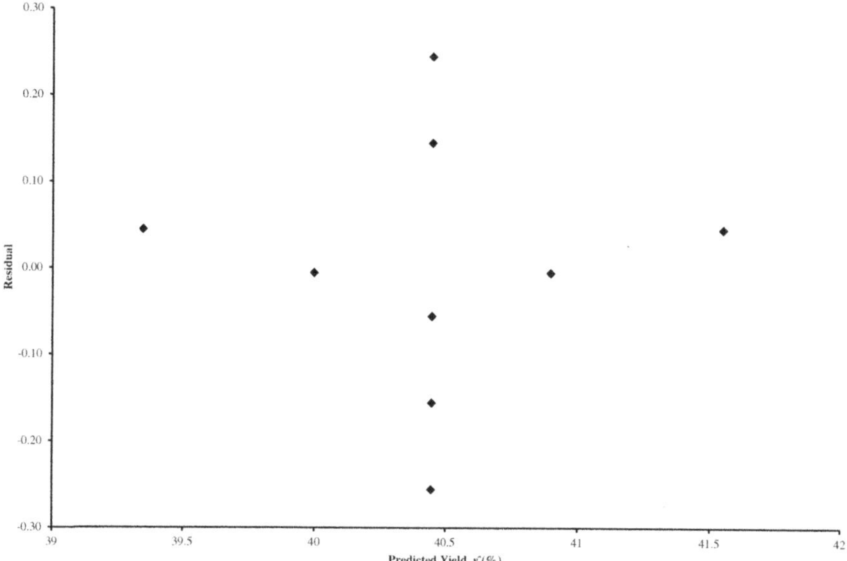

Fig. 4.13 Residuals for the reduced model as a function of $\hat{y}$

4.9 Response Surface Design

Another approach for designing second-order models to be used for process optimization, without using higher-order factorial design for optimization, is **response surface design**. Many different approaches have been developed, many of which require the use of specialized computer software to obtain a tractable solution.

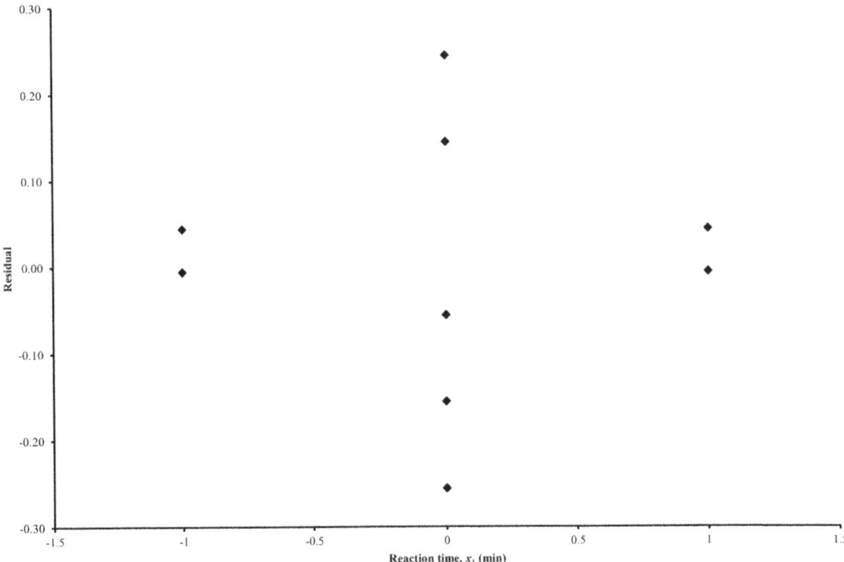

Fig. 4.14 Residuals for the reduced model as a function of x_1

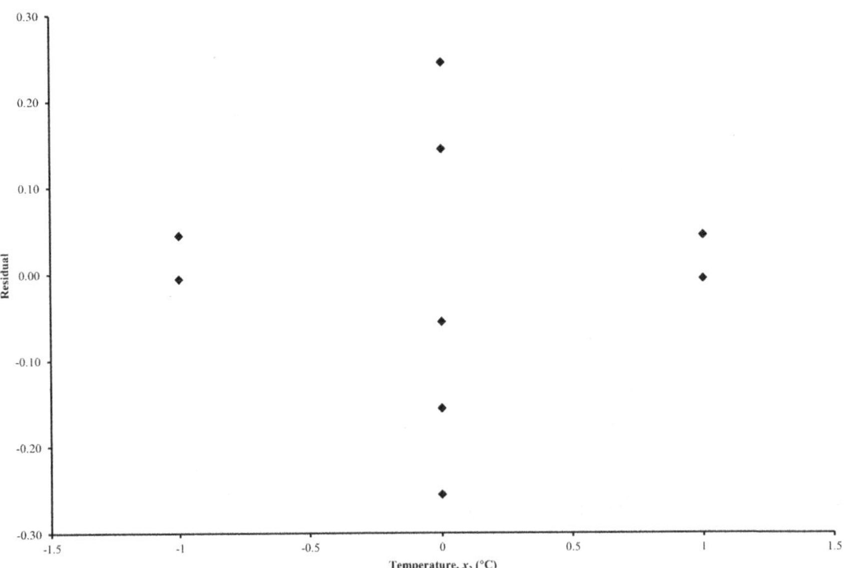

Fig. 4.15 Residuals for the reduced model as a function of x_2

4.9.1 Central Composite Design

The **central composite design** avoids the problem of higher-order factorial designs, by judiciously selecting additional parameter points. The generic regression matrix, $\mathcal{A}$, for k factors consists of three different parts:

(1) The regression matrix obtained from the 2^k-factorial experiment. This will be denoted by $\mathcal{F}$.

(2) The central point of the system denoted by $(0, 0, \ldots, 0, 0)$, where there are k zeros. This part can be repeated as often as is desired or required. This part will be denoted by $\mathcal{C}$.

(3) A matrix which consists of each factor at some value $\pm\alpha$, where α is a value determined by the designer. Thus, this part, denoted by $\mathcal{E}$, will have the following form:

$$
\mathcal{E} =
\begin{bmatrix}
\alpha & 0 & 0 & \cdots\cdots & 0 \\
-\alpha & 0 & 0 & \cdots\cdots & 0 \\
0 & \alpha & 0 & \cdots\cdots & 0 \\
0 & -\alpha & 0 & \cdots\cdots & 0 \\
\vdots & & & & \vdots \\
0 & 0 & 0 & \cdots\cdots & \alpha \\
0 & 0 & 0 & \cdots\cdots & -\alpha
\end{bmatrix}.
\tag{4.99}
$$

Thus, the regression matrix will have the following form:

$$
\mathcal{A} =
\begin{bmatrix}
\mathcal{F} \\
\mathcal{C} \\
\mathcal{E}
\end{bmatrix}.
\tag{4.100}
$$

The model that can be fit to this type of experiment is given as

$$
y = \beta_0 + \sum_{i=1}^{k} \beta_i x_i \underbrace{+ \ldots +}_{\substack{\text{remaining} \\ \text{factorial} \\ \text{terms}}} \sum_{i=1}^{k} \beta_{ii} x_i^2
\tag{4.101}
$$

which is basically the standard 2^k-factorial experiment with the addition of the pure quadratic terms due to each of the factors themselves.

4.9.1.1 Determining the Value of α

There are many different methods to determine the value of α. Let $F = 2^k$, the number of points due to the factorial design, and $T = 2k + n_C$, the number of additional points, where n_C is the number of central points in the design. Common values are as follows (Myers 1971):

(1) **Rotatable design:** $\alpha = F^{0.25}$, which is the design implemented by the `ccdesign(k)` function in MATLAB®. Rotatability implies that the variance for points equidistant from the centre is equal. This is useful if the region of interest is spherical.

(2) **Face-centred cubical design:** $\alpha = 1$, which is useful if the region of interest is a cube. This approach will place the values in the centre of the plane defined by the factorial experiment.

(3) **Orthogonal design:** $\alpha = (0.25QF)^{0.25}$, where $Q = \left(\sqrt{F+T} - \sqrt{F} \right)^2$, which minimizes the correlation between the different parameter estimates.

4.9.2 Optimal Design

Optimal design is the process of designing an experiment so that it is optimal with respect to some criteria. All such designs are computer generated in that some complex formula must be used in order to perform the relevant optimization. The most common forms of optimal design are

(1) **A-optimal design**, which seeks to minimize the trace of $(A^T A)^{-1}$. This can be interpreted as the minimization of the **average** value of the variance of the regression coefficients.

(2) **D-optimal design**, which seeks to minimize the determinant of $(A^T A)^{-1}$. This can be interpreted as the minimization of the volume of the joint confidence region of the regression coefficients.

(3) **E-optimal design**, which seeks to minimize the largest eigenvalue of $(A^T A)^{-1}$.

(4) **G-optimal design**, which seeks to minimize the largest value of $N \mathrm{var}(\hat{y})/\sigma^2$, where N is the number of design points in the experiment, for example, for a full factorial design, $N = l^k$. This can be interpreted as the minimization of the scaled prediction variance. This approach is useful if the goal of the model is to use it for future predictions.

(5) **V-optimal design**, which seeks to minimize the average prediction variance for a set of prediction points. This approach is useful if the goal of the model is to use it for future predictions over multiple different points.

4.9.3 Response Surface Procedure

The response surface method is an iterative procedure for determining the optimal point of a process. Starting from an initial large set of variables, a **screening experiment** can be performed to determine which of the variables have the largest influence on the desired variable. The number of variables selected at this stage can be a factor of the time available to perform the detailed experiments and the influence that the given variables have on the system. Once the screening experiment has been completed, then a factorial or response surface method experiment can be designed to obtain an initial estimate of the optimal point. Based on this estimate of the optimal point, a new experiment could be designed to determine a new optimal point. This procedure can be repeated until either the procedure stabilizes about a fixed point or after a fixed number of experiments. It can be noted that the above procedure does not guarantee that the true optimal point will be reached.

In seeking the optimal point, it is common to use the **method of steepest ascent** or the **gradient method**. In this approach, based on the resulting relationship, the new process operating point will be determined based on the direction with the steepest gradient. This approach ensures that the optimal point can be reached the fastest.

4.10 Further Reading

The following are references that provide additional information about the topic:

(1) **General Design and Analysis of Experiments**: Most of these references contain information about factorial and fractional factorial design.

    a. Box, G. E., Hunter, W. G., & Hunter, J. S. (1978). *Statistics for Experimenters: An Introduction to Design, Data Analysis, and Model Building.* New York, New York, United States of America: John Wiley & Sons, Inc.

    b. Hinkelmann, K., & Kempthorne, O. (2007). *Design and Analysis of Experiments* (Vols. I, II, III). Hoboken, New Jersey, United States of America: John Wiley & Sons, Inc.

    c. Montgomery, D. C. (1991). *Design and Analysis of Experiments* (3rd ed.). New York, New York, United States of America: John Wiley and Sons.

(2) **Response Surface Methodology**:

    a. Myers, R. H. (1971). *Response Surface Methodology.* Boston, Massachusetts, United States of America: Allyn and Bacon, Inc.

4.11 Chapter Problems

Problems at the end of the chapter consist of three different types: (a) Basic Concepts (True/False), which seek to test the reader's comprehension of the key concepts in the chapter; (b) Short Exercises, which seek to test the reader's ability to compute the required parameters for a simple data set using simple or no technological aids. This section also includes proofs of theorems; and (c) Computational Exercises, which require not only a solid comprehension of the basic material, but also the use of appropriate software to easily manipulate the given data sets.

4.11.1 Basic Concepts

Determine if the following statements are true or false and state why this is the case.

(1) Confounding of variables implies that $(A^T A)$ is uninvertible.

(2) Randomization is important, since it can minimize the effect of unknown variables on the regression results.

(3) Blocking seeks to minimize the effect of unknown variables on the regression results.

(4) A model defined based on an energy balance over the system is a black-box model of the system.

(5) A model developed for analysing the performance of the system can always be used to predict the future behaviour of the system under new conditions.

(6) A 3^4 full factorial experiment has four levels and three factors.

(7) For a factorial experiment with no replicates, the variance can be determined.

(8) If the design is orthogonal, after a parameter is removed, then the remaining parameters need to be recalculated.

(9) The factor $x_1 x_3 x_4$ is a third-order interaction.

(10) Factorial design experiments with a large number of levels and factors can be expensive to run.

(11) For fractional factorial experiment, it is useful if higher-order interactions are known to be unimportant.

(12) If the complete defining relationship is I = ABCD = ADE = ABF, then the resolution of this design is IV.

(13) The more letters (factors) in a defining relationship the larger the potential resolution of the design.

(14) Two interactions that are confounded can be individually estimated.

(15) If the generator is F = ABD, then the defining relationship for this generator is I = ABDF.

(16) Adding central points to a factorial design experiment allows for the testing of curvature (or second-order terms, such as A^2 in the design).

(17) If it is desired to use a model for future prediction, then an *A*-optimal design is best.

(18) If there are only a few factors to consider, then designing a screening experiment is profitable.
(19) The response surface methodology is an iterative procedure that requires multiple experiments and modelling exercises.
(20) A rotatable central composite design is useful if it is desired to have a common variance for points equidistant from the centre.

4.11.2 Short Exercises

These questions should be solved using only a simple, nonprogrammable, nongraphical calculator combined with pen and paper.

(21) Consider the problem of trying to fit data to the following model

$$y_t = -\alpha_1 y_{t-1} + \beta_1 u_{t-1} + e_t, \tag{4.102}$$

where α and β are coefficients to be determined, t is a subscript representing the time of measurement, that is t represents the current time, $t-1$ represents the time one sampling unit in the past. For this model, it can be shown that (Ljung, 1999)

$$\mathcal{A}^T \mathcal{A} = \begin{bmatrix} E(y_t y_t) & -E(y_t u_t) \\ -E(y_t u_t) & E(u_t u_t) \end{bmatrix}. \tag{4.103}$$

In order to obtain parameter estimates, the matrix given by Equation (103) must be invertible for all nonzero values of α and β. If the input signal is of the following form, under what conditions can the process be identified:

(a) $u_t = -Ky_t$.
(b) $u_t = -Ky_{t-2}$.

(Further information about fitting such processes can be found in Chap. 6).

(22) Consider the following factorial design whose regression matrix is shown in Table 4.11. Determine:

(a) What are the independent (or basic) factors and what are the dependent factors?
(b) What are the generators?
(c) What is the complete defining relationship?
(d) What are the aliases of C (or what is the confounding relationship for C) and for AF?
(e) What is the resolution?
(f) What type of factorial design it is?

Table 4.11 Design for the fractional factorial experiment (for Question 22)

Run	A	B	C	D	E	F
1	−	−	+	+	−	+
2	+	−	+	−	−	−
3	−	+	−	+	−	−
4	+	+	−	−	−	+
5	−	−	−	−	+	+
6	+	−	−	+	+	−
7	−	+	+	−	+	−
8	+	+	+	+	+	+

(23)   You have to run a fractional factorial experiment where there are six (A, B, C, D, E, and F) factors each at two different levels. You have decided that a 2^{6-2} fractional factorial experiment will be performed with the following two generators:

$$D = AC$$

$$F = CD.$$

Answer the following questions:

(a)   What is the complete defining relation?
(b)   What is the confounding pattern for all the first-order interactions?
(c)   What is the resolution of this experiment?
(d)   If a resolution of IV is desired, how could the above generators be changed to achieve this? Give the complete defining relation to prove that the design is indeed resolution IV.

(24)   You ran a 4^2 full factorial experiment. Not wanting to go through the hassle of developing an orthogonal basis for such an experiment, you decided to analyse this experiment as a two-level factorial experiment. Answer the following questions:

(a)   What model can you fit with the original 4^2 experiment?
(b)   Clearly explain how you could analyse this experiment as a two-level factorial experiment.
(c)   Can you fit the original model using this type of analysis?

(25)   As the plant engineer, you have been asked to optimize the performance of a chemical reactor. Using the information below, suggest a potential experimental design (including all defining relationships, generators, resolution, and the regression matrix in terms of the factors). Clearly justify your choices based on the requirements. The situation is as follows:

(a)   There are five factors (A, B, C, D, and E).

(b)   You have managed to obtain two days (48 h) to do the complete experiment.

(c)   Each run takes 2 h, that is you have enough time to complete 24 runs.

(d)   All third-order and higher interactions are known to be minimal.

(e)   It is expected that only some of the five factors are significant.

(f)   It is desired to run some of the runs at the centre point in order to test for curvature and variability in the results.

(26)   Consider the following 3^{3-1} factorial experiment with the generator given as

$$x_3^2 = x_1 x_2^2.$$

Answer the following questions.

(a)   What is the complete defining relationship for this experiment?

(b)   Determine the complete confounding pattern? (*Hint: You will need to confound eighteen variables.*)

(c)   What type of model could be estimated using this type of experiment? Give all estimable terms.

(27)   Propose a 2^{6-2}-factorial experiment so that it has resolution IV.

(28)   Determine a suitable design for a 3^{4-1} factorial experiment.

4.11.3 Computational Exercises

The following problems should be solved with the help of a computer and appropriate software packages, such as MATLAB® or Excel®.

(29)   In the article "In the Soup: A Case Study to Identify Contributors to Filling Variability," Hare (1988) considers the problem of trying to determine the variables which affect the filling variability of dry soup mix. Five different variables where considered: (A) number of filling ports being used in the mixer; (B) the operating temperature; (C) the mixing time in seconds; (D) the batch weight in pounds (lbs.); and (E) the delay between mixing and packaging. The operating temperature was maintained using a cooling jacket: either the cooling jacket was on (denoted by C) or the process was operated at ambient conditions (denoted by A). The results are shown in Table 4.12. Run 7 is the normal operating conditions. The results are presented in the order in which they were run. Perform the following analysis of this data set:

(a)   Determine an appropriate coding for this experiment.

(b)   What is the generator for this design?

(c)   What is the resolution of this design?

(d)   Estimate the factors? Which ones are significant?

(e)   Using the reduced-order model, analyse the residuals and determine if the design assumptions are met?

Table 4.12 Dry soup variability data (for Question 29)

Run	$\hat{\sigma}_p$ (lbs.)	A	B	C (s)	D (lbs.)	E (days)
1	0.78	1	C	60	2,000	7
2	1.10	3	C	80	2,000	7
3	1.70	3	A	60	1,500	1
4	1.28	3	C	80	1,500	1
5	0.97	1	A	60	1,500	7
6	1.47	1	C	80	1,500	7
7	1.85	1	A	60	2,000	1
8	2.10	3	A	80	2,000	1
9	0.76	1	A	80	2,000	7
10	0.62	3	A	60	2,000	7
11	1.09	1	C	80	2,000	1
12	1.13	1	C	60	1,500	1
13	1.25	3	C	60	1,500	7
14	0.98	3	A	80	1,500	7
15	1.36	3	C	60	2,000	1
16	1.18	1	A	80	1,500	1

(*Data from Lynne B. Hare, "In the Soup: A Case Study to Identify Contributors to Filling Variability"* in the Journal of Quality Technology ©*1988 Taylor & Francis.*)

(30)   Consider the problem of trying to determine which conditions impact the life (in hours) of a machine tool. The variables of interested have been selected as cutting speed (A), tool geometry (B), and cutting angle (C). Consider the following full factorial design whose regression matrix and results are shown in Table 4.13. Perform all analysis at the 95% level. Answer the following questions:

(a)   Determine the model for the full factorial experiment.

(b)   Fit the model and obtain confidence intervals for the parameter estimates. Determine which parameter estimates should be kept.

(c)   Calculate the F-score for each parameter estimate. Determine which parameters should be kept.

(d)   Are the results from (b) and (c) the same? Do you think that this is a coincidence or will this always be the case?

(e)   Based on your results from (b) and (c) what model would you suggest for the data? Which interactions are significant? Why?

(f)   Examine the residuals for the full model and determine if there are any issues with the distribution of the residuals. (*Hint: Plot the residuals for each replicate in a different colour or on separate graphs*)

(31)   Consider the data shown in Table 4.14, which is for the optimization of crystal growth. In crystal growth optimization, it is desired to produce the heaviest

Table 4.13 Tool life data (for Question 30)

Run	A	B	C	Replicate		
				y_1	y_2	y_3
1	−	−	−	22	31	25
2	+	−	−	32	43	29
3	−	+	−	35	34	50
4	+	+	−	55	47	46
5	−	−	+	44	45	38
6	+	−	+	40	37	36
7	−	+	+	60	50	54
8	+	+	+	39	41	47

(*Data taken from D. Montgomery (2007),* Design and Analysis of Experiments, 6th *Edition (2007)*)

Table 4.14 Crystal optimisation data (for Question 31)

Y (grammes)	A	B	C
66	−1	−1	−1
80	1	−1	−1
78	−1	1	−1
90	1	1	−1
75	−1	−1	1
70	1	−1	1
60	−1	1	1
75	1	1	1
90	−1.682	0	0
86	1.682	0	0
68	0	−1.682	0
63	0	1.682	0
75	0	0	−1.682
75	0	0	1.682
105	0	0	0
100	0	0	0
103	0	0	0
95	0	0	0
100	0	0	0
96	0	0	0

(*Data taken from D. Montgomery (2007),* Design and Analysis of Experiments, *6th Edition*)

crystal. Using the concept presented in this and previous chapters, analyse the provided data to answer the following questions:

(a)   What is the best model for the data provided?
(b)   At what optimal point should the process be operated?
(c)   Criticize the experimental design and suggest ways to improve it.

Appendix A4: Nonmatrix Approach to the Analysis of 2^k-Factorial Design Experiments

It will be assumed that a 2^k-factorial experiment has been designed with n_R full replicates. Furthermore, it will be assumed that all the factors have been coded so that -1 and 1 represent the upper and lower levels in the experiment. The same notation as used in this chapter will be used. Thus, instead of calculating inverses and transposes, the following simplifications work for a 2^k-factorial experiment:

$$A^T A = 2^k \mathcal{I}_k, \tag{4.104}$$

where $\mathcal{I}_k$ is the $k \times k$ identity matrix,

$$\left(A^T A\right)^{-1} = 2^{-k}\mathcal{I}_k \tag{4.105}$$

$$\hat{\vec{\beta}} = 2^{-k} A^T \vec{y}. \tag{4.106}$$

If $\overline{A}$ is used, then the results are

$$\overline{A}^T\overline{A} = 2^k n_R \mathcal{I}_k \tag{4.107}$$

$$\left(\overline{A}^T\overline{A}\right)^{-1} = 2^{-k}(n_R)^{-1}\mathcal{I}_k. \tag{4.108}$$

The sum of squares due to errors, SSE, can be computed using the following formula

$$SSE = (n_R - 1)\sum_{i=1}^{2^k} s_i^2, \tag{4.109}$$

where s_i is the standard deviation for the replicates for treatment i. Thus, the standard deviation, $\hat{\sigma}$, can be determined as

$$\hat{\sigma} = \sqrt{\frac{SSE}{l^k(n_R - 1)}} = \sqrt{\frac{\sum_{i=1}^{2^k} s_i^2}{l^k}}. \qquad (4.110)$$

The effect due to each variable can be determined from

$$\text{effect} = 2\hat{\beta}. \qquad (4.111)$$

Chapter 5
Modelling Stochastic Processes with Time-Series Analysis

So far with a few minor exceptions, the data used for regression analysis have been assumed to be independent of time, that is, the same values would be obtained irrespective of the time or sequence of events. In many process engineering examples, such an assumption is not necessarily true. The process values change from sampling point to sampling point, and the individual values strongly depend on adjacent process values, for example, the growth of a crystal is a time-dependent process. In such cases, analysing the data set using standard regression analysis may not be appropriate. A data set where the time element is important is often called a **time series**. Time series are found in many different fields including economics, business, social sciences, and of course, science and engineering. The development of time-series analysis methods has often been done in parallel in many different fields. Three different approaches to the analysis of time series can be considered (Shumway and Stoffer 2011). The first approach, which will be referred to here as the **transfer-function-based approach** seeks to develop a class of suitable models that describe the observable system using the available information. The internal dynamics of the system are not considered in this approach. The transfer-function-based approach finds wide application in process engineering, as it forms one of the main forms of modelling in process control applications. The foundations of this approach were laid by Box and Jenkins (Box and Jenkins 1970) in the middle part of the twentieth century. The second approach, which will be referred to here as the **state-space-based approach**, seeks to develop an understanding of the internal process dynamics that can be used to predict the future behaviour of the system. This approach incorporates the Kalman filter into developing appropriate parameter estimates. This approach although relatively new is finding more widespread implementation in process engineering due to its ability to handle complex process dynamics. The approach is based on the work of Kalman in developing a new method for dealing with time-series data (Kalman 1960; Kalman and Bucy 1961). Applications of this approach to time-series analysis are detailed in Harvey (1991). The third and final approach, which will be referred to here as the **frequency-domain-based approach**, seeks to develop methods for understanding

Y. A. W. Shardt, *Statistics for Chemical and Process Engineers*,
https://doi.org/10.1007/978-3-030-83190-5_5

the process in the frequency domain, that is, what kind of periodic signals are present in the given data set. In chemical and process engineering, transfer-function- and frequency-domain-based approaches are the most common ways of examining time series. Frequency-domain analysis often uses a Fourier transformation of the data set to highlight the key frequencies present. A good overview of these methods can be found in Bloomfield (2000) and Priestley (1981).

Given the pre-eminence of the transfer-function- and frequency-domain-based approaches in process and chemical engineering, these two approaches will be discussed in greatest detail in this chapter. Nevertheless, information about the state-space-based approach will also be considered. This chapter will present the basic, univariate approach to time-series analysis, which will be extended in Chap. 6 to consider the multivariate case containing both stochastic and deterministic components in order to model complex processes for process control, economic analysis, and simulation development.

5.1 Fundamentals of Time-Series Analysis

A time series is a data set whose individual entries are ordered chronologically, that is, let the time series Y be defined as follows:

$$Y = \{y_0, y_1, y_2, \ldots, y_m\}, \tag{5.1}$$

where y_0 represents the data sample at the initial sampling point, y_1 the data sample at the first sampling point, and m the total number of samples taken. It should be noted that the time interval between each of the samples is constant, that is, each sample is obtained every S seconds or other appropriate time unit.

For a time series, it is possible to compute such statistical properties as mean and standard deviation. However, these values can easily depend on time, that is, different subsets will have completely different means, for example, a growing crystal will initially have a small mean value, which will increase as the crystal grows. Since dealing with changing values can be difficult in many statistical applications, some assumptions are made about the properties of the time series.

A time series is said to be **strictly stationary** if the probabilistic behaviour of every subset of values, $\{x_t, x_{t+1}, \ldots x_{t+j}\}$ is identical to the time-shifted set $\{x_{t+k}, x_{t+k+1}, \ldots x_{t+k+j}\}$, that is,

$$P\left(x_t < c_1, \ldots, x_{t+j} < c_j\right) = P\left(x_{t+k} < c_1, \ldots, x_{t+k+j} < c_j\right) \tag{5.2}$$

for all $j = 1, 2, \ldots$; all time points $t = t_1, t_2, \ldots$; all constants $c_1, c_2 \ldots, c_j$, and all shifts, $k = 0, 1, 2, \ldots$. Strict stationarity implies that the distribution of variables is the same at every time instant and that all statistical properties are constant. In practice, it can be difficult to determine if a single data set satisfies this strict stationary condition.

Therefore, a weaker version that only constrains the first two moments has been developed.

A time series is said to be **weakly stationary**, if

1. The time series has a finite variance.
2. The mean of the time series is constant and independent of time, that is, $\mu = E(x_t)$.
3. The autocovariance function, defined by Eq. (5.3) is independent of time.

Unless otherwise specified, references to a stationary signal will imply that we are dealing with the weakly stationary definition.

In order to compute various properties, it is useful to treat each entry in a time series as a random variable. Therefore, the **autocovariance** of the time series, $\gamma(t, \tau)$, at time t and lag τ can be given as:

$$\gamma(t, \tau) = E\big((y_t - E(y_t))(y_{t+\tau} - E(y_{t+\tau}))^T\big). \tag{5.3}$$

The autocovariance of a signal shows the degree of correlation between the signal at different time periods. It can be noted that the autocovariance at a lag of zero is always equal to the covariance of the time series. However, since the values of the autocovariance are not normalised, comparing different time series can be difficult. Instead, the normalised autocovariance, or the **autocorrelation**, $\rho(t, \tau)$, is more useful. Autocorrelation is defined as

$$\rho(t, \tau) = \frac{\gamma(t, \tau)}{\gamma(t, 0)} = \frac{\gamma(t, \tau)}{\sigma_t^2}. \tag{5.4}$$

The autocorrelation, unlike the autocovariance, varies only between -1 and 1. From the definition of a weakly stationary signal, the autocovariance and autocorrelation will not depend on t. Therefore, the t can be dropped from the computation to give

$$\gamma(\tau) = E\big((y_t - \mu_Y)(y_{t+\tau} - \mu_Y)^T\big) \tag{5.5}$$

$$\rho(\tau) = \frac{\gamma(\tau)}{\gamma(0)} = \frac{\gamma(\tau)}{\sigma_y^2}, \tag{5.6}$$

where μ_Y is the mean value of the given time series. For a stationary signal, both the autocovariance and autocorrelation functions are even functions, that is,

$$\rho(\tau) = \rho(-\tau). \tag{5.7}$$

For two stationary time series, X and Y, the cross-covariance, $\gamma_{XY}(\tau)$, is defined as

$$\gamma_{XY}(t, \tau) = E\big((x_t - \mu_X)(y_{t+\tau} - \mu_Y)^T\big). \tag{5.8}$$

Similar to the autocovariance, this measure is not very useful as it is not bounded. Instead, the cross-correlation, $\rho_{XY}(\tau)$, which is a measure of the correlation between the two signals, is more useful. The cross-correlation is defined as

$$\rho_{XY}(t, \tau) = \frac{E\left((x_t - \mu_X)(y_{t+\tau} - \mu_Y)^T\right)}{\sqrt{\sigma_X^2 \sigma_Y^2}}. \tag{5.9}$$

Two time series are said to be **jointly stationary** if:

1.   Both series are stationary.
2.   The cross-covariance function for the two time series does not depend on t.

Joint stationarity implies that the cross-correlation function can be written as

$$\rho_{XY}(\tau) = \frac{E\left((x_t - \mu_X)(y_{t+\tau} - \mu_Y)^T\right)}{\sqrt{\sigma_X^2 \sigma_Y^2}}. \tag{5.10}$$

In such cases, the cross-correlation function is an odd function, that is,

$$\rho_{XY}(\tau) = \rho_{YX}(-\tau). \tag{5.11}$$

The final useful type of correlation is the **partial correlation**, which represents the amount of correlation between two variables after accounting for any mutual variables, that is, it is a conditional correlation defined as

$$\rho_{XY|Z} = \frac{\mathrm{cov}(x, y|z)}{\sqrt{\mathrm{cov}(x|z)\mathrm{cov}(y|z)}}, \tag{5.12}$$

where Z represents the variables on which conditioning is performed (Franke et al. 2011). In time-series analysis, the most common form of partial correlation is the **partial autocorrelation** defined as

$$\rho_{X|Z}(\tau) = \frac{\mathrm{cov}(x_t, x_{t+\tau}|x_{t+1}, \ldots, x_{t+\tau-1})}{\sqrt{\mathrm{cov}(x_t|x_{t+1}, \ldots, x_{t+\tau-1})\mathrm{cov}(x_{t+\tau}|x_{t+1}, \ldots, x_{t+\tau-1})}}. \tag{5.13}$$

In partial autocorrelation, the conditioning is performed on all values located between the two variables of interest. This implies that the effect of the intermediate variables is removed from the computed autocorrelation between the two variables. In order to compute this function, it is first necessary to develop the appropriate models for time-series analysis. Further information about computing the values are provided in Sect. 5.4.1.2.

5.1.1 *Estimating the Auto- and Cross-Covariance and Correlation Functions*

The autocovariance and cross-covariance can be estimated using the following formulae:

$$\hat{\gamma}(\tau) = \frac{1}{m} \sum_{t=1}^{m-\tau} y_t \, y_{t+\tau}^T$$

$$\hat{\gamma}_{XY}(\tau) = \frac{1}{m} \sum_{t=1}^{m-\tau} x_t \, y_{t+\tau}^T. \tag{5.14}$$

The parameter estimates given above are biased but more useful. On the other hand, to obtain unbiased estimates, m can be replaced by $m - \tau$.

The estimated autocorrelation function is defined as

$$\hat{\rho}(\tau) = \frac{\hat{\gamma}(\tau)}{\hat{\gamma}(0)} \tag{5.15}$$

while the estimated cross-correlation function is defined as

$$\hat{\rho}_{XY}(\tau) = \frac{\hat{\gamma}_{XY}(\tau)}{\hat{\gamma}_{XY}(0)}. \tag{5.16}$$

The large-sample properties for both the estimated autocorrelation and the estimated cross-correlation functions are similar. The large-sample distribution of $\hat{\rho}$ (h) is normal with zero mean and a standard deviation given as

$$\sigma_{\hat{\rho}} = \frac{1}{\sqrt{m}}, \tag{5.17}$$

where m is the number of data points used to estimate the correlation function. This relationship applies both for the auto- and cross-correlation functions. The confidence interval for the estimated correlation values can then be written as

$$\mathrm{CI}_{\hat{\rho}} = \frac{t_{1-0.5\alpha,\, m}}{\sqrt{m}}. \tag{5.18}$$

In practice, since we have assumed that m is large, then $t_{1-0.5\alpha,\, m} \approx Z_{1-0.5\alpha} = 1.96 \approx 2$. This confidence interval is useful in determining which correlation values are significant and which ones could be equal to zero.

5.1.2 Obtaining a Stationary Time Series

Since the results presented in this chapter require that a series be stationary, it is necessary to consider the procedure for obtaining a stationary series. A nonstationary signal can be made stationary by taking the difference between two adjacent values. This procedure is called **differencing**. If the differences themselves are not stationary, then they can be differenced until a stationary differenced signal is obtained. However, it should be noted that differencing will lead to a loss of information in the signal and can introduce correlations where there are none.

There are two types of differencing: **true** and **periodic**. In **true differencing**, the difference between adjacent values is taken, that is,

$$\Delta y_t = y_t - y_{t-1}, \tag{5.19}$$

where Δ represents the difference operator. The value is an approximation to the derivative at this point. Therefore, to obtain the final model, it will be necessary to integrate the differenced model.

On the other hand, in **periodic differencing**, the difference is computed between values that are some constant distance apart, that is,

$$\Delta_p y_t = y_t - y_{t-p}, \tag{5.20}$$

where p is the period for differencing and Δ_p is the periodic difference operator. Such differencing can be useful if the data set shows cyclic or periodic behaviour. This is especially common in econometric or meteorological data series, where a period of one year or twelve months (depending on the sampling) is commonly observed.

5.1.3 Edmonton Weather Data Series Example

Consider the case of a chemical engineer who is involved in the optimisation of a plastics plant located in Refinery Row in Edmonton, Alberta, Canada. After analysing the available data, it was determined that the summer temperature had an overall effect on profitability of the plant. For this reason, it was desired to model the mean summer temperature so as to be able to predict the temperature in the future.

After examining all the available weather data, the data set presented in Table 5.3 for Edmonton was compiled using original data from Environment Canada. Further information regarding this data set, including all the original data points, can be found in Appendix A5. The challenge is to use this data set to develop a model of the mean summer temperature.

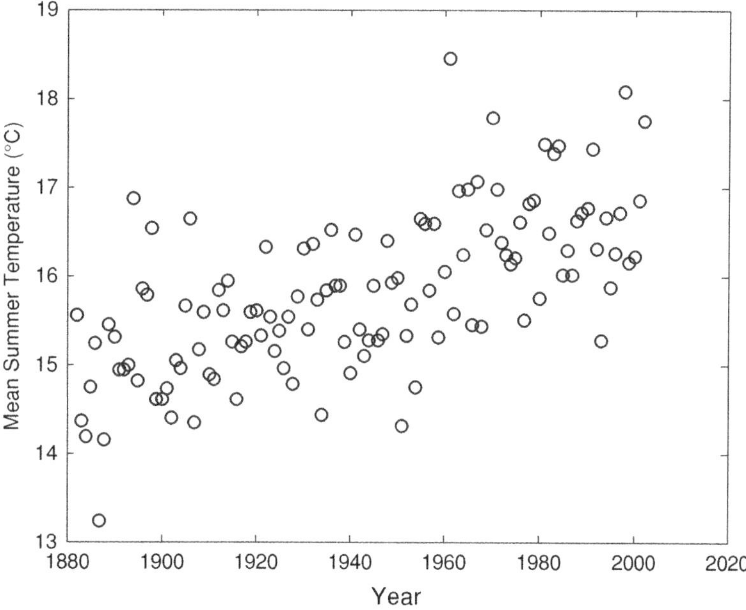

Fig. 5.1 Time-series plot of the mean summer temperature in Edmonton

A time-series plot of the mean summer temperature is shown in Fig. 5.1. In this example, the autocorrelation of the mean summer temperature will be determined. As well, the cross-correlation between the mean summer temperature and the mean spring temperature will be examined. The methods used to compute these plots can be found in either Chap. 7 for MATLAB® or Chap. 8 for Excel®.

The autocorrelation plot is shown in Fig. 5.2, the partial autocorrelation plot in Fig. 5.3, and the cross-correlation between the mean summer and spring temperatures in Edmonton in Fig. 5.4. For the autocorrelation plot shown in Fig. 5.2, there are some salient features that need to be considered. First, it can be noted that at a lag of zero, the autocorrelation is, as expected, 1. Second, it can be seen that all of the autocorrelations are located above the 95% confidence interval for significance. Note that the confidence intervals are equal to $2 / \sqrt{121} = 0.\overline{18}$. This suggests that all of the observed correlations are significant. Finally, there seems to be a weak, but noticeable, eight-lag oscillation.

The partial autocorrelation plot for the mean summer temperature in Edmonton, shown in Fig. 5.3, has the same format as the autocorrelation plot. Unlike in the autocorrelation plot, here, there are values located both inside and outside of the confidence region. A similar pattern to that previously observed can be seen here, that is, the values are significant at multiples of some constant. In this case, the significant partial autocorrelation values are located at lags of 1, 2, 3, and 8. This suggests a potential 2-year seasonal component (with values at 2, 4, 6, and 8).

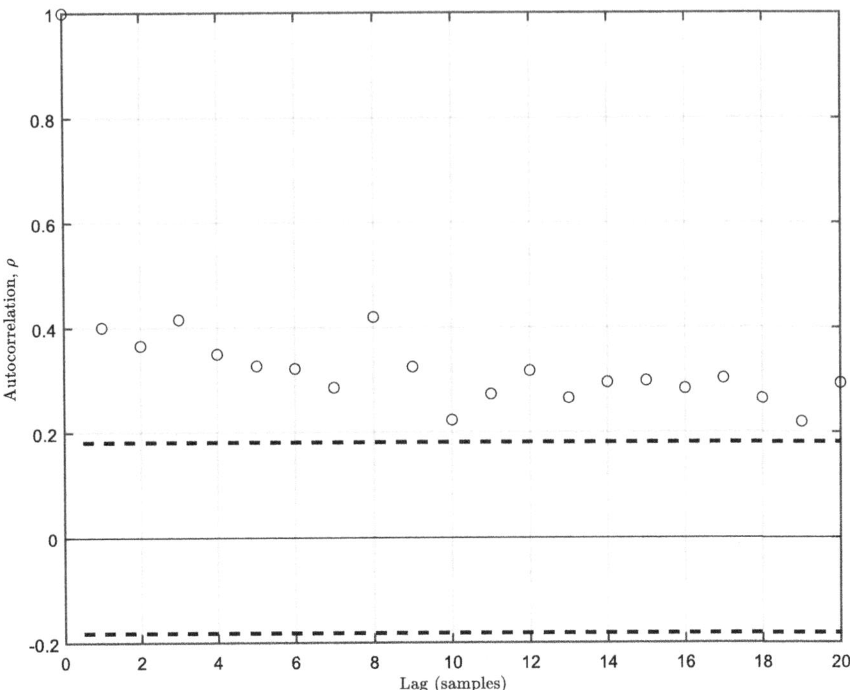

Fig. 5.2 Autocorrelation plot for the mean summer temperature in Edmonton. The thick dashed lines show the 95% confidence intervals for the given data set

The cross-correlation plot shown in Fig. 5.4 between the mean summer and spring temperatures has a similar format to the previously considered autocorrelation plot. The confidence interval is, as was previously noted, the same as for the autocorrelation plot. The salient feature is the four largest lags at -20, -16, 3, and 4. At this point, it would be useful to comment briefly about the meaning of these values. Since the formula for computing the cross-covariance can be written as $y_{t+\tau} = x_t$ or equivalently, $y_t = x_{t-\tau}$, we can see that positive values correspond to a relationship between past values of x (or in our case, the mean spring temperature) and current values of y (the mean summer temperature). Negative lags correspond to future values of x impacting a current value of y. Since we are interested in developing a model for y, a future lag relationship is not too useful, as it would imply that we will need to know values about the mean spring temperature in years that have not yet occurred. Since this requires either a good crystal ball or another model with its own inherent imperfections, the negative lag correlations are not terribly useful for the stated purpose. Therefore, the only two lags of interest are the positive ones at three and four years.

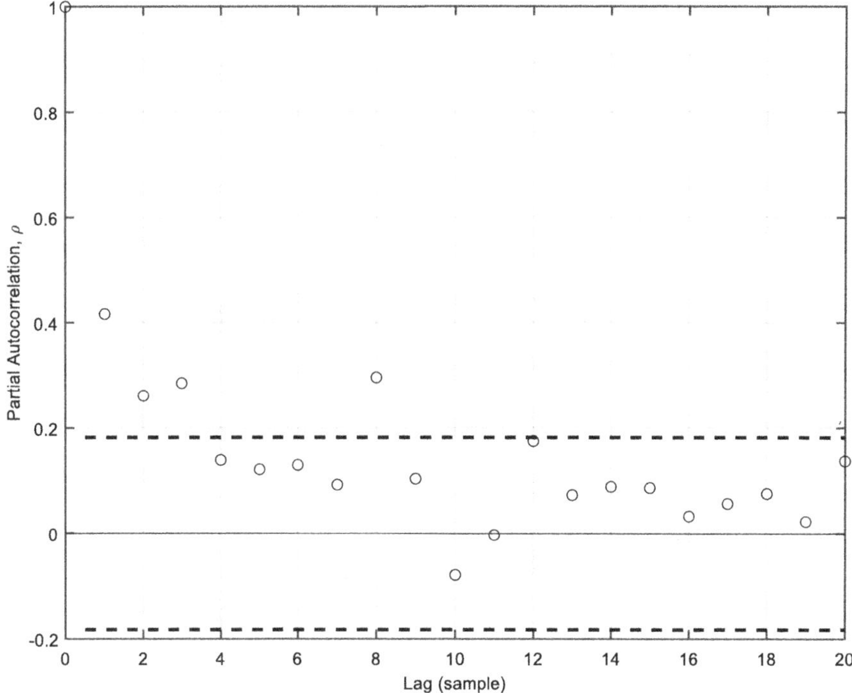

Fig. 5.3 Partial autocorrelation plot for the mean summer temperature in Edmonton. The thick dashed lines show the 95% confidence intervals for the given data set

5.2 Common Time-Series Models

In order to describe different time-series models compactly, it is necessary to introduce the *z*- or **forward shift operator**.[1] It is defined as

$$y_{t+1} = z\, y_t \tag{5.21}$$

that is, the forward shift operator shifts the time forward by one sample. Similarly, z^{-1} or the **backshift operator** performs the opposite job of the forward shift operator, that is,

$$y_{t-1} = z^{-1} y_t \tag{5.22}$$

or a backward shift of one sample.

---

[1] In the literature, different definitions can be found leading to slightly different overall forms, especially when it comes to the analysis of the model properties. The approach taken here is the most common, especially in the process control field.

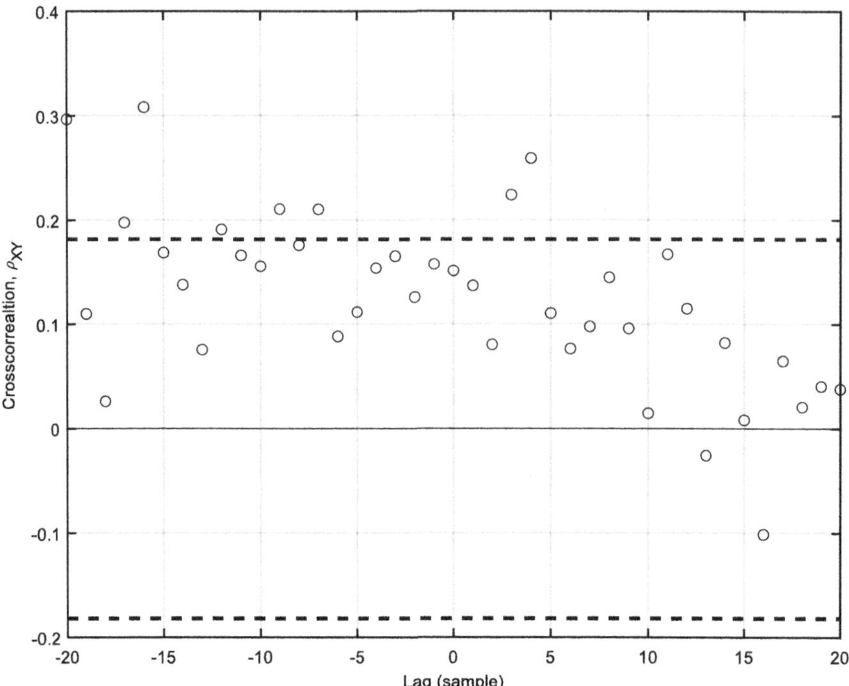

Fig. 5.4 Cross-correlation between the mean summer temperature (y) and the mean spring temperature (x) in Edmonton. The thick dashed lines show the 95% confidence intervals for the given data set

This implies that the differencing formulae can be rewritten as

$$\Delta y_t = y_t - z^{-1} y_t = \left(1 - z^{-1}\right) y_t \tag{5.23}$$

and

$$\Delta_{py_t} = y_t - z^{-P} y_t = \left(1 - z^{-P}\right) y_t. \tag{5.24}$$

The most general, time-series model called a **seasonal, autoregressive, integrated, moving-average (SARIMA) model** of order $(p, d, q) \times (P, D, Q)_s$ has the form

$$A_P\left(z^{-s}\right) A_p(z^{-1})\left(1 - z^{-s}\right)^D \left(1 - z^{-1}\right)^d y_t = B_Q\left(z^{-s}\right) B_q\left(z^{-1}\right) e_t, \tag{5.25}$$

where $A_p(z^{-1})$ and $B_q(z^{-1})$ are polynomials in z^{-1} of the form

$$A_p\left(z^{-1}\right) = 1 + \sum_{i=1}^{p} \alpha_i z^{-i}$$

$$B_q\left(z^{-1}\right) = 1 + \sum_{i=1}^{q} \beta_i z^{-i} \tag{5.26}$$

α_i and β_i are the parameters of the corresponding polynomial, and $A_P(z^{-s})$ and $B_Q(z^{-s})$ are defined as

$$A_P\left(z^{-s}\right) = 1 + \sum_{i=1}^{P} \alpha_{si} z^{-si}$$

$$B_Q\left(z^{-s}\right) = 1 + \sum_{i=1}^{Q} \beta_{si} z^{-si} \tag{5.27}$$

The process to be modelled is denoted by y, and e represents an independent, random variable drawn from a Gaussian distribution at each time instance, t. The model can be split into two components: the **seasonal component** and the **conventional component**. The **seasonal component** concerns those polynomials that contain an s in their power representation and have orders given by capital letters, that is, polynomials A_P, B_Q, and $(1 - z^{-s})^D$. The remaining terms (A_p, B_q, and $(1 - z^{-1})^d$) are called the **conventional component**. In practice, various simplifications are made to this model. The following are the most common simplified models:

1. **Autoregressive Model**: In this model, all polynomials except the $A_p(z^{-1})$-polynomial are assumed to have zero order. This gives a model of the form

$$y_t = \frac{1}{A_p\left(z^{-1}\right)} e_t = \frac{1}{A\left(z^{-1}\right)} e_t. \tag{5.28}$$

It is called an autoregressive model because the model solely depends on the past values of the process itself, that is, Eq. (5.28) can be written as

$$y_t = \alpha_1 y_{t-1} + \alpha_2 y_{t-2} + \ldots + \alpha_p y_{t-p} + e_t. \tag{5.29}$$

The pure seasonal autoregressive model is similarly defined but solely considers the seasonal component given by $A_P(z^{-s})$.

2. **Moving-Average Model**: In this model, all polynomials except the $B_q(z^{-1})$-polynomial are assumed to have zero order. This gives a model of the form

$$y_t = B_q\left(z^{-1}\right)e_t = B\left(z^{-1}\right)e_t. \tag{5.30}$$

It is called a moving-average model, since it computes a (weighted) average of the past random values, Eq. (5.30) can be written as

$$y_t = e_t + \beta_1 e_{t-1} + \beta_2 e_{t-2} + \ldots + \beta_q e_{t-q}. \tag{5.31}$$

The pure seasonal autoregressive model is similarly defined but solely considers the seasonal component given by $B_Q(z^{-s})$, that is,

$$y_t = B_Q(z^{-s})e_t. \tag{5.32}$$

3. **Integrating Model**: In this model, all polynomials except $(1 - z^{-1})^d$ are assumed to have zero order. This gives a model of the form

$$y_t = \frac{1}{\left(1 - z^{-1}\right)^d}e_t. \tag{5.33}$$

It is called an integrating model because the error is continually added to the previous values, that is, for the case, $d = 1$, Eq. (5.33) can be written as

$$y_t = y_{t-1} + e_t. \tag{5.34}$$

The integrating model is often also called the random walk model, since the time-series plot of the data can resemble a person walking randomly along the xy-plane. The pure seasonal integrating model is similarly defined but solely considers the seasonal component given by $(1 - z^{-s})^D$.

4. **Autoregressive, Moving-Average Model (ARMA)**: This model combines the autoregressive and moving average models but ignores the seasonal and integrating components to give a model of the form

$$y_t = \frac{B_q\left(z^{-1}\right)}{A_p\left(z^{-1}\right)}e_t = \frac{B\left(z^1\right)}{A\left(z^{-1}\right)}e_t. \tag{5.35}$$

Another type of model that can be used to describe a time series is the **infinite-order moving average**, also known as the **causal form of the model**, which is defined as

$$y_t = \sum_{j=0}^{\infty} h_j e_{t-j}, \tag{5.36}$$

where h is the impulse response coefficient that can be obtained by performing long division with the polynomials in the general model. The first term of the error impulse response model, h_0, is traditionally equal to 1. This model is commonly encountered in many theoretical applications. The corresponding form in terms of the noise term is called the infinite-order autoregressive model or the **invertible form of the model** and is defined as

$$e_t = \sum_{j=0}^{\infty} k_j y_{t-j}, \tag{5.37}$$

where k is defined analogously to h in the causal form.

In order to understand the different properties of these models, the next section will examine the theoretical behaviour of the different models. This will provide insight into methods that can be used to estimate some of the required parameters.

5.3 Theoretical Examination of Time-Series Models

The theoretical examination of time-series models considers the computation of theoretical properties and their interpretation. These results will allow us to develop different methods for modelling and understanding time series.

For the analysis of time-series models, two concepts need to be introduced: **causality** and **invertibility**. A process is said to be causal, if and only if, the current value of the process can be determined solely using past or current values of the process. This means that no unavailable, future values of the process are required. A process is said to be causal if and only if all roots of the denominator (that is, the A-polynomials) lie inside the unit circle in the complex domain, that is, $|z| < 1$. Under such circumstances, a causal process is also **stationary**. Furthermore, for a causal process, the infinite-order moving-average model will converge to a finite value.

A time-series process is said to be **invertible** if and only if the inverse process is also causal, that is, if $y_t = L e_t$, then if $e_t = L^{-1} y_t$ is causal, it is said that the original process is invertible. Furthermore, this implies that an invertible form of the model converges to a defined value. Clearly, if the denominator of the model determines causality, then invertibility will be determined by roots of the numerator of the original process (that is, the B-polynomials) lying inside the complex unit circle, that is, $|z| < 1$. Invertibility arises in the analysis of moving-average processes, since it can be impossible to distinguish between two different moving-average processes by solely examining their output, that is, y_t.

5.3.1 Properties of a White-Noise Process

Before considering the different types of models mentioned above, it is instructive to first examine a white-noise signal. A white noise signal is defined as

$$y_t = e_t, \tag{5.38}$$

where $e_t \sim \mathfrak{N}(0, \sigma^2)$. Furthermore, it is assumed that the individual values are identical (same distribution) and independent. The mean of this signal is

$$\mu_y = E(y_t) = E(e_t) = 0. \tag{5.39}$$

The autocovariance infinite-order moving is

$$\gamma_y(\tau) = E(y_t y_{t-\tau}) = E(e_t e_{t-\tau}) = \begin{cases} \sigma^2 & \tau = 0 \\ 0 & \text{otherwise} \end{cases}. \tag{5.40}$$

The result follows from the fact that the signal values are independent of each other. This implies that for two values e_t and $e_{t-\tau}$, $\tau \neq 0$, the expected value will be zero. This is a very useful property of a white-noise signal. The autocorrelation will then be zero for all $|\tau| > 0$ and one for $\tau = 0$. This implies that a white-noise signal will have a single peak on a autocorrelation plot at $\tau = 0$. All other values will be zero. A pure, white-noise signal is always invertible and causal.

5.3.2 Properties of a Moving-Average Process

This section will examine the properties of a moving-average process, MA(q), defined as

$$y_t = e_t + \beta_1 e_{t-1} + \beta_2 e_{t-2} + \ldots + \beta_q e_{t-q} = \sum_{i=0}^{q} \beta_i e_{t-i}, \tag{5.41}$$

where, as before, $e_t \sim \mathfrak{N}(0, \sigma^2)$ and the individual values are identical (same distribution) and independent. It should be noted that, by convention, $\beta_0 = 1$. The mean of the process is easy to compute as being equal to 0.

Theorem 5.1 *Autocovariance of a Moving-Average Process.* The autocovariance of a moving-average process can be written as

$$\gamma(\tau) = \begin{cases} \sum_{i=0}^{q-\tau} \beta_i \beta_{i+\tau} \sigma^2 & \tau \leq q \\ 0 & \text{otherwise} \end{cases}.$$

(5.42)

Proof Substituting the definition of a moving-average series (Eq. 5.41) into the definition of the autocovariance (Eq. 5.3) gives

$$\gamma(\tau) = E\left(\sum_{i=0}^{q} \beta_i e_{t-i} \sum_{j=0}^{q} \beta_j e_{t-j+\tau} \right).$$

(5.43)

Multiplying through Eq. (5.43) gives

$$\gamma(\tau) = \left(\sum_{i=0}^{q} \beta_i E\left(e_{t-i} \sum_{j=0}^{q} \beta_j e_{t-j+\tau} \right) \right)$$

$$= \left(\sum_{i=0}^{q} \beta_i E\left(\beta_0 e_{t-i} e_{t+\tau} + \beta_1 e_{t-i} e_{t+\tau-1} + \cdots + \beta_q e_{t-i} e_{t-q+\tau} \right) \right).$$

(5.44)

From the definition of a white-noise signal, the autocovariance will be nonzero if and only if the two white-noise realisations have the same subscripts. Therefore, solving for the subscripts gives

$$t - i = t - j + \tau$$
$$j = i + \tau.$$

(5.45)

Thus, the only nonzero terms inside the expectation sign will be the terms such that $j = i + \tau$. This implies that Eq. (5.44) can be rewritten as

$$\gamma(\tau) = \left(\sum_{i=0}^{q} \beta_i \beta_{i+\tau} \sigma^2 \right).$$

(5.46)

Now, it should be noted that the values of β_j for $j > q$ will be equal to zero. Therefore, setting $j = q$ in the subscript equation gives that $i \leq q - \tau$. Furthermore, for lags $\tau > q$, there will not be any nonzero parameters (at least one of the two parameters will be zero), and so the sum will be zero. Therefore, Eq. (5.42) is obtained.

$$Q.E.D.$$

Corollary 5.1 *Variance of a Moving-Average Process. The variance of a moving-average process can be written as*

$$\gamma(0) = \sum_{i=0}^{q} \beta_i^2 \sigma^2. \tag{5.47}$$

Proof Setting $\tau = 0$ into Eq. (5.42) will quickly give the above solution.

Q.E.D.

Corollary 5.2 *Autocorrelation of a Moving Average Process. The autocorrelation of a moving average process can be written as*

$$\rho(\tau) = \begin{cases} \dfrac{\sum_{i=0}^{q-\tau} \beta_i \beta_{i+\tau}}{\sum_{i=0}^{q} \beta_i^2} & \tau \leq q \\ 0 & \text{otherwise} \end{cases}. \tag{5.48}$$

Proof From Eq. (5.42) and the definition of autocorrelation.

Q.E.D.

For a moving-average process, it can be seen that the autocorrelation will decrease from 1 at a lag of zero to a value of zero for all lags greater than q. This suggests that plotting the (estimated) autocorrelation function of an unknown moving-average process can reveal the underlying order q.

Example 5.1 *Example of a Moving-Average Process*

Consider the following moving-average process

$$y_t = e_t + 0.5e_{t-1} + 0.1e_{t-2},$$

where $e_t \sim \mathfrak{N}(0, 4)$ is a Gaussian, white-noise process. Compute the mean, variance, and general autocorrelation function for the process. Compare with the results from Corollary 5.2.

Solution

The mean value can be computed as follows:

$$\begin{aligned} \mu_y = E(y_t) &= E(e_t + 0.5e_{t-1} + 0.1e_{t-2}) \\ &= E(e_t) + 0.5E(e_{t-1}) + 0.1E(e_{t-2}) \\ &= 0, \end{aligned}$$

since $E(e_t) = E(e_{t-1}) = E(e_{t-2}) = 0$. The variance can be computed as follows

$$\begin{aligned}
\sigma_y^2 = E(y_t y_t) &= E((e_t + 0.5e_{t-1} + 0.1e_{t-2})(e_t + 0.5e_{t-1} + 0.1e_{t-2})) \\
&= E(e_t e_t) + 0.5E(e_{t-1}e_t) + 0.1E(e_{t-2}e_t) + \\
&\quad 0.5E(e_{t-1}e_t) + 0.5^2 E(e_{t-1}e_{t-1}) + 0.1(0.5)E(e_{t-2}e_{t-1}) + \\
&\quad 0.1E(e_{t-2}e_t) + 0.5(0.1)E(e_{t-1}e_{t-2}) + 0.1^2 E(e_{t-2}e_{t-2}) \\
&= \sigma^2 + 0.5^2\sigma^2 + 0.1^2\sigma^2 \\
&= 1.26(4) \\
&= 5.04.
\end{aligned}$$

Note that only the terms where the subscripts are equal have a nonzero value. All other values are zero. The autocovariance function for $\tau > 0$ can be computed as follows

$$\begin{aligned}
\gamma(\tau) = E(y_t y_{t+\tau}) &= E((e_t + 0.5e_{t-1} + 0.1e_{t-2})(e_{t+\tau} + 0.5e_{t-1+\tau} + 0.1e_{t-2+\tau})) \\
&= E(e_t e_{t+\tau}) + 0.5E(e_{t-1}e_{t+\tau}) + 0.1E(e_{t-2}e_{t+\tau}) + \\
&\quad 0.5E(e_{t-1+\tau}e_t) + 0.5^2 E(e_{t-1+\tau}e_{t-1}) + 0.1(0.5)E(e_{t-2}e_{t-1+\tau}) + \\
&\quad 0.1E(e_{t-2+\tau}e_t) + 0.5(0.1)E(e_{t-1}e_{t-2+\tau}) + 0.1^2 E(e_{t-2+\tau}e_{t-2}).
\end{aligned}$$

Again, the only significant terms will be for those whose subscripts are equal at the given lag. For $\tau = 1$, we get,

$$\begin{aligned}
\gamma(1) &= 0.5E(e_t e_t) + 0.5(0.1)E(e_{t-1}e_{t-1}) = 0.55\sigma^2 = 0.55(4) \\
&= 2.2.
\end{aligned}$$

Similarly, for $\tau = 2$, we get

$$\begin{aligned}
\gamma(2) &= 0.1E(e_t e_t) = 0.1\sigma^2 = 0.1(4) \\
&= 0.4.
\end{aligned}$$

For $\tau \geq 3$, the autocovariance will be zero. Since $\rho(\tau) = \gamma(\tau) \,/\, \gamma(0)$, the autocorrelation function becomes

$$\rho(\tau) = \begin{cases} 1 & \tau = 0 \\ 55/126 & \tau = 1 \\ 5/63 & \tau = 2 \\ 0 & \text{otherwise} \end{cases}.$$

This is identical to the results obtained using the formula in Corollary 5.2.

Example 5.2 *Simulation of a Moving-Average Process*

Consider the same moving-average process as in Example 5.1, which has been simulated for 2,000 samples. Examine the provided autocorrelation plot and compare it with the values obtained previously. The results are shown in Fig. 5.5.

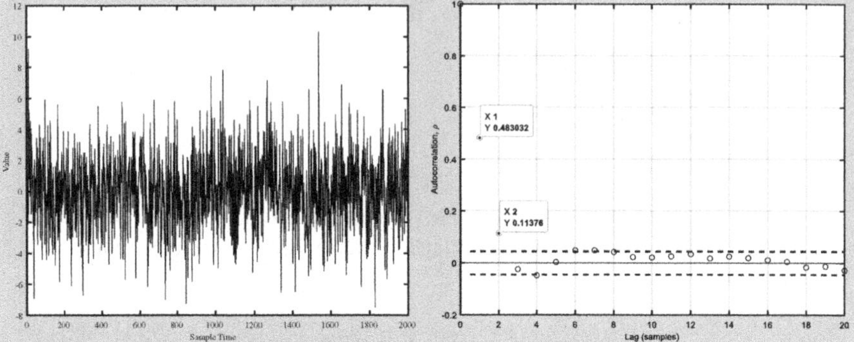

Fig. 5.5 (left) Time-series plot of the given moving-average process and (right) Autocorrelation plot for the same process

Solution

As expected, the autocorrelation plot has only three significant peaks (at $\tau = 0$, 1, and 2). The estimated values are close to the theoretical values. Notice that even though the values for $\tau \geq 3$ should be zero, we see some of them being on the boundary or slightly over. This will always be the case with estimated values, which makes the exact determination of the order slightly more complicated. Nevertheless, for a moving-average process, the autocorrelation plot does allow for the process order to be estimated.

Since a moving-average process does not contain a denominator, it is always causal. However, it may not always be invertible. Consider a simple MA(1) process where $\beta_1' = 2$ and $\beta_1'' = 2^{-1} = 0.5$. It is easy to see that both will have the same autocorrelation function. In fact, this is true for any pair of values, β_1 and β_1^{-1}. Furthermore, the autocovariance function for both $\beta_1' = 2$ with $\sigma^2 = 1$ and $\beta''_1 = 0.5$ with $\sigma^2 = 4$ will be the same. This implies that it will be impossible to distinguish these two pairs solely by examining y_t. Unfortunately, since y_t is the only information about the process that we have, it is necessary to break the tie somehow. In this case, the best choice is to take the one that is invertible, that is, its inverse process is causal. Invertibility can be useful when dealing with process analysis.

5.3.3   *Properties of an Autoregressive Process*

This section will examine the properties of an autoregressive process, AR(p), defined as

$$y_t = e_t - \alpha_1 y_{t-1} - \alpha_2 y_{t-2} - \ldots - \alpha_p y_{t-p} = e_t - \sum_{i=0}^{p} \alpha_i y_{t-i}$$

$$= \frac{1}{1 + \alpha_1 z^{-1} + \alpha_2 z^{-2} + \cdots + \alpha_p z^{-p}} e_t, \tag{5.49}$$

where, as before, $e_t \sim \mathfrak{N}(0, \sigma^2)$ and the individual values are identical (same distribution) and independent. In order to simplify the computation of the various properties of an autoregressive process, it is useful to recast the process into its equivalent infinite-order moving-average form, that is,

$$y_t = e_t - \sum_{i=0}^{p} \alpha_i y_{t-i} = \sum_{i=0}^{\infty} h_i e_{t-i}, \tag{5.50}$$

The coefficients h_i can be obtained by performing long division on the original polynomial expression given in Eq. (5.49).

The mean value of the autoregressive process can be computed using Eq. (5.50), that is,

$$\mu_y = E(y_t) = \sum_{i=0}^{\infty} h_i E(e_{t-i}) = 0, \ \text{if the series converges.} \tag{5.51}$$

Clearly, if the series does not converge, then the mean value cannot exist.

Theorem 5.2 *Autocovariance of an Autoregressive Process. The autocovariance of an autoregressive process can be written as*[2]

$$\gamma(\tau) = \sum_{i=1}^{p} \sum_{j=1}^{p} \frac{(-\theta_j)^\tau \phi_i \phi_j}{1 - \theta_i \theta_j} \sigma^2, \tag{5.52}$$

---

[2] The presented formula assumes that there are no repeated roots in the decomposition of the function. If there are repeated roots, then the value can be obtained by either taking the limit of the above equation as two of the roots approach each other or looking at Appendix 2 of (Shardt 2012), which presents a detailed method for the symbolic computation of the cross-covariance for two arbitrary time series.

where θ are the distinct, possibly complex, roots of the A-polynomial and ϕ are the partial fraction coefficients, that is,

$$y_t = \frac{1}{1 + \alpha_1 z^{-1} + \alpha_2 z^{-2} + \cdots + \alpha_p z^{-p}} e_t = \sum_{i=1}^{p} \frac{\phi_i}{1 + \theta_i} e_t. \tag{5.53}$$

Proof Consider a causal, autoregressive process that can be written as follows

$$y_t = \frac{1}{\prod_{i=1}^{p} \left(1 + \theta_i z^{-1}\right)} e_t, \tag{5.54}$$

where θ_i are the distinct, possibly complex, roots of the A-polynomial. By partial fractioning, Eq. (5.54) can be rewritten as

$$y_t = \sum_{i=1}^{p} \frac{\phi_i}{1 + \theta_i z^{-1}} e_t. \tag{5.55}$$

Noting that, for any causal autoregressive process where $|\theta| < 1$,

$$\frac{1}{1 + \theta_i z^{-1}} = \sum_{i=1}^{\infty} (-\theta_i)^i z^{-i}. \tag{5.56}$$

Equation (5.55) can be rewritten as

$$y_t = \sum_{i=1}^{p} \phi_i \sum_{j=1}^{\infty} (-\theta_i)^j e_{t-j}. \tag{5.57}$$

Since we are interested in the autocovariance, we can write the autocovariance as

$$\gamma(\tau) = E\left(\left(\sum_{i=1}^{p} \phi_i \sum_{j=1}^{\infty} (-\theta_i)^j e_{t-j}\right)\left(\sum_{i=1}^{p} \phi_i \sum_{k=1}^{\infty} (-\theta_i)^k e_{t-k+\tau}\right)\right). \tag{5.58}$$

Similar to the solution for the moving-average process, the only terms that are significant are those whose subscripts are the same, that is,

$$t - j = t - k + \tau$$
$$k = j + \tau. \tag{5.59}$$

Noting and keeping only those terms that are feasible as defined by the subscripts and substituting this into Eq. (5.58) gives

$$\gamma(\tau) = \sum_{l=1}^{p}\sum_{i=1}^{p}\phi_i\phi_l e_{t-j}\sum_{j=1}^{\infty}(-\theta_i)^j(-\theta_l)^{j+\tau}E\left(e_{t-j}e_{t-j}\right). \tag{5.60}$$

Using Eq. (5.56) in reverse and relabelling the indices gives

$$\gamma(\tau) = \sum_{i=1}^{p}\sum_{j=1}^{p}\frac{\left(-\theta_j\right)^{\tau}\phi_i\phi_j}{1-\theta_i\theta_j}\sigma^2. \tag{5.61}$$

Q.E.D.

Corollary 5.3 *Variance of an Autoregressive Process. The variance of an autoregressive process can be written as*

$$\sigma_y^2 = \gamma(0) = \sum_{i=1}^{p}\sum_{j=1}^{p}\frac{\phi_i\phi_j}{1-\theta_i\theta_j}\sigma^2. \tag{5.62}$$

Proof Setting $\tau = 0$ into Eq. (5.52) will quickly give the above solution.

Q.E.D.

Corollary 5.4 *Autocorrelation of an Autoregressive Process. The autocorrelation of an autoregressive process can be written as*

$$\rho(\tau) = \frac{\sum_{i=1}^{p}\sum_{j=1}^{p}\frac{\left(-\theta_j\right)^{\tau}\phi_i\phi_j}{1-\theta_i\theta_j}}{\sum_{i=1}^{p}\sum_{j=1}^{p}\frac{\phi_i\phi_j}{1-\theta_i\theta_j}}. \tag{5.63}$$

Proof From Eq. (5.52) and the definition of autocorrelation.

Q.E.D.

For an autoregressive process, it can be seen that the autocorrelation does not have an abrupt endpoint; rather, it continues to decay and slowly approach 0. This means that it is not possible to determine the order of an autoregressive process from the autocorrelation plot . Instead, we need to consider the **partial autocorrelation**

function (PACF). It can be shown (see Sect. 5.4.1.2 for the details) that the PACF of an autoregressive process stops after p lags. This makes the PACF analogous to the ACF for a moving-average process. In fact, for a moving-average process, the PACF will decay for all lags.

Example 5.3 *Example of an Autoregressive Process*
  Consider the following autoregressive process

$$y_t = 0.5 y_{t-1} + e_t,$$

where $e_t \sim \mathfrak{N}(0, 4)$ is a Gaussian, white-noise process. Compute the mean, variance, and general autocorrelation function for the process.

Solution

Before computing any of the required values, let us first rewrite this into the infinite-order moving-average form. Rewriting the process into the transfer-function form gives

$$y_t = \frac{1}{1 - 0.5z^{-1}} e_t.$$

From the derivation for the autocovariance, Eq. (5.56) gives that the infinite-order moving-average form will be

$$y_t = \sum_{i=0}^{\infty} 0.5^i e_{t-i}.$$

The mean value can be computed as follows:

$$\mu_y = E(y_t) = E\left(\sum_{i=0}^{\infty} 0.5^i e_{t-i}\right)$$

$$= \sum_{i=0}^{\infty} 0.5^i E(e_{t-i})$$

$$= 0,$$

since $E(e_{t-i}) = 0 \; \forall i$. The variance can be computed using Corollary 5.3

$$\sigma_y^2 = \sum_{i=1}^{p} \sum_{j=1}^{p} \frac{\phi_i \phi_j}{1 - \theta_i \theta_j} \sigma^2 = \sum_{i=1}^{1} \sum_{j=1}^{1} \frac{\phi_i \phi_j}{1 - \theta_i \theta_j} \sigma^2$$

$$= \frac{1}{1-(0.5)^2} \quad (4)$$

$$= 5\frac{1}{3}.$$

Note that only the terms where the subscripts are equal have a nonzero value. All other values are zero. The autocorrelation function for $\tau > 0$ can be computed using Corollary 5.4 to give

$$\rho(\tau) = \frac{\sum_{i=1}^{p}\sum_{j=1}^{p}\frac{(-\theta_j)^{\tau}\phi_i\phi_j}{1-\theta_i\theta_j}}{\sum_{i=1}^{p}\sum_{j=1}^{p}\frac{\phi_i\phi_j}{1-\theta_i\theta_j}} = \frac{\sum_{i=1}^{1}\sum_{j=1}^{1}\frac{(-\theta_j)^{\tau}\phi_i\phi_j}{1-\theta_i\theta_j}}{\sum_{i=1}^{1}\sum_{j=1}^{1}\frac{\phi_i\phi_j}{1-\theta_i\theta_j}} = \frac{\frac{0.5^{\tau}}{1-0.5^2}}{\frac{1}{1-0.5^2}} = 0.5^{\tau}.$$

This result clearly shows the behaviour of the autocorrelation function for an autoregressive process and its difference from the moving-average process.

Example 5.4 *Simulation of an Autoregressive Process*

Consider the same autoregressive process as in Example 5.3, which has been simulated for 2,000 samples. Examine the provided autocorrelation plot and compare it with the values obtained previously. The results are shown in Fig. 5.6.

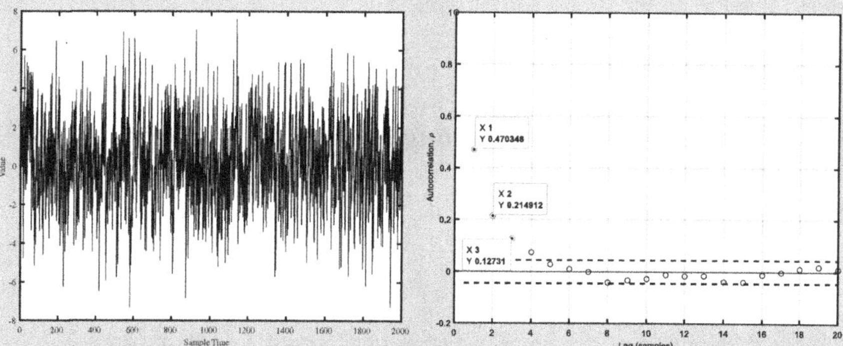

Fig. 5.6 (left) Time-series plot of the given autoregressive process and (right) Autocorrelation plot for the same process

Solution

As expected, the autocorrelation plot decays slowly to zero. The first three values are close to the theoretical values of 0.5, 0.25, and 0.125. As well, note

that the estimated autocorrelation values for large lags are relatively impre-
cise, since, in reality, the value could easily be close to zero. For comparison
purposes, the partial autocorrelation plot is shown in Fig. 5.7 for both the
AR(1) process considered in this example and the MA(1) process considered
in Example 5.1. Here, it is quite obvious that the autoregressive process has
a single spike at a lag of 1, while the moving-average process has at least
two significant points. The difference between the expected and observed
behaviours can be attributed to the fact that these are simulation examples,
for which there will be a wide range of possible outcomes.

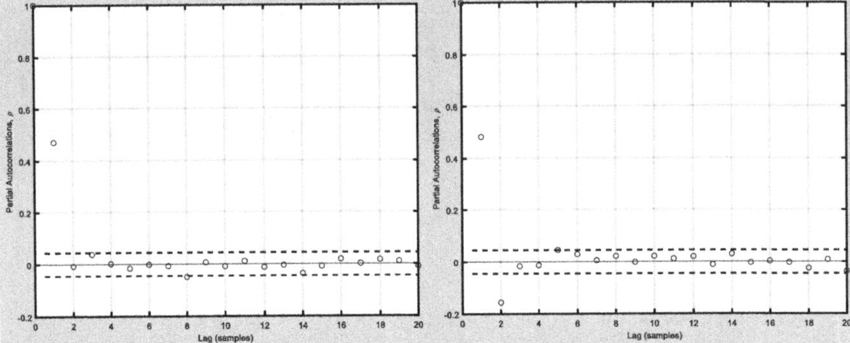

Fig. 5.7　Partial autocorrelation plot for (left) AR(1) and (right) MA(1) processes

5.3.4　Properties of an Integrating Process

The integrating process, also known as the random walk, is defined as

$$y_t = e_t + y_{t-1} = \frac{1}{1 - z^{-1}} e_t. \tag{5.64}$$

It is easy to see that an integrating process is unstable, since $z = 1$. Therefore, the
mean value for this process is undefined. The theoretical autocorrelation will be a
uniform one for all lags. Practically, when estimating the autocorrelation, it will very
slowly decrease as a function of the lags (Wichern 1973; Hassler 1994). This makes
it difficult to distinguish from a true autoregressive model with $\alpha \approx 1$. In practice,
if the data do not look stationary, then it is quite likely that the process contains an
integrator rather than a slowly varying autoregressive component.

Example 5.5 *Simulation of an Integrating Process*

Simulate an integrating process for 2,000 samples and compare it with an AR(1) process with $\alpha_1 = -0.98$. Compute the sample autocorrelation and partial autocorrelation functions. Compare and suggest ways to distinguish the two cases. The simulation results are shown in Fig. 5.8 The Gaussian noise for both processes is the same.

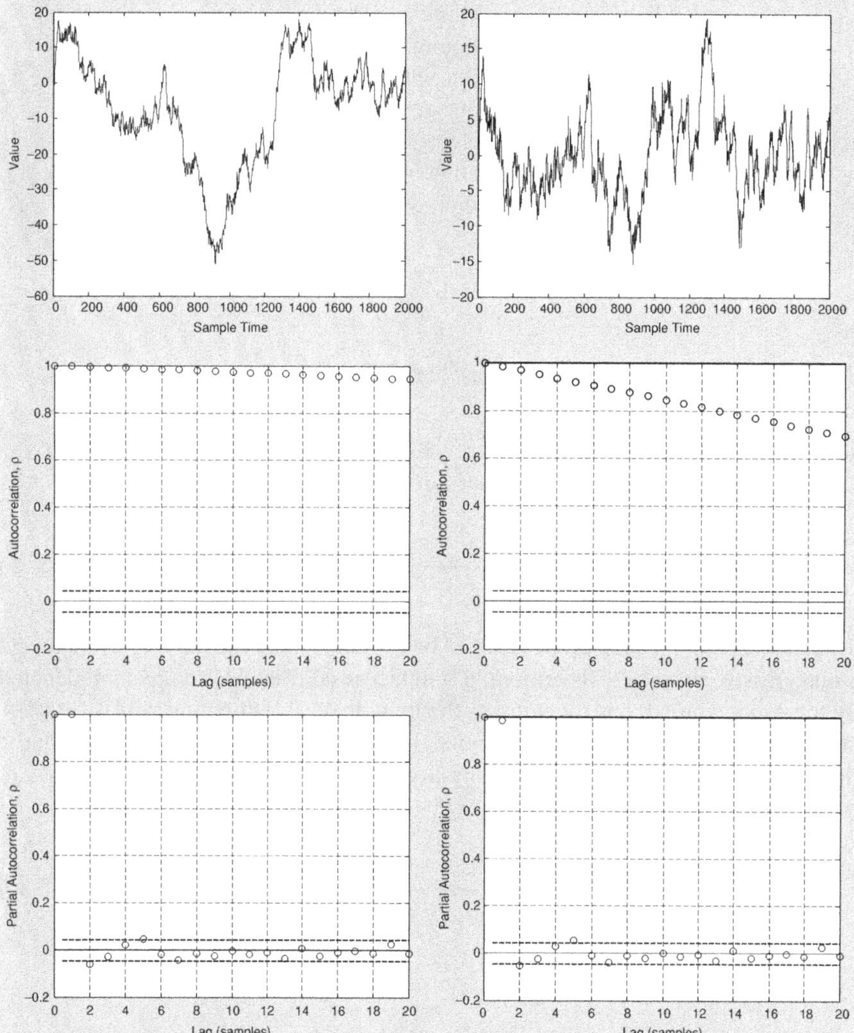

Fig. 5.8 (top) Time-series plot, (middle) Autocorrelation plot, and (bottom) Partial autocorrelation plot for (left) Integrating and (right) AR(1) with $\alpha = -0.98$ processes

Solution

First, it can be noted that the integrating process has a much larger deviation from the mean value than the causal autoregressive process with uneven distribution about the mean. The autocorrelation plot shows that the value for both decays slowly. However, for the autoregressive process, it is much faster than for the integrating process. On the other hand, for the partial autocorrelation plot, the overall behaviour is quite similar. Both have two peaks (at zero and one), but the values of the peaks are quite different. For the integrating process, the value is exactly one for both peaks, while for the autoregressive process, the value is, as expected, less than one (≈ 0.9856). This suggests that in addition to identifying the order of the autoregressive component, the partial autocorrelation plot can be useful in separating integrating processes from other types of autoregressive processes based on the value of the lag-one term.

5.3.5 *Properties of ARMA and ARIMA Processes*

The autoregressive, moving-average process denoted as ARMA(p, q) is one of the most common times series models that can be used. It has the general form given as

$$y_t = \frac{B(z^{-1})}{A(z^{-1})} e_t = \frac{1 + \beta_1 z^{-1} + \cdots + \beta_q z^{-q}}{1 + \alpha_1 z^{-1} + \cdots + \alpha_p z^{-p}} e_t. \tag{5.65}$$

This model combines the properties and behaviour of both the moving-average and autoregressive processes. Therefore, it will be causal if its AR component, denoted by the A-polynomial, has roots inside the unit circle. It will be invertible if its MA component, denoted by the B-polynomial, has roots inside the unit circle. Similarly, the autoregressive, integrating, moving-average process, denoted as ARIMA(p, d, q), has the general form given as

$$y_t = \frac{B(z^{-1})}{A(z^{-1})(1 - z^{-1})^d} e_t = \frac{1 + \beta_1 z^{-1} + \cdots + \beta_q z^{-q}}{\left(1 + \alpha_1 z^{-1} + \cdots + \alpha_p z^{-p}\right)(1 - z^{-1})^d} e_t. \tag{5.66}$$

The autocorrelation function for an ARMA process can be computed exactly (for details, see Appendix A3 of Shardt 2012). In general, the determination of the orders can be estimated by examining the autocorrelation and partial autocorrelation plots. In most cases, it is desired to obtain an approximate value for these parameters to be used as an initial estimate for the identification procedure.

Example 5.6 *Simulation of an ARMA Process*

Consider the following ARMA(2, 2) process

$$y_t = \frac{1 + 0.2z^{-1} - 0.5z^{-2}}{1 - 0.4z^{-1} + 0.5z^{-2}} e_t.$$

Determine the causality and invertibility of this process. Simulate it for 2,000 samples and obtain estimates for the autocorrelation and partial autocorrelation functions. Can the true orders be determined?

Solution

The causality of this process is determined by considering the roots of the denominator, that is, $1 - 0.4z^{-1} + 0.5z^{-2}$. Using the quadratic formula, gives two roots, $z = 0.2 \pm 0.678i$. Computing the absolute value (or modulus) of the roots gives $|z| = 0.7071 < 1$. Since this value is less than 1, it can be concluded that the process is causal. Likewise, for invertibility, considering the numerator of the process, we get that the roots are $z = 0.61414, -0.81414$. Since the absolute value of both roots is less than 1, it can be concluded that the process is invertible.

The simulation results are shown in Figs. 5.9 and 5.10. First, it can be noted that both the autocorrelation and partial autocorrelation plots do not show any clear behaviour or cut-off points. The autocorrelation plot does not decay exponentially to zero; rather around a lag of 4, there is some unexpected behaviour. Furthermore, both plots show values that alternate in sign. This behaviour is common if any of the roots are complex numbers. Since we are dealing with a real process, this observation implies that there must be at least two (or a similar *even* number of) such complex roots, that is, the order is at least two. For both plots, the largest peaks occur at a lag of 2. This reflects well on what we know about the process.

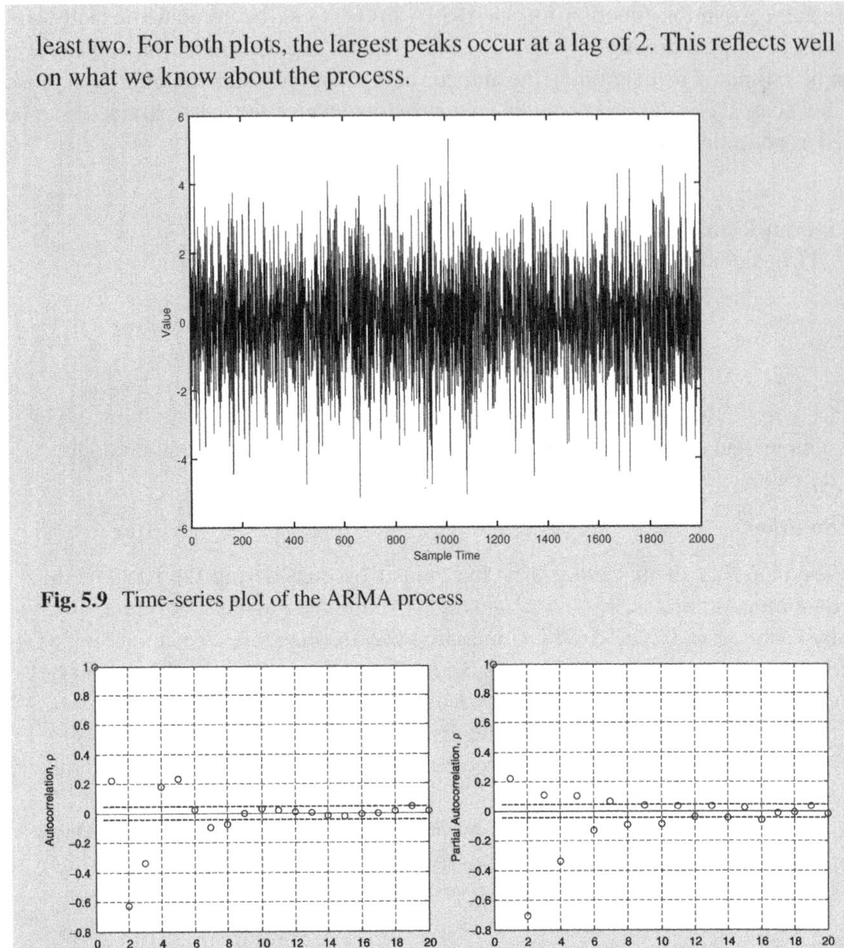

Fig. 5.9 Time-series plot of the ARMA process

Fig. 5.10 (left) Autocorrelation plot and (right) Partial autocorrelation plot for the ARMA process

5.3.6 *Properties of the Seasonal Component of a Time-Series Model*

There are three seasonal components: the seasonal autoregressive component defined as

$$y_t = \frac{1}{A_P(z^{-s})}e_t = \frac{1}{1 + \alpha_1 z^{-s} + \cdots + \alpha_P z^{-sP}}e_t, \tag{5.67}$$

the seasonal moving-average component defined as

$$y_t = B_Q(z^{-s})e_t = \left(1 + \beta_1 z^{-s} + \cdots + \beta_Q z^{-Qs}\right)e_t \tag{5.68}$$

and, finally, the seasonal integrating process defined as

$$y_t = \frac{1}{(1 - z^{-s})^D}e_t. \tag{5.69}$$

The properties of the seasonal components are similar to those of the corresponding base components after taking into consideration the seasonal component. This can be accomplished by replacing the z^{-1} in the base polynomials by z^{-s} to yield the seasonal forms. The mean and variance will stay the same. The autocovariance and autocorrelation can be computed by replacing τ in the base formulae by τs. At all other points, the values will be zero. A similar transformation applies to the partial autocorrelation. This implies that the seasonal component can be identified by noting consistent gaps in the process between significant values. For example, the autocorrelation for the seasonal autoregressive component can be computed as follows:

$$\rho(\tau) = \begin{cases} \dfrac{\sum_{i=1}^{P}\sum_{j=1}^{P}\frac{(-\theta_j)^{ks}\phi_i\phi_j}{1-\theta_i\theta_j}}{\sum_{i=1}^{P}\sum_{j=1}^{P}\frac{\phi_i\phi_j}{1-\theta_i\theta_j}} & \tau = ks \;\forall k \in \mathbb{N} \\[4mm] 0 & \text{otherwise} \end{cases}. \tag{5.70}$$

Example 5.7 *Simulation of the Seasonal Component*
  Simulate the following seasonal processes for 2,000 samples and comment on their autocorrelation and partial autocorrelation plots:

$$y_1 = \frac{1}{1 - 0.6z^{-4}}e_t, \quad y_2 = \left(1 + 0.5z^{-3} - 0.2z^{-6}\right)e_t, \quad y_3 = \frac{1}{1 - z^{-5}}e_t.$$

Solution

The simulation results are shown in Figs. 5.11, 5.12, and 5.13. In all cases, the autocorrelation and partial autocorrelation functions show only significant values at multiples of a seasonal component. In all other cases, the results are very similar to those previously considered with the autocorrelation function being useful to show the moving-average values and the partial autocorrelation function useful for the autoregressive and integrating processes.

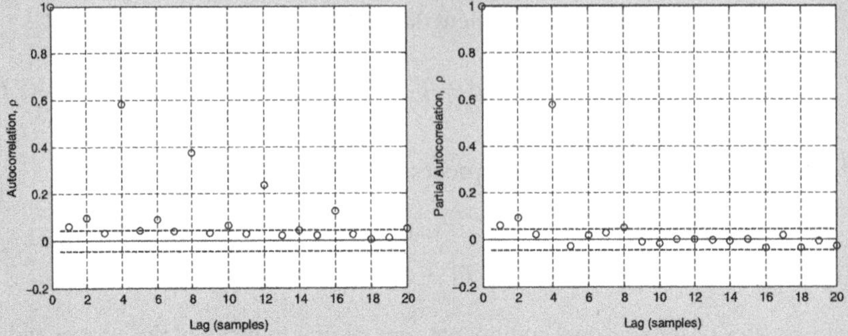

Fig. 5.11 (left) Autocorrelation plot and (right) Partial autocorrelation plot for the seasonal autoregressive process

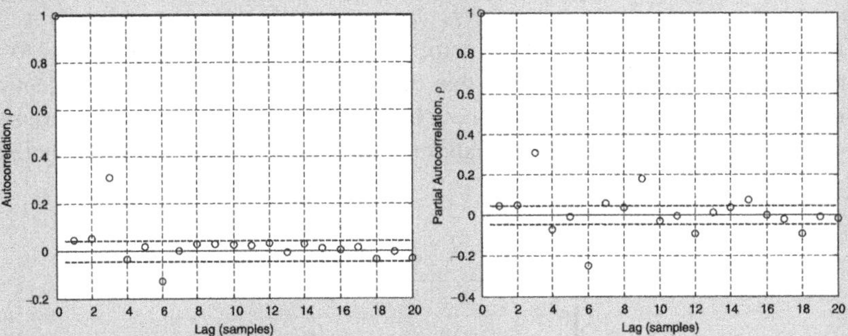

Fig. 5.12 (left) Autocorrelation plot and (right) Partial autocorrelation plot for the seasonal moving-average process

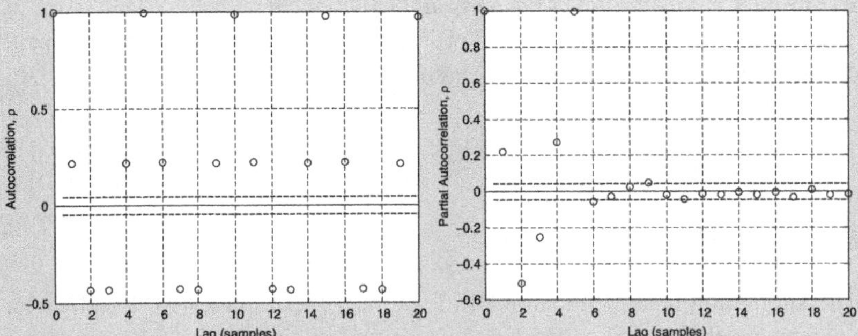

Fig. 5.13 (left) Autocorrelation plot and (right) Partial autocorrelation plot for the seasonal integrating process

5.3.7 *Summary of the Theoretical Properties for Different Time-Series Models*

A useful summary of the different properties of the common time-series models is shown in Table 5.1. This summary is very useful when trying to determine the initial orders for the data set.

5.4 Time-Series Modelling

Having examined the different theoretical properties of time-series models in the previous sections, it is now possible to consider the modelling of the time series given some data set. There are two separate steps in this procedure: determining the model orders and determining the model parameters. The model orders determine the type of model that will be used, while the model parameters provide the actual values for the model. The general, time-series modelling procedure can be summarised into the following steps:

1. **Stationarity Testing**: Determine if the data set is stationary, by examining the data set itself, its autocorrelation, and partial autocorrelation plots. If there is evidence of an integrator, difference the data, and repeat the procedure with the differenced data until the data are stationary. This will give the value of d and D. Note that it may sometimes be necessary to transform the data by applying a nonlinear transformation, for example, $\tilde{y}_t = \log(y_t)$.

Table 5.1 Summary of the theoretical properties of different time-series models

Model	Form	Autocorrelation $\rho(\tau)$	Partial Autocorrelation
Pure White Noise	$y_t = e_t$	Single peak at $\tau = 0$	Exponential decay
Moving Average, MA(q)	$y_t = B(z^{-1})$	q significant peaks	Exponential decay
Autoregressive Model, AR(p)	$y_t = \frac{1}{A(z^{-1})} e_t$	Exponential decay	p significant peaks
Integrating Process, I(d)	$y_t = \frac{1}{(1-z^{-1})^d} e_t$	Theoretically: always 1 Practically: Very slowly decaying values	Two peaks at $\tau = 0$ and 1, both with value 1
Autoregressive, Moving-Average Model, ARMA(p, q)	$y_t = \frac{B(z^{-1})}{A(z^{-1})} e_t$	A combination of the MA and AR graphs from which an estimate of the orders can be obtained	
Seasonal Component	Replace z^{-1} by z^{-s} in the above formulae	The graphs are the same as above, except that there is now a space of s between each of the peaks	

2. **Model Order Determination**: Using the (differenced) data, determine the model orders for the process. Model orders are determined by examining the autocorrelation and partial autocorrelation plots for the data set combined with the information presented in Table 5.1. This will give the value of p, P, q, and Q.

3. **Model Parameter Estimation**: Using the selected model orders and an appropriate method, estimate the model parameters.

4. **Model Validation**: Validate the model by examining the residuals. If the residuals satisfy the assumptions, then consider the model to be sufficient. Otherwise, change the model orders (including, if necessary, the value of d and D) or try a nonlinear transformation of the data. One can continue as long as one wants on this step.

5.4.1 Estimating the Time-Series Model Parameters

Estimating the model parameter values is, in general, performed using one of two methods: the method of moments leading to the Yule–Walker equations or the maximum likelihood method. Although the Yule–Walker equations are simpler, they only provide an efficient estimator for autoregressive models. As well, the Yule–Walker equations are useful for estimating the partial autocorrelation function. Least-square estimates are also possible, but they are difficult to solve analytically due to the complex nature of the models.

5.4.1.1 Yule–Walker Equations for Estimating an Autoregressive Model

Consider the standard autoregressive model given by Eq. (5.49)

$$y_t = e_t - \alpha_1 y_{t-1} - \alpha_2 y_{t-2} - \cdots - \alpha_p y_{t-p}. \tag{5.71}$$

Multiply this equation by $y_t, y_{t-1}, \ldots, y_{t-p}$ and take the expectation of the resulting $p+1$ equations to give

$$\gamma(0) = \sigma^2 - \alpha_1 \gamma(1) - \alpha_2 \gamma(2) - \cdots - \alpha_p \gamma(p)$$
$$\gamma(1) = -\alpha_1 \gamma(0) - \alpha_2 \gamma(-1) - \cdots - \alpha_p \gamma(1-p)$$
$$\vdots$$
$$\gamma(p) = -\alpha_1 \gamma(p-1) - \alpha_2 \gamma(p-2) - \cdots - \alpha_p \gamma(0). \tag{5.72}$$

Re-arranging this system of equations into matrix form gives

$$\sigma^2 = \gamma(0) + \vec{\alpha}^T \vec{\gamma}$$
$$\vec{\gamma} = -\Gamma \vec{\alpha}, \tag{5.73}$$

where $\vec{\alpha}$ is the $p \times 1$-row vector $\langle \alpha_1, \alpha_2, \ldots, \alpha_p \rangle^T$, Γ is the $p \times p$-matrix of autocovariances defined as

$$
\Gamma = \begin{bmatrix}
\gamma(0) & \gamma(1) & \cdots & \cdots & \gamma(p-1) \\
\gamma(1) & \gamma(0) & \gamma(1) & \cdots & \gamma(p-2) \\
\vdots & & \ddots & & \vdots \\
\vdots & & & \ddots & \vdots \\
\gamma(p-1) & \cdots & \cdots & \gamma(1) & \gamma(0)
\end{bmatrix}
\tag{5.74}
$$

and $\vec{\gamma}$ is the $p \times 1$-row vector $\langle \gamma(1), \gamma(2), \ldots, \gamma(p) \rangle^T$. Using the method of moments approach and using the estimated autocovariances in place of the true values, it is possible to obtain an estimate for $\vec{\alpha}$ and σ^2. The resulting equations are[3]

$$
\boxed{
\begin{aligned}
\hat{\sigma}^2 &= \hat{\gamma}(0) - \hat{\vec{\gamma}}^{\,T} \hat{\Gamma}^{-1} \hat{\vec{\gamma}} \\
\hat{\vec{\alpha}} &= -\hat{\Gamma}^{-1} \hat{\vec{\gamma}}
\end{aligned}
}
\tag{5.75}
$$

An equivalent formulation in terms of the autocorrelation function gives

$$
\hat{\vec{\rho}} = -\hat{P}\hat{\vec{\alpha}},
\tag{5.76}
$$

where $\hat{\vec{\alpha}}$ is the $p \times 1$-row vector $\langle \alpha_1, \alpha_2, \ldots, \alpha_p \rangle^T$, P is the $p \times p$-matrix of autocorrelations defined as

$$
\hat{P} = \begin{bmatrix}
1 & \hat{\rho}(1) & \cdots & \cdots & \hat{\rho}(p-1) \\
\hat{\rho}(1) & 1 & \hat{\rho}(1) & \cdots & \hat{\rho}(p-2) \\
\vdots & & \ddots & & \vdots \\
\vdots & & & \ddots & \vdots \\
\hat{\rho}(p-1) & \cdots & \cdots & \hat{\rho}(1) & 1
\end{bmatrix}
\tag{5.77}
$$

and $\hat{\vec{\rho}}$ is the $p \times 1$-row vector $\langle \hat{\rho}(1), \hat{\rho}(2), \ldots, \hat{\rho}(p) \rangle^T$. Parameter estimates obtained with either method will give the same result.

Although it is possible to solve this system of equations directly, computationally, it can be better to solve them using an iterative approach. The most common such method is the Durbin–Levinson algorithm (see Sect. 5.4.4.1 for an explanation of this method).

The Yule–Walker estimators can be shown to be asymptotically normally distributed (Shumway and Stoffer 2011), such that

---

[3] Since Γ is symmetric, $\Gamma^T = \Gamma$.

$$\hat{\vec{\alpha}} - \vec{\alpha} \sim \mathfrak{N}\left(0, \frac{\sigma^2 \Gamma^{-1}}{m}\right)$$

$$\hat{\sigma}^2 \approx \sigma^2,$$

(5.78)

where m is the number of data points used in the estimation. This implies that approximate $100(1 - \alpha)\%$ confidence intervals for the parameters $\hat{\vec{\alpha}}$ can be constructed as

$$\hat{\alpha}_i \pm t_{1-\frac{\alpha}{2}, m-p} \hat{\sigma} \sqrt{\frac{\Gamma_{ii}^{-1}}{m}}.$$

(5.79)

Example 5.8 *Fitting an AR(2) Process Using the Yule–Walker Equations*
The following AR(2) process

$$y_t = 1.4 y_{t-1} - 0.5 y_{t-2} + e_t$$

was simulated for 100 samples (provided in Sect. A5.2). Using the Yule–Walker equations, fit an AR(2) model to the data and comment on the resulting parameters. Compute the 95% confidence intervals.

Solution

Using the sample autocovariance formula, the sample autocovariances are

$$\hat{\gamma}(0) = 13.6002, \ \hat{\gamma}(1) = 12.5275, \ \hat{\gamma}(2) = 10.7976.$$

The required matrices then become

$$\hat{\vec{\gamma}} = \begin{bmatrix} 12.5275 & 10.7976 \end{bmatrix}^T$$

$$\hat{\Gamma} = \begin{bmatrix} 13.6002 & 12.5275 \\ 12.5275 & 13.6002 \end{bmatrix}.$$

Solving the Yule–Walker equations gives

$$\hat{\vec{\alpha}} = -\begin{bmatrix} 13.6002 & 12.5275 \\ 12.5275 & 13.6002 \end{bmatrix}^{-1} \begin{bmatrix} 12.5275 \\ 10.7976 \end{bmatrix}$$

$$= \frac{-1}{13.6002^2 - 12.5275^2} \begin{bmatrix} 13.6002 & -12.5275 \\ -12.5275 & 13.6002 \end{bmatrix} \begin{bmatrix} 12.5275 \\ 10.7976 \end{bmatrix}.$$

$$= \begin{bmatrix} -1.2527 \\ 0.3599 \end{bmatrix}$$

$$\hat{\sigma}^2 = 13.6002 - \begin{bmatrix} 12.5275 \\ 10.7976 \end{bmatrix}^T \begin{bmatrix} 13.6002 & 12.5275 \\ 12.5275 & 13.6002 \end{bmatrix}^{-1} \begin{bmatrix} 12.5275 \\ 10.7976 \end{bmatrix}$$

$$= 1.7937$$

The 95% confidence bound, which, in this case, will be the same for both parameters, can be computed as

$$\delta = 1.96(1.7937)^{0.5} \sqrt{\frac{0.485\,25}{100}} = 0.182\,86.$$

This implies that the confidence intervals for the two parameters are.

$$\hat{\hat{\alpha}} = \begin{bmatrix} -1.25 \pm 0.18 \\ 0.36 \pm 0.18 \end{bmatrix}.$$

We can see that the confidence interval for both parameters covers the true values.

5.4.1.2 Computing the Partial Autocorrelation Function

In order to compute the partial autocorrelation function, assume that the process of interest can be modelled as a τ-order autoregressive process. Note that it does not matter what model the true process has. Set up the τ-order Yule–Walker equation in the form given by Eq. (5.76). It can be shown that the partial autocorrelation of lag τ is equal to $-\alpha_\tau$, that is, the final parameter that is estimated (Franke et al. 2011).[4] Practically, rather than computing all the parameters, it is easier to simply compute the final desired value using Cramer's Rule, that is,

$$\rho_{X|X_{t+1},\ldots,X_{t+\tau-1}}(\tau) = \frac{\|\mathbf{P}_\tau^*\|}{\|\mathbf{P}_\tau\|}, \tag{5.80}$$

where $\|\cdot\|$ is the determinant of a matrix and $\mathbf{P}_\tau^*$ is the $\mathbf{P}_\tau$ matrix with the τth column replaced by $\vec{\rho}$. Equation (5.80) gives that

$$\rho_{X|X_{t+1},\ldots,X_{t+\tau-1}}(1) = \rho(1) \tag{5.81}$$

---

[4] The negative sign arises from the way the model has been defined.

and

$$\rho_{X|X_{t+1},\ldots,X_{t+\tau-1}}(2) = \frac{\left\| \begin{matrix} 1 & \rho(1) \\ \rho(1) & \rho(2) \end{matrix} \right\|}{\left\| \begin{matrix} 1 & \rho(1) \\ \rho(1) & 1 \end{matrix} \right\|} = \frac{\rho(2) - \rho(1)^2}{1 - \rho(1)^2}. \tag{5.82}$$

Theoretical values for higher-order partial autocorrelation values can be similarly computed. In practice, the Durbin–Levinson iterative method is used to compute the required partial autocorrelations (see Sect. 5.4.4.1 for an explanation of this method).

Example 5.9 *Partial Autocorrelation for an AR(1) Process*

Consider the standard first-order autoregressive process and compute its partial autocorrelation values.

Solution

From Corollary 5.4, the autocorrelation for a first-order process can be written as

$$\rho(\tau) = (-\alpha_1)^\tau.$$

Therefore, the partial autocorrelation function can be written as

$$\rho_{X|X_{t+1},\ldots,X_{t+\tau-1}}(1) = \rho(1) = (-\alpha)^1 = -\alpha$$

$$\rho_{X|X_{t+1},\ldots,X_{t+\tau-1}}(2) = \frac{\rho(2) - \rho(1)^2}{1 - \rho(1)^2} = \frac{(-\alpha)^2 - (-\alpha)^2}{1 - (-\alpha)^2} = 0$$

$$\rho_{X|X_{t+1},\ldots,X_{t+\tau-1}}(\tau) = 0 \; for \; \tau \geq 3.$$

This shows that, as previously mentioned, the partial autocorrelation function can be useful in identifying the order of the autoregressive function.

The partial autocorrelation function can be estimated by replacing the true autocorrelations with the estimated ones. The statistical properties of the estimated partial autocorrelation function are identical to those of the estimated autocorrelation function, that is,

$$\sigma_{\hat{\rho}_{X|X_{t+1},\ldots,t+\tau-1}} = \frac{1}{\sqrt{m}}, \tag{5.83}$$

where m is the number of data points used in estimating the partial autocorrelation function.

5.4.2 Maximum Likelihood Parameter Estimates for ARMA Models

Assume that the ARMA process of interest can be written as an infinite-order autoregressive process, that is,

$$e_t = \frac{A\left(z^{-1}\right)}{B\left(z^{-1}\right)} y_t = y_t + \sum_{i=1}^{\infty} w_i y_{t-i}, \tag{5.84}$$

where w is the coefficient obtained upon performing long division on the term A/B, which can be rewritten as

$$y_t = \sum_{i=1}^{\infty} w_i y_{t-i} + e_t. \tag{5.85}$$

Assuming that the residuals (or e_t) are normally distributed, then $\vec{y} = \langle y_1, y_2, \ldots, y_m \rangle^T$ will have a multivariate normal distribution with a probability density

$$p\left(\vec{y}|\vec{\theta}\right) = (2\pi)^{-m/2} \|\Gamma\|^{-0.5} \exp\left(-\frac{1}{2\sigma^2} \vec{y}^T \Gamma^{-1} \vec{y}\right), \tag{5.86}$$

where Γ is the matrix of autocovariances defined by Eq. (5.74) and $\vec{\theta}$ is the model parameter vector defined as

$$\vec{\theta} = \left\langle \alpha_1, \alpha_2, \ldots, \alpha_p, \beta_1, \beta_2, \ldots \beta_q, \sigma_\varepsilon^2 \right\rangle^T \tag{5.87}$$

and σ_ε^2 is the variance of the white-noise process. Let the likelihood function for this problem be defined as

$$L\left(\vec{\theta}|\vec{y}\right) = p\left(\vec{y}|\vec{\theta}\right) = \left(2\pi\sigma_\varepsilon^2\right)^{-m/2} \|\Gamma\|^{-0.5} \exp\left(-\frac{1}{2\sigma_\varepsilon^2} \vec{y}^T \Gamma^{-1} \vec{y}\right). \tag{5.88}$$

The log-likelihood function, $\ell\left(\vec{\theta}|\vec{y}\right)$, is given as

$$\ell\left(\vec{\theta}\,|\,\vec{y}\right) = \log L\left(\vec{\theta}\,|\,\vec{y}\right) = -\frac{m}{2}\log 2\pi\sigma_\varepsilon^2 - \frac{1}{2}\log\|\Gamma\| - \frac{1}{2\sigma_\varepsilon^2}\vec{y}^{\,T}\Gamma^{-1}\vec{y}. \qquad (5.89)$$

The parameter estimates are then obtained by maximising either Eq. (5.88) or Eq. (5.89). The result given by Eq. (5.89) is called the **exact log-likelihood function**. However, when dealing with a long time series with many data points, it will be computationally expensive to compute the determinant and inverse required by the exact log-likelihood function at every iteration. Instead, the exact probability can be replaced by the conditional probability, that is, $p\left(\vec{y}\,|\,\vec{\theta}\right) \approx p\left(y_t\,|\,y_{t-1}, \ldots, y_1, \vec{\theta}\right)$. This approximation holds well for large data sets. With such an approximation, Eq. (5.88) can be written as

$$L\left(\vec{\theta}\,|\,\vec{y}\right) = \prod_{t=1}^{m} p\left(y_t\,|\,y_{t-1}, \ldots, y_1, \vec{\theta}\right). \qquad (5.90)$$

The expected value of the conditional probability can be written as

$$E\left(y_t\,|\,y_{t-1}, \ldots, y_1, \vec{\theta}\right) = \sum_{j=1}^{\infty} w_j y_{t-j} \approx \sum_{j=1}^{t-1} w_j y_{t-j}. \qquad (5.91)$$

In such cases, the conditional log-likelihood function can be written as

$$\ell\left(\vec{\theta}\,|\,\vec{y}\right) = \log L\left(\vec{\theta}\,|\,\vec{y}\right) = -\frac{m}{2}\log 2\pi\sigma_\varepsilon^2 - \frac{1}{2}\log\sigma_\varepsilon^2 - \frac{1}{2\sigma_\varepsilon^2}\sum_{i=1}^{m}\left(y_i - \sum_{j=1}^{t-1} w_j y_{i-j}\right)^2. $$
$$(5.92)$$

In the literature, this is often called the **innovation-based approximation**, since the difference between the measured and estimated values, called an **innovation**, is used in the computation of the log-likelihood function. This approach has the advantage that no inverses or determinants need to be computed. In general, since most data series are quite long, with many data points, this approach provides sufficiently accurate values. Various other simplifications can be made to improve the computational aspect of the result. One common assumption is that values in the past with $t < 0$ are equal to zero.

The required noise variance is then computed using the following formula

$$\sigma_\varepsilon^2 = \frac{1}{m - p - q} \sum_{i=1}^{m} \left(y_i - \hat{y}_i\right)^2,\tag{5.93}$$

where $\hat{y}_i$ is the predicted time-series value based on the model and m is the number of data points in the time series.

It can be noted that, irrespective of the approach taken, these equations will generally have to be solved numerically using some form of an optimisation algorithm. The required initial guess can be obtained based on either the Yule–Walker parameter estimates or some other approach. For autoregressive models, a closed-form solution to the above equations is available. The final result is identical to the Yule–Walker parameter estimates for an autoregressive model.

Example 5.10 *Exact Solution of the Maximum Likelihood Equation for an Autoregressive Model*

Consider an autoregressive model

$$y_t + \sum_{j=1}^{p} \alpha_j y_{t-j} = e_t.$$

Derive the corresponding maximum likelihood estimates for this model using the conditional probability density approach.

Solution

First, we can note that Eq. (5.91) can be made exact if we consider that there are exactly p past values that need to be considered. This gives

$$E\left(y_t | y_{t-1}, \ldots, y_1, \vec{\theta}\right) = \sum_{j=1}^{\infty} w_j y_{t-j} = -\sum_{j=1}^{p} \alpha_j y_{t-j}.$$

Substituting this relationship into the log-likelihood function (Eq. 5.92) gives

$$\ell\left(\vec{\theta}|\vec{y}\right) = \log L\left(\vec{\theta}|\vec{y}\right) = -\frac{m}{2}\log 2\pi\sigma_\varepsilon^2 - \frac{1}{2}\log\sigma_\varepsilon^2 - \frac{1}{2\sigma_\varepsilon^2}\sum_{i=1}^{m}\left(y_i + \sum_{j=1}^{p}\alpha_j y_{i-j}\right)^2.$$

If it is assumed that σ_ε is known, then the solution to the above equation can be determined analytically by considering solely the last, quadratic term. Thus, taking the derivative of the quadratic term with respect to the parameters α_j and setting the result equal to zero gives

$$\frac{1}{\sigma_\varepsilon^2} \sum_{t=1}^{m} \left[y_t + \sum_{i=1}^{p} \alpha_i y_{t-i} \right] y_{t-j} = 0 \,\forall j = 1, 2, \ldots, p.$$

In order to solve the resulting system of equations, it can be helpful to note that $y_t y_{t-j} \approx \hat{\gamma}(j)$, an estimated autocovariance. Therefore, the resulting system of equations can be written as

$$\hat{\vec{\gamma}} = -\hat{\Gamma}\hat{\vec{\alpha}},$$

where $\hat{\vec{\alpha}}$ is the $p \times 1$-row vector $\left(\hat{\alpha}_1, \hat{\alpha}_2, \ldots, \hat{\alpha}_p \right)^T$, which is identical to the Yule–Walker equation.

As with all maximum likelihood methods, the parameter estimates are asymptotically normally distributed, and approximate confidence intervals can be obtained using the following formulae:

$$\hat{\vec{\theta}} - \vec{\theta} \sim \mathfrak{N}\left(0, \frac{\sigma^2 \Gamma_{pq}^{-1}}{m} \right) \tag{5.94}$$

$$\hat{\sigma}_\varepsilon^2 \approx \sigma_\varepsilon^2,$$

where m is the number of data points used in the estimation,

$$\Gamma_{pq} = \begin{bmatrix} \Gamma_{\alpha\alpha} & \Gamma_{\alpha\beta} \\ \Gamma_{\beta\alpha} & \Gamma_{\beta\beta} \end{bmatrix}, \tag{5.95}$$

$\Gamma_{\alpha\alpha}$ is the $p \times p$ matrix whose (i, j)th-entry is the $\gamma(i - j)$ of the process $A(z^{-1})y_t = e_t$, $\Gamma_{\beta\beta}$ is the $q \times q$ matrix whose (i, j)th-entry is the $\gamma(i - j)$ of the process $y_t = B(z^{-1})e_t$, $\Gamma_{\alpha\beta}$ is the $p \times q$ matrix whose (i, j)th-entry is the $\gamma_{XY}(i - j)$ of the above two processes, that is, each entry is the cross-covariance between the AR and MA components at lag $(i - j)$, and $\Gamma_{\beta\alpha} = \Gamma_{\alpha\beta}^T$. This implies that approximate $100(1 - \alpha)\%$ confidence intervals for the parameters $\hat{\vec{\theta}}$ can be constructed as

$$\hat{\theta}_i \pm t_{1-\frac{\alpha}{2}, m-p-q} \hat{\sigma}_\varepsilon \sqrt{\frac{\left(\Gamma_{pq} \right)_{ii}^{-1}}{m}}. \tag{5.96}$$

Example 5.11 *Modelling the Mean Summer Temperature in Edmonton*

Consider the previously examined Edmonton temperature data series detailed in Sect. A5.1. For the purposes of this example, consider the problem of estimating a model for the mean summer temperature. The autocorrelation and the partial autocorrelation plots have already been shown and analysed previously (see Figs. 5.2 and 5.3). Using the results from there, obtain an initial model for the data.

Solution

Before a model can be fit, it is necessary to determine the orders of the model. From the autocorrelation plot, it would seem that some type of autoregressive component is present, since the values do not decay to zero. It can be noted that there are pronounced spikes at lags 1, 2, and 8. In fact, all of the values are significant and located above the confidence bounds for zero. On the other hand, the partial autocorrelation plot shows only a few significant points, namely lags 1, 2, 3, and 8. For the purposes of this example, it will be assumed that $p = q = 8$, that is, an autoregressive moving-average model will be fit to the data set. However, any potential seasonal component will be ignored at this point.

In general, when fitting a model, the data should be detrended, that is, the mean value should be removed from the data set. Using appropriate computer software, the model parameters with their standard deviation are

$$A\left(z^{-1}\right) = 1 - 0.948(\pm 0.2)z^{-1} + 0.758(\pm 0.1)z^{-2} - 0.612(\pm 0.2)z^{-3}$$
$$- 0.045(\pm 0.155)z^{-4} - 0.451(\pm 0.2)z^{-5} + 0.508(\pm 0.2)z^{-6}$$
$$- 0.643(\pm 0.2)z^{-7} - 0.446(\pm 0.1)z^{-8}$$

$$B\left(z^{-1}\right) = 1 - 0.877(\pm 0.1)z^{-1} + 0.840(\pm 0.06)z^{-2} - 0.493(\pm 0.1)z^{-3}$$
$$- 0.137(\pm 0.09)z^{-4} - 0.598(\pm 0.08)z^{-5} + 0.700(\pm 0.1)z^{-6}$$
$$- 0.936(\pm 0.06)z^{-7} + 0.760(\pm 0.1)z^{-8}.$$

The 95% confidence interval would be approximately twice (exactly 1.96) the given standard deviation. It can be seen that, in general, the estimated coefficients corresponding to the values between z^{-4} and z^{-6} have large confidence intervals that could cover zero. This agrees well with the observed results that suggest that only the first few lags and a lag of 8 are significant. Model validation will be performed in the subsequent example.

5.4.3 *Model Validation for Time-Series Models*

The basic principles of model validation, testing the residuals and the overall model, are the same as for regression analysis. The goal of this validation is to confirm that the residuals obtained are independent, normally distributed, white-noise values and that the model captures a significant portion of the overall variability. The main tools for model validation are:

1. **Tests for Normal Distribution**: In time-series analysis, there are three common approaches that can be used to test for normality:

    a. **Normal Probability Plot**: The most common method to test normality is to plot a normal probability plot of the residuals. The points should lie along a straight line. Examples of good and bad normal probability plots are shown in Table 3.2.

    b. **Autocorrelation Plot of the Residuals**: The autocorrelation plot of the residuals provides a useful visual aid in determining if the residuals are white noise. If the 95% confidence intervals for a zero autocorrelation are included in the plot, then at least 95% of the computed autocorrelations should lie inside the plotted confidence intervals.

    c. **Ljung–Box–Pierce Q-Statistic**: This is a general test statistic that seeks to determine if the observed autocorrelation is that of a white-noise signal, that is, do all nonzero lags have an autocorrelation of zero. The statistic is given as

$$Q = m(m+2) \sum_{h=1}^{H} \frac{\hat{\rho}_e(h)}{m-h}, \tag{5.97}$$

where $\hat{\rho}_e$ is the autocorrelation of the residuals and H is an arbitrary value, normally selected to be about 20. The critical value, $Q_{critical}$, is computed using the χ^2-distribution with $H - p - q$ degrees of freedom. If $Q > Q_{critical}$, then it can be concluded that the autocorrelations are not indicative of white noise.

2. **Tests for Independence and Homoscedasticity**: These two aspects are most commonly tested together using various types of scatter plots. The most common scatter plots to examine are:

    a. Plotting a time-series plot of the residuals.
    b. Plotting the residuals against the measured values, y.

In all cases, there should not be any discernible patterns in the plots. Common scatter plots are shown in Table 3.3 (see Sect. 3.3.5 for details on how to interpret and analyse such plots).

3. **Using the confidence interval for each of the parameters, θ_i:** If the confidence interval includes 0, then the parameter can be removed from the model. Ideally, a new regression analysis excluding that parameter would need to be performed and continued until there are no more parameters to remove or add.
4. **Calculating Pearson's coefficient of regression:** The closer the value is to 1, the better the regression is. This coefficient gives what fraction of the observed behaviour can be explained by the given variables.

When performing model validation, it is important to bear in mind the final goal for which the model will be used. In time-series analysis, the majority of the time, such models are used to forecast or predict future values of the system. In such cases, it is very important to not only test the performance of the model using the initial data set but also use another model validation data set. This validation data set can be obtained by splitting the original data set into parts. The first part is used for model estimation, while the second part is used for model validation. The residuals obtained using the data from the second part would then be used for model validation. The data set is often split $\frac{1}{3}$ for estimation and $\frac{2}{3}$ for validation.

Another approach to model validation is to consider various information criteria that assess the trade-off between the number of parameters selected and the variance of the model. These criteria can be useful for automating the estimation of initial process parameters. However, any model obtained using such an approach still needs to be validated for normality and purpose. The most common information criterion is **Akaike's Information Criterion**, which for any time-series model can be written as

$$\eta_{\mathrm{AIC}} = \ln \hat{\sigma}_\varepsilon^2 + \frac{m + 2k}{m}, \tag{5.98}$$

where m is the number of data points and k is the total number of parameters, that is, $k = p + P + q + Q$. Another commonly used criterion is the **Bayesian** or **Schwarz Information Criterion** (BIC) defined as

$$\eta_{BIC} = \ln \hat{\sigma}_\varepsilon^2 + \frac{k \log m}{m}. \tag{5.99}$$

Example 5.12 *Validating the Initial Mean Summer Temperature in Edmonton Model*

Consider the model fit in Example 5.11 and perform model validation to determine the overall quality of the model. Figure 5.14 shows the normal probability plot and the autocorrelation plot of the residuals.

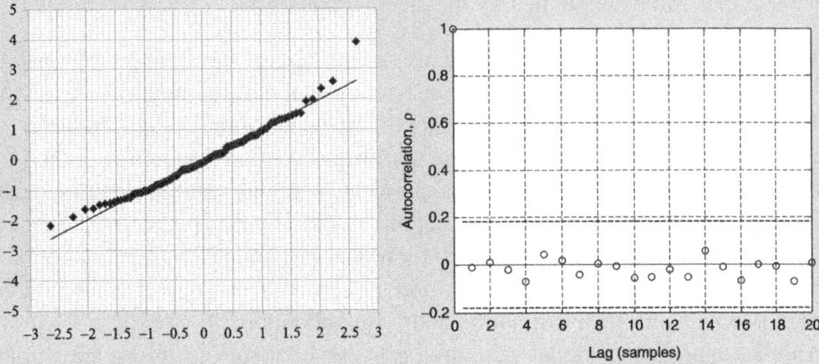

Fig. 5.14 (left) Normal probability plot and (right) autocorrelation plot for the residuals

Solution

The Ljung–Box–Pierce Q-statistic has a value of 4.35 (for $H = 20$). The 95% critical value is 31.41. Since $Q_{computed} < Q_{critical}$, the null hypothesis cannot be rejected, and it can be concluded that the residuals are white noise.

Figure 5.15 shows the measured and predicted mean summer temperatures as a function of time. It can be seen that the model does follow the trends in the data well. However, it does not predict well the extreme values. Looking at this plot, it seems quite clear that there is a seasonal component to the occurrence of extreme values. The fit as determined by Pearson's coefficient of regression is 32.04%.

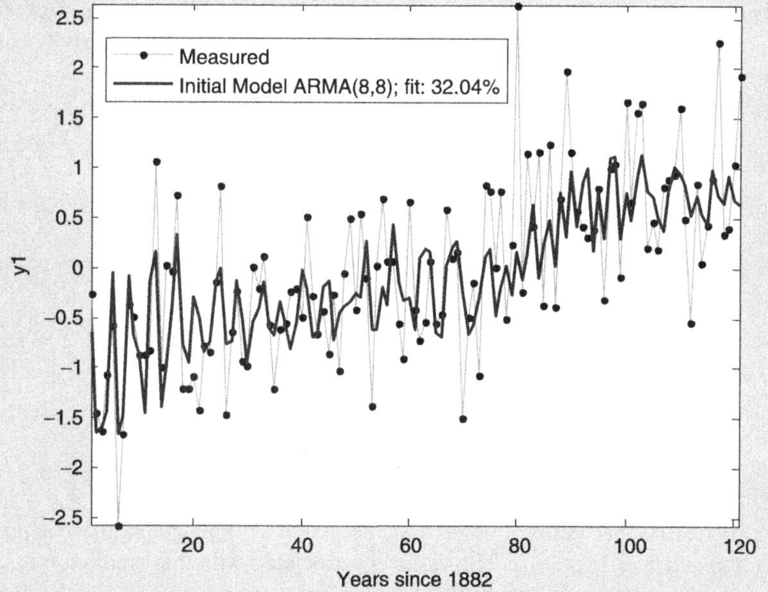

Fig. 5.15 Measured and one-step-ahead forecast temperatures as a function of years since 1882

Finally, we may note that $\eta_{AIC} = \ln(0.395\,884) + (120 + 2 \times 16)\,/\,120 = 0.340$.

Based on the results obtained here, it can be seen that although the model passes all the assumption tests, its predictive capability is not the best. There seems to be some component in the data that is not being reflected in the model. Furthermore, some of the parameter estimate confidence intervals are close to zero, suggesting that those estimates should be excluded from the model.

5.4.4 Model Prediction and Forecasting Using Time-Series Models

One of the primary purposes for a model is to determine future values of the process. For time series, it is useful to distinguish between two cases: **prediction** and **forecasting**. Prediction refers to determining future values using a model whose parameters are known and represent the true system values, while forecasting refers to determining future values using a model whose parameters have been obtained using some form of modelling. The results in this section will be presented without necessarily going into great detail regarding the derivation of the forms.

For predicting a value into the future given a known model, let $y_{t+\tau|t}$ be the τ-step-ahead prediction given all available data up to the current time point t. The τ-step-ahead prediction can be obtained by solving the following system of equations (Shumway and Stoffer 2011):

$$y_{t+\tau|t} = \vec{\phi}_t^T \vec{y}_t$$
$$\Gamma_t \vec{\phi}_t = \vec{\gamma}_t,$$

$$(5.100)$$

where

$$\Gamma_t = \begin{bmatrix} \gamma(0) & \cdots & \gamma(t-1) \\ \vdots & & \vdots \\ \gamma(t-1) & \cdots & \gamma(0) \end{bmatrix}, \vec{\gamma}_t = \begin{bmatrix} \gamma(\tau) \\ \vdots \\ \gamma(\tau+t-1) \end{bmatrix}, \vec{y}_t = \begin{bmatrix} y_t \\ \vdots \\ y_1 \end{bmatrix}, \vec{\phi}_t = \begin{bmatrix} \phi_{t1} \\ \vdots \\ \phi_{tt} \end{bmatrix}$$

$$(5.101)$$

and $\vec{\phi}_t$ represents the vector of prediction coefficients. For an autoregressive process with $t \geq p$ and $\tau = 1$, $\vec{\phi}_t = \vec{\alpha}$. The variance associated with this prediction is

$$\sigma^2_{t+\tau|t} = \gamma(0) - \vec{\gamma}_t^T \Gamma_t^{-1} \vec{\gamma}_t.$$

$$(5.102)$$

For large data sets, the solution of these equations directly can be difficult. Instead, iterative methods are used. The most common such iterative method is the Durbin–Levinson algorithm, which is described in Sect. 5.4.4.1.

For forecasting a value into the future (or filtering an already measured value[5]) given an estimated model, the situation is a bit more complicated. Consider the infinite-order moving-average representation for an ARMA process, that is,

$$y_{t+\tau} = \sum_{j=0}^{\infty} h_j e_{t+\tau-j},$$

$$(5.103)$$

which can be partitioned into parts: one which considers the future errors between $t + 1$ and $t + \tau$ and the second component that solely considers the remaining terms, that is,

---

[5] This is also called filtering because one reason for forecasting is to remove (filter) the noise from the (already made) measurements.

$$y_{t+\tau} = \sum_{j=0}^{\tau-1} h_j e_{t+\tau-j} + \sum_{j=\tau}^{\infty} h_j e_{t+\tau-j}. \tag{5.104}$$

Since the best linear predictor in the mean-squared sense is obtained by conditioning the estimate on all available past information, we get that the τ-step-ahead forecast, $\hat{y}_{t+\tau|t}$, is

$$\hat{y}_{t+\tau|t} = E(y_{t+\tau}|y_t, y_{t-1}, y_{t-2}, \ldots) = \sum_{j=\tau}^{\infty} h_j e_{t+\tau-j}. \tag{5.105}$$

It should be noted that

$$E\left(e_{t+\tau-j}|y_t, y_{t-1}, y_{t-2}, \ldots\right) = \begin{cases} 0 & j < \tau \\ e_{t+\tau-j} & j \geq \tau \end{cases}. \tag{5.106}$$

Therefore, the error associated with the forecast can be written as

$$\varepsilon_{t+\tau|t} = y_{t+\tau} - \hat{y}_{t+\tau|t} = \sum_{j=0}^{\tau-1} h_j e_{t+\tau-j}. \tag{5.107}$$

The forecast error variance then can be determined as

$$\hat{\sigma}_{t+\tau|t}^2 = \sigma_e^2 \sum_{j=0}^{\tau-1} h_j^2. \tag{5.108}$$

The $100(1 - \alpha)\%$ confidence interval can then be computed using the standard formula, that is,

$$y_{t+\tau} = \sum_{j=\tau}^{\infty} h_j e_{t+\tau-j} \pm t_{1-\frac{\alpha}{2},m-n} \hat{\sigma}_e \sum_{j=0}^{\tau-1} h_j^2, \tag{5.109}$$

where n is the total number of parameters in the model and α is α-error value.

It should be noted that, in practice, the above equations for the forecast value contain two issues, that is, they require an infinite amount of process data to be available and need to know the past values of the white noise. The first issue of requiring an infinite amount of data is not an issue for large data sets. This implies that truncation after m data points will not impact the final result. The second issue of requiring past white-noise values can be resolved by computing estimated white-noise values using the following formula

$$\hat{e}_t = y_t - \left(\sum_{j=1}^{q} \beta_i \hat{e}_{t-i} - \sum_{j=i}^{p} \alpha_i y_{t-i} \right). \qquad (5.110)$$

The algorithm is initialised by setting all e_t for which $t < p + 1$ to zero and then iterating forward. For pure moving-average or autoregressive processes, the forecast equation has a very simple form.

Example 5.13 *Forecasting a MA(3) Process*

Provide one-, two-, and four-step-ahead forecasts and 95% confidence intervals for the estimated model

$$y_t = e_t + 0.436e_{t-1} - 0.293e_{t-2} - 0.763e_{t-3}$$

using the data provided in Sect. A5.3. The noise variance, σ_e^2, is 1.0807.

Solution

In order to compute the required forecasts, it is first necessary to obtain estimates of the errors. Computing the errors using Eq. (5.110) gives the following results for the final five errors: $\hat{e}_{96} = -0.5949$, $\hat{e}_{97} = -1.3221$, $\hat{e}_{98} = -0.4402$, $\hat{e}_{99} = 0.5044$, and $\hat{e}_{100} = 0.8410$.

One-Step-Ahead Predictor

The one-step-ahead predictor can be obtained from Eq. (5.105)

$$\begin{aligned}
\hat{y}_{101|100} &= 0.436e_{100} - 0.293e_{99} - 0.763e_{98} \\
&= 0.436(0.8410) - 0.293(0.5044) - 0.763(-0.4402) \\
&= 0.5548.
\end{aligned}$$

The variance associated with this prediction is obtained from Eq. (5.108)

$$\hat{\sigma}_{101|100}^2 = 1.0807(1) = 1.0807.$$

Using $t_{0.975,\,97} = 1.96$ gives a 95% confidence interval as 0.6 ± 2.1.

Two-Step-Ahead Predictor

The two-step-ahead predictor can be obtained from Eq. (5.105)

$$\begin{aligned}
\hat{y}_{102|100} &= -0.293e_{100} - 0.763e_{99} \\
&= -0.293(0.8410) - 0.763(0.5044) \\
&= -0.6312.
\end{aligned}$$

The variance associated with this prediction is obtained from Eq. (5.108)

$$\hat{\sigma}^2_{102|100} = 1.0807\left(1 + 0.436^2\right) = 1.286.$$

Using $t_{0.975,\,97} = 1.96$ gives a 95% confidence interval as -0.6 ± 2.5.

Four-Step-Ahead Predictor

The four-step-ahead predictor can be obtained from Eq. (5.105). However, note that, in this case, there are no nonzero terms for h_j. Therefore, the predicted value is $\hat{y}_{104|100} = 0$ (or the mean value). The variance obtained from Eq. (5.108) will then be

$$\hat{\sigma}^2_{104|100} = 1.0807\left(1 + 0.436^2 + (-0.293)^2 + (-0.763)^2\right) = 2.0081.$$

Using $t_{0.975,\,97} = 1.96$ gives a 95% confidence interval as 0 ± 3.9. In fact, for a $MA(q)$ process, the prediction will be the same for all $\tau > q$.

5.4.4.1 Durbin–Levinson Algorithm

The Durbin–Levinson algorithm is a useful iterative method for inverting and hence solving the prediction equations given by Eq. (5.100) for a large data set. The procedure can be summarised as follows:

1. **Initialisation**: Set $\phi_{00} = 0$, $\sigma^2_{1|0} = \gamma(0)$, and $n = 1$.
2. **Diagonal Terms**: For $n \geq 1$, compute

$$\phi_{nn} = \frac{\rho(n) - \sum_{k=1}^{n-1} \phi_{n-1,k}\rho(n - k)}{1 - \sum_{k=1}^{n-1} \phi_{n-1,k}\rho(k)}, \quad \sigma^2_{n+1|n} = \sigma^2_{n|n-1}\left(1 - \phi_{nn}^2\right). \quad (5.111)$$

3. **Off-Diagonal Terms**: For $n \geq 2$, compute

$$\phi_{nk} = \phi_{n-1,k} - \phi_{nn}\phi_{n-1,n-k}, \quad k = 1, 2, \ldots, n - 1. \quad (5.112)$$

4. **Increment**: Increment n by 1 and go to Step 2. Stop when the desired value of n has been reached.

The Durbin–Levinson Algorithm can be used to also compute the Yule–Walker parameters and the partial autocorrelation function, since both have a similar matrix form.

Example 5.14 *Using the Durbin–Levinson Algorithm to Obtain the Predictor*
    Use the Durbin–Levinson algorithm to obtain the one-step-ahead predictors for the case where there are one, two, and three past data points.

Solution

Since each subsequent step requires the information from the previous step, the values will first be computed and then the predictors assembled.

$n = 0$: Set $\phi_{00} = 0$, $\sigma^2_{1|0} = \gamma(0)$, and $n = 1$.

$n = 1$: In this step, there will only be a single diagonal term,

$$\phi_{11} = \frac{\rho(1) - \sum_{k=1}^{0} \phi_{0,k}\rho(1-k)}{1 - \sum_{k=1}^{0} \phi_{0,k}\rho(k)} = \rho(1), \sigma^2_{n+1|n} = \sigma^2_{1|0}(1 - \rho(1))$$

$$= \gamma(0)(1 - \rho(1)).$$

$n = 2$: In this step, both diagonal and off-diagonal terms will be computed

$$\phi_{22} = \frac{\rho(2) - \sum_{k=1}^{1} \phi_{1,k}\rho(2-k)}{1 - \sum_{k=1}^{1} \phi_{1,k}\rho(k)} = \frac{\rho(2) - \rho(1)^2}{1 - \rho(1)^2}$$

$$\sigma^2_{3|2} = \sigma^2_{2|1}\left(1 - \phi^2_{22}\right) = \gamma(0)(1 - \rho(1))\left(1 - \frac{\rho(2) - \rho(1)^2}{1 - \rho(1)^2}\right)$$

$$= \gamma(0)(1 - \rho(1))\left(\frac{1 - \rho(2)}{1 - \rho(1)^2}\right)$$

$$\phi_{21} = \phi_{11} - \phi_{22}\phi_{11}, k = 1$$

$$= \rho(1) - \rho(1)\frac{\rho(2) - \rho(1)^2}{1 - \rho(1)^2} = \rho(1)\left(\frac{1 - \rho(2)}{1 - \rho(1)^2}\right).$$

$n = 3$: Like for $n = 2$, both diagonal and off-diagonal terms will be computed

$$\phi_{33} = \frac{\rho(3) - \sum_{k=1}^{2} \phi_{2,k}\rho(3-k)}{1 - \sum_{k=1}^{2} \phi_{2,k}\rho(k)} = \frac{\rho(3) - \phi_{21}\rho(2) - \phi_{22}\rho(1)}{1 - \phi_{21}\rho(2) - \phi_{22}\rho(1)}$$

$$\sigma^2_{4|3} = \sigma^2_{3|2}\left(1 - \phi^2_{33}\right)$$

$$= \gamma(0)(1 - \rho(1))\left(\frac{1 - \rho(2)}{1 - \rho(1)^2}\right)\left(1 - \frac{\rho(3) - \phi_{21}\rho(2) - \phi_{22}\rho(1)}{1 - \phi_{21}\rho(2) - \phi_{22}\rho(1)}\right)$$

$$= \gamma(0)(1 - \rho(1))\left(\frac{1 - \rho(2)}{1 - \rho(1)^2}\right)\left(\frac{1 - \rho(3)}{1 - \phi_{21}\rho(2) - \phi_{22}\rho(1)}\right)$$

$$\phi_{31} = \phi_{21} - \phi_{33}\phi_{22}, \ k = 1$$
$$\phi_{32} = \phi_{22} - \phi_{33}\phi_{21}, \ k = 2.$$

It should be noted that if the process model is known, then it is relatively easy to obtain the required auto- and cross-correlations and use them to compute the predictor.

5.5 Frequency-Domain Analysis of Time Series

Frequency-domain analysis of a time series is a useful tool for analysing and determining the presence of periodic signals in a given signal. This can be useful in determining and confirming the presence of seasonal components in the data set. It may not always be obvious what seasonal components are present purely by examining the auto- and partial autocorrelation plots.

5.5.1 Fourier Transform

The **Fourier transform** is the decomposition of an original signal into its periodic components. Formally, the Fourier transform, represented by $\mathfrak{F}$, converts a function, $f(t)$, defined in the time domain, to its corresponding frequency-domain function, $\mathfrak{f}(\omega)$. The relationship between the two functions can be written as

$$\mathfrak{f}(\omega) = \mathfrak{F}(f(t)) = \int_{-\infty}^{\infty} f(t)e^{-i\omega t}dt. \tag{5.113}$$

The function $\mathfrak{f}(\omega)$ is called the **spectral density** or **power spectrum** of $f(t)$. It can be noted that this definition of the Fourier transform does not explicitly show its connection with a periodic signal. However, we can note that by Euler's formula

$$e^{-i\omega t} = \cos(\omega t) + i\sin(\omega t) \tag{5.114}$$

the definition can be converted into one containing more obvious periodic terms. For this reason, ω is called the frequency. This clearly shows that the underlying system is periodic.

Furthermore, it should be noted that the resulting spectral density function will be a complex function. Since, in frequency-domain analysis, it is the strength of the individual frequencies that are of interest, it is common to convert the imaginary numbers into a more useful form. Consider an imaginary number $C = x + yi$, where

x and y are real. The magnitude of C, $|C|$, is defined as

$$|C| = \sqrt{x^2 + y^2} \qquad (5.115)$$

and the angle, $\angle C$, is defined as

$$\angle C = \arctan 2(x, y). \qquad (5.116)$$

The *arctan2* function is equivalent to $\arctan(yx^{-1})$ with the resulting angle placed in the correct quadrant based on the signs of x and y. Based on this analysis, the magnitude plotted as a function of the frequency will provide information about which frequencies are most prevalent in the signal.

However, in practice, the signal of interest is discrete rather than continuous. In such a case, there is a need to modify the above results to take into consideration the effect that discretisation has on the signal properties. Before considering the changes, it is instructive to consider the effect of discretisation on the signal properties. The largest effect is that information about the original signal is lost. For the purposes of frequency-domain analysis, the important result is the Nyquist frequency, f_{Nyquist}

$$f_{\text{Nyquist}} = 0.5 f_{\text{sampling}}, \qquad (5.117)$$

where f_{sampling} is the sampling rate. Frequencies above the Nyquist frequency cannot be recovered given a sampling rate. This implies that frequencies above this cut-off cannot be estimated.

The discrete Fourier transform is defined as

$$f\left(\frac{k}{n}\right) = \mathfrak{F}(X_n) = \sum_{t=0}^{n-1} x_t e^{-\frac{2\pi i k t}{n}}, \qquad (5.118)$$

where k is an integer and n is the length of the signal. The frequency is given as k/n cycles per sample. The amplitude, or strength, of the given frequency can be obtained using Eq. (5.115). Although it is possible to compute the discrete Fourier transform using a number of different methods, the most popular and efficient method for computer implementation is the fast Fourier transform method, which computes the required values efficiently and quickly (Bloomfield 2000; Welch 1967).

The following theorems will show some further relationships between the Fourier transform and various time-series properties.

Theorem 5.3 (Wiener–Khinchin Theorem) *For a discrete signal, the autocovariance and the spectral density function are related as follows (Khintchine 1934):*

$$\mathfrak{f}(\omega) = \sum_{\tau=-\infty}^{\infty} \gamma(\tau)e^{-i\omega\tau} \tag{5.119}$$

and

$$\gamma(\tau) = \frac{1}{2\pi} \int_{-\pi}^{\pi} \mathfrak{f}(\omega)e^{-i\omega\tau}\,d\omega. \tag{5.120}$$

Proof A detailed proof is presented in Priestley (1981, p. 219).

$$Q.E.D.$$

Corollary 5.5 (Spectral Density of White Noise) *The spectral density of white noise is a constant function equal to the white-noise variance, that is,*

$$\mathfrak{f}(\omega) = \sigma_w^2. \tag{5.121}$$

Proof For white noise, $\gamma(0) = \sigma_w^2$ and all other autocovariances are zero. From the first part of Theorem 5.3, the stated result follows.

$$Q.E.D.$$

Corollary 5.6 (Parseval's Theorem) *The variance of a discrete signal can be computed from its spectral density using the following equation*

$$\sigma^2 = \gamma(0) = \frac{1}{2\pi} \int_{-\pi}^{\pi} \mathfrak{f}(\omega)d\omega. \tag{5.122}$$

Proof Setting $\tau = 0$ in the second part of Theorem 5.3 and simplifying produces the given result.

$$Q.E.D.$$

Theorem 5.4 (Filtering Theorem) *If two signals are related by*

$$y(t) = \frac{B(z^{-1})}{A(z^{-1})}u(t) \tag{5.123}$$

then the spectral density functions are related as

$$\mathfrak{f}_y(\omega) = \left|\frac{B(e^{-j\omega})}{A(e^{-j\omega})}\right|^2 \mathfrak{f}_u(\omega). \tag{5.124}$$

Proof A detailed proof is presented in Stoica and Moses (2005).

Q.E.D.

Corollary 5.7 (Filtering White Noise) *If* u(t) *is white noise and is being filtered by a function of the form in Theorem 5.4 , then the filtered spectral density will be*

$$f_y(\omega) = \left| \frac{B\left(e^{-j\omega}\right)}{A\left(e^{-j\omega}\right)} \right|^2 \sigma_w^2. \tag{5.125}$$

Proof This results directly follows from combining the definition of the spectral density of white noise into Theorem 5.4.

Q.E.D.

5.5.2   *The Periodogram and Its Use in the Frequency-Domain Analysis of Time Series*

Since it has been shown that the Fourier transform contains the same information as the autocorrelation function, one may wonder what is the advantage of using it. Basically, the Fourier transform provides a different perspective on the same information allowing for different features to be more prominent. In the case of the Fourier transform, the periodicities are made clear, while in the (partial) autocorrelation plots, the different orders are emphasised.

The most common way to use the Fourier transform is to construct a **periodogram** that shows all the identifiable frequencies and their amplitudes. When using the fast Fourier transform to obtain the periodogram, only half of the values are plotted, since the other half is a mirror image (about $f = 0$).[6] The periodogram is constructed as follows:

(1)   Obtain the complex spectral density function, (ω), using any appropriate method (most often the fast Fourier method). Let n be the number of data points in the signal.

(2)   Compute the amplitude $q = |(\omega)|^2 = \mathrm{Re}((\omega))^2 + \mathrm{Im}((\omega))^2$, where *Re* represents the real component and *Im* the imaginary component of (ω).

---

[6] The formatting and layout of a periodogram vary greatly from source to source. The form presented here is the most convenient for time-series analysis. Appropriate code for creating such a periodogram is presented in Chap. 7 for MATLAB® and Chap. 8 for Excel®.

(3)   Set the centre point to be $C = \lfloor 0.5n \rfloor + 1$, where $\lfloor \cdot \rfloor$ is the floor or round down to the nearest integer function.

(4)   Compute the frequency range, F, as follows:

    a.   For a half-frequency periodogram, set $F = \langle 0, 1, \ldots, C \rangle\,/\,n$.

    b.   For a full-frequency periodogram, set $F = \langle 0, 1, \ldots, n-1 \rangle\,/\,n$.

(5)   To plot the full periodogram, set the x-axis equal to F and the y-axis to $q\,/\,n$.

(6)   To plot half of the periodogram, set the x-axis equal to F and the y-axis to $2q(1:C-1)/n$. The coefficient of 2 augments the amplitude to take into consideration the fact that only half of the periodogram was plotted.

(7)   The units of the graph will be cycles per sample for the x-axis and the original units of the signal for the y-axis.

(8)   To plot the frequency in the original units, multiple F by the sampling rate to give cycles per unit time, that is $F' = F\, f_{\text{sampling}}$. F' would then be used in place of F when plotting the periodogram.

(9)   In some applications, it may be desired to plot a full, zero-centred periodogram. In this case, there is a need to re-arrange both the F and q vectors obtained above in order to account for the differences. Basically, the second half of the original vector must be placed at the beginning. The following steps can be followed:

    a.   Let $q' = \langle q(C\!:\!\text{end}), q(1\!:\!C-1) \rangle$.

    b.   If n is even, let $F = \langle -0.5n, -0.5n+1, \ldots, 0, 1, \ldots, 0.5n-1 \rangle\,/\,n$.

    c.   If n is odd, let $F = \langle -0.5(n-1), -0.5(n-1)+1, \ldots, 0, 1, \ldots, 0.5(n-1) \rangle\,/\,n$.

    d.   Plot F on the x-axis and q' on the y-axis. The axis labels and interpretation will be the same as before.

The resulting half periodogram represents a decomposition unto either a series of cosines or a series of sines of the form

$$y = \sum_{k=0}^{\lfloor 0.5n \rfloor} A \cos(2\pi f k)$$

$$y = \sum_{k=0}^{\lfloor 0.5n \rfloor} A \sin(2\pi f k), \tag{5.126}$$

where A is the amplitude, defined as $|A| = 2q$, and f is the corresponding frequency, that is, $f = F(k)/2$. This can be seen in Fig. 5.16, where three very simple periodograms are shown: one single cosine, one single sine, and both sine and cosine. It should be noted that the periodogram ignores signs, that is, both a positive and a negative amplitude will appear as the same positive value.

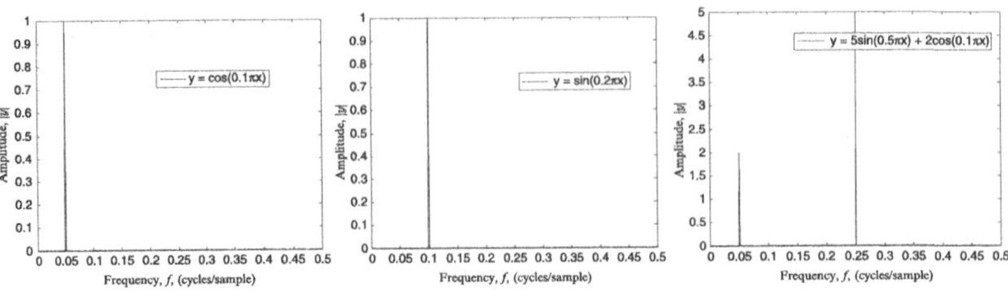

Fig. 5.16 Periodograms for three simple cases: (left) Single cosine, (middle) Single sine, and (right) Both cosine and sine together

The interpretation of more complex cases follows naturally from simple cases. Figure 5.17 shows periodograms for a seasonal process with a seasonal component of three samples and that of white noise. The first two periodograms show, as expected, a strong peak around $f = \frac{1}{3}$, which corresponds more or less to the seasonal component. It should be noted that the integrator has the strongest and cleanest peak, while the autoregressive example has a less clean peak. The last example, that of white noise, shows what the expected flat spectrum looks like in practice. We can see that the values are quite jagged fluctuating about some mean point. As α_1 approaches 0, it is expected that the overall graph will approach a white-noise graph. Furthermore, the peaks may appear at some multiple of the period, which can make identifying the true value a bit more difficult. Finally, rather than seeing a peak, a trough may occur at a given point. This is very common with seasonal moving-average processes, as these can remove certain frequencies from the signal. Figure 5.18 shows different moving-average processes all with a seasonal component of 3. Both the trough and multiple frequencies can be clearly seen in this figure.

Example 5.15 *Periodograms for the Edmonton Temperature Series*
Consider the Edmonton temperature series that is fully described in Sect. A5.1. Plot the periodograms for the spring, summer, and winter mean

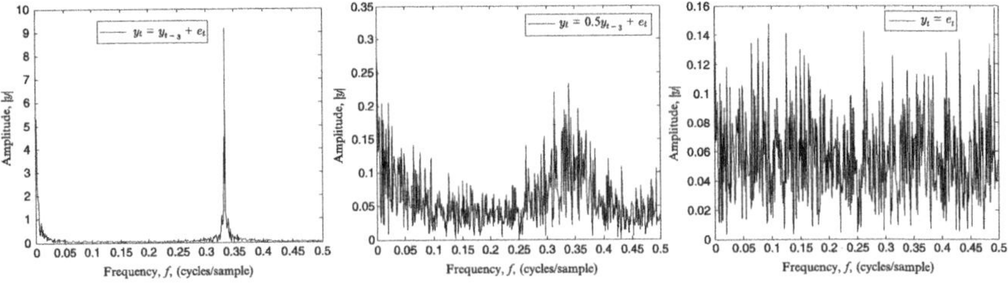

Fig. 5.17 Process with a seasonal component of three samples: (left) Integrator, (middle) Autoregressive, and (right) White noise

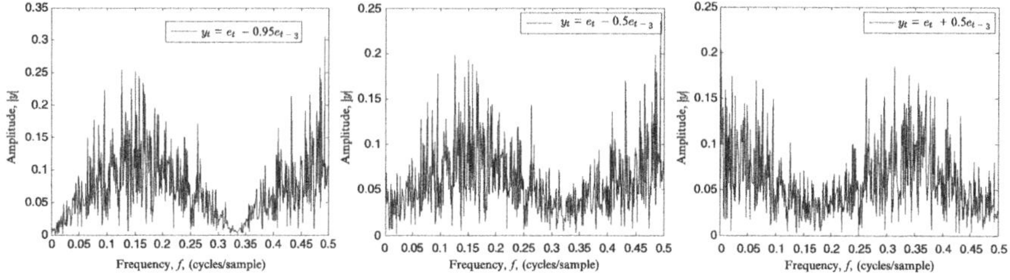

Fig. 5.18 A seasonal moving-average process with a seasonal component of 3 and (left) $\beta_1 = -0.95$, (middle) $\beta_1 = -0.5$, and (right) $\beta_1 = 0.5$

temperature series. As well, plot the periodogram for the differenced summer temperature series. What are some of the salient features?

Solution

The three undifferenced periodograms are shown in Fig. 5.19. Before running the Fourier transform, it is necessary to remove the mean value from the data set. It can be noted that both the spring and summer graphs have a peak close to the zero value (> 60 years/cycle). This can suggest that an integrator is present in the signal, since an integrator will have peaks at zero and one. On the other hand, the winter series shows a single large peak at 3.3 years/cycle and multiple smaller peaks throughout the spectrum.

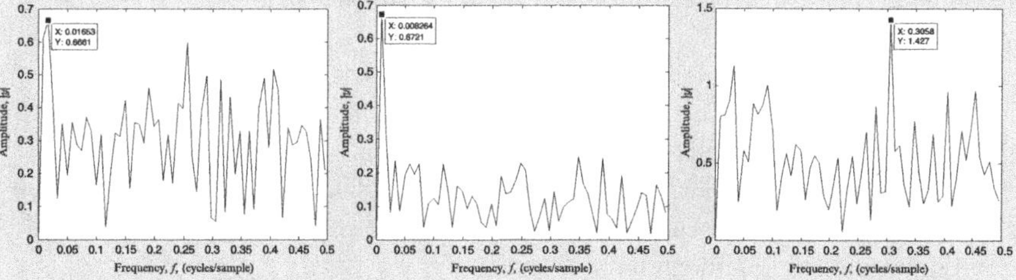

Fig. 5.19 Periodograms for (left) Spring, (middle) Summer, and (right) Winter of the Edmonton temperature series

Figure 5.20 shows the periodogram for the once-differenced summer temperature time series. Unlike in the previous periodogram, there are now a series of peaks clustered in the area around 2.5 to 3 years/cycle. As well, there is a secondary peak around 4 years/cycle followed by a rather weak peak in the 8 years/cycle region. All these values seem to be multiples of each other suggesting that they represent a single feature rather than separate features.

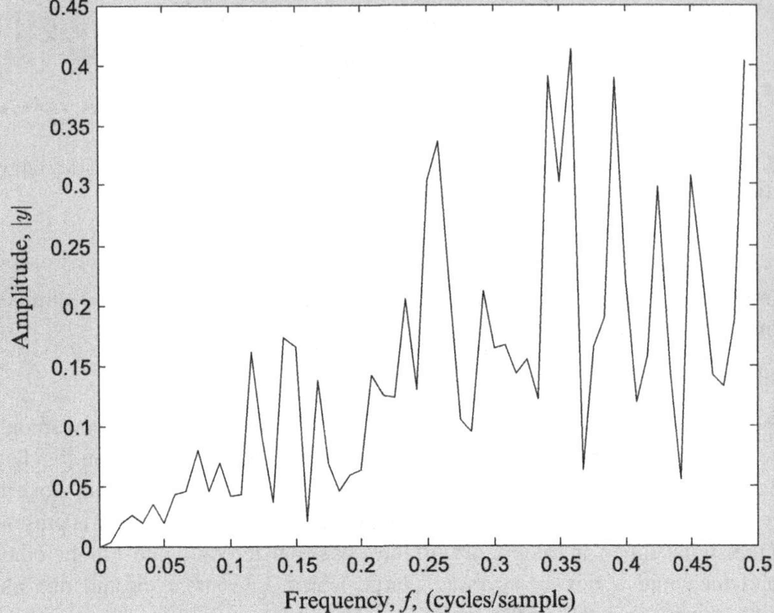

Fig. 5.20 Periodogram for the differenced summer temperature series

It can be noted that the spring and summer temperature series have a similar undifferenced behaviour, which suggests that an integrator could be present. Differencing the summer temperature series reveals the potential of seasonal components at three, four, and eight years. On the other hand, the winter temperature series has a different behaviour with a single peak at 3.3 years/cycle and no suggestions of an integrator.

5.6   State-Space Modelling of Time Series

State-space modelling is a useful, generalised approach to modelling a wide range of different systems under many different situations, including such cases as missing observations, outliers, or changing process parameters. Such cases are difficult, if not impossible, to incorporate into a transfer-function-based approach to modelling. The disadvantage of a state-space-based approach is that the models can be complex, without necessarily having a direct, physical meaning.

5.6.1 State-Space Model for Time Series

The complete state-space model can be written as a set of two coupled equations

$$\vec{x}_{k+1} = \mathcal{A}\vec{x}_k + \mathcal{B}\vec{u}_k + \vec{\omega}_k$$
$$\vec{y}_k = \mathcal{C}\vec{x}_k + \mathcal{D}\vec{u}_k + \vec{e}_k, \tag{5.127}$$

where $\vec{x}$ is the $n \times 1$ vector of **states**; $\vec{y}$ is the $p \times 1$ vector of **observations** (or **outputs**); $\vec{u}_k$ is the $m \times 1$ vector of **inputs**; ω_k is the $n \times 1$ vector of independent, white-noise random variables drawn from a Gaussian distribution with zero mean and covariance Σ_ω; e_k is the $p \times 1$ vector of independent, white noise random variables drawn from a Gaussian distribution with zero mean and covariance Σ_e; $\mathcal{A}, \mathcal{B}, \mathcal{C}$ and $\mathcal{D}$ are appropriately sized matrices; and the subscript k is an integer representing the current sample.

It can be noted that, in Eq. (5.127), an additional term, called the input, has been included in the model. This term allows the model to consider deterministic or other stochastic signals that have an impact on the overall process. When dealing with state-space models, it is traditional to consider the full form from the start rather than separating it out. The principles and ideas of such a term are developed further in Chap. 6. As well, it can be noted that in state-space models, only the input and output are normally measured directly. The remaining terms will need to be inferred using an appropriate method.

Finally, a state-space model is said to be causal (or stable) if the absolute value of the eigenvalues of the state matrix ($\mathcal{A}$-matrix) lie inside the unit circle.

5.6.2 The Kalman Equation

One of the most commonly used methods for predicting and forecasting new values using a state-space model is the **Kalman filter** developed by Kalman *et al.* in the early 1960s (Kalman 1960; Kalman and Bucy 1961). The power of the Kalman equation lies in its ability to deal with a wide range of situations including missing data or time-varying parameters.

Before defining the Kalman filter itself, it is necessary to define some notation that will make the formulation and interpretation of the results simpler. First, let the vector of available observations be defined as

$$\vec{Y}_s = \langle \vec{y}_1, \ldots, \vec{y}_s \rangle. \tag{5.128}$$

Next, let the time point at which the prediction is to be made be denoted by t. If $s < t$, then the problem to be solved is called a **prediction** or **forecasting** problem. If $s = t$, then the problem is called **filtering**. If $s > t$, then the problem is called **smoothing**.

Irrespective of the time horizon, a general term for this problem is (state) **estimation**. Finally, define the conditional estimate of the state given the information currently available as

$$\vec{x}_{t|s} = E\left(\vec{x}_t | \vec{Y}_s\right) \tag{5.129}$$

and the prediction covariance as

$$\Sigma_{t_1,t_2|s} = E\left(\left(\vec{x}_{t_1} - \vec{x}_{t_1|s}\right)\left(\vec{x}_{t_2} - \vec{x}_{t_2|s}\right)^T\right). \tag{5.130}$$

If $t_1 = t_2 = t$, then only a single t will be written to simplify the notation.

Theorem 5.5 (Basic Kalman Filter) *For the state-space model previously defined with initial conditions $\vec{x}_{0|0} = \mu_0$ and $\Sigma_{0|0}$, for $t = 0, \ldots, n$, the Kalman filter can be defined as*

$$\vec{x}_{t|t-1} = \mathcal{A}\vec{x}_{t-1|t-1} + \mathcal{B}\vec{u}_t \tag{5.131}$$

and the covariance is given as

$$\Sigma_{t|t-1} = \mathcal{A}\Sigma_{t-1|t-1}\mathcal{A}^T + \Sigma_\omega. \tag{5.132}$$

The update equations can be written as

$$\begin{aligned}
\vec{x}_{t|t} &= \vec{x}_{t|t-1} + \mathcal{K}_t\left(\vec{y}_t - \mathcal{C}\vec{x}_{t|t-1} - \mathcal{D}\vec{u}_t\right) \\
\Sigma_{t|t} &= [\mathcal{I} - \mathcal{K}_t\mathcal{C}]\Sigma_{t|t-1}
\end{aligned}, \tag{5.133}$$

where $\mathcal{K}_t$ is the Kalman gain, defined as

$$\mathcal{K}_t = \Sigma_{t|t-1}\mathcal{C}^T\left(\mathcal{C}\Sigma_{t|t-1}\mathcal{C}^T + \Sigma_e\right)^{-1}. \tag{5.134}$$

Proof The proof for these equations can be found in Kalman (1960).

Q.E.D.

Using the basic Kalman filter, the **prediction error** (or **innovation**), ε_t,[7] can be defined as

$$\vec{\varepsilon}_t = \vec{y} - \hat{\vec{y}} = \vec{y}_t - E\left(\vec{y}|\vec{Y}_{t-1}\right) = \vec{y}_t - \mathcal{C}\vec{x}_{t|t-1} - \mathcal{D}\vec{u}_t \tag{5.135}$$

with a covariance matrix, Σ_ε, defined as

---

[7] Strictly speaking, this is a one-step-ahead prediction error.

$$\Sigma_\varepsilon = \text{var}(\vec{\varepsilon}_t) = C\Sigma_{t|t-1}C^T + \Sigma_e. \tag{5.136}$$

Theorem 5.6 (Kalman Smoother) *For the state-space model previously defined with initial conditions $\vec{x}_{n|n}$ and $\Sigma_{n|n}$, for $t = n, n-1, \ldots, 1$, the Kalman smoother can be defined as*

$$\vec{x}_{n|t-1} = \vec{x}_{t-1|t-1} + \mathcal{J}_{t-1}\left(\vec{x}_{t|n} - \vec{x}_{t|t-1}\right) \tag{5.137}$$

and the covariance given as

$$\Sigma_{t-1|n} = \Sigma_{t-1|t-1} + \mathcal{J}_{t-1}\left(\Sigma_{t-1|n} - \Sigma_{t-1|t-1}\right)\mathcal{J}_{t-1}^T, \tag{5.138}$$

where $\mathcal{J}_t$ is the Kalman smoother gain, defined as

$$\mathcal{J}_{t-1} = \Sigma_{t-1|t-1}A^T\Sigma_{t|t-1}^{-1}. \tag{5.139}$$

Proof The proof for these equations can be found in Shumway and Stoffer (2011).

$$Q.E.D.$$

The estimated values and their corresponding bounds depend on the application. For each of the three previously considered cases, the corresponding formula is:

(1) **Smoothing**: For smoothing, the mean response interval for a Kalman smoothed value can be written as

$$\boxed{\left(\vec{x}_{t-1|r}\right)_i \pm t_{1-\frac{\alpha}{2},m}\sqrt{\left(\Sigma_{t-1|r}\right)_{ii}}}, \tag{5.140}$$

where i refers to the ith entry in the state vector, r is the data length used, and m is the total number of data points available. In most circumstances, the t-value can be replaced by the Z-value of 1.96.

(2) **Filtering**: For filtering, the mean response interval for a Kalman filtered value can be written as

$$\boxed{\left(\vec{x}_{t|t}\right)_i \pm t_{1-\frac{\alpha}{2},m}\sqrt{\left(\Sigma_{t|t}\right)_{ii}}}, \tag{5.141}$$

where i refers to the ith entry in the state vector and m is the total number of data points available. In most circumstances, the t-value can be replaced by the Z-value of 1.96.

(3) **Prediction**: The (one-step-ahead) prediction of future values using the Kalman filter can be accomplished by setting the initial conditions to be $\vec{x}_{0|0} = \vec{x}_{n|n}$

and $\Sigma_{n|n}$ and then using Eqs. (5.131) and (5.132) to obtain the prediction. The mean response interval for this prediction would be

$$\boxed{\left(\vec{x}_{t|t-1}\right)_i \pm t_{1-\frac{\alpha}{2},m}\sqrt{\left(\Sigma_{t|t-1}\right)_{ii}}},\tag{5.142}$$

where m is the total number of data points available.

The above equations can be simply extended to the case where the $\mathcal{A}$-, $\mathcal{B}$-, $\mathcal{C}$-, and $\mathcal{D}$-matrices, as well as the Σ_ω- and Σ_e-matrices, are time varying, that is, their values change with time. In such cases, one simply needs to simply make the corresponding changes in the above equations.

5.6.3 Maximum Likelihood State-Space Estimates

The estimation of the state-space parameters can be challenging given the complexities of the resulting equations. Consider the case where all the parameters, $\vec{\theta} = \langle \Sigma_0, \mathcal{A}, \mathcal{B}, \mathcal{C}, \mathcal{D}, \Sigma_\omega, \Sigma_e \rangle$, are to be estimated. It will be assumed that the initial state $\vec{x}_0$ is normally distributed with zero mean and a covariance matrix Σ_0, and the errors (ω and e) are jointly normally distributed and uncorrelated. Assume that the time series contains m data points and that there are a total of n unknown parameters.

In order to compute the maximum likelihood estimates, compute the one-step-ahead prediction error, ε_t, using Eq. (5.135) and the corresponding covariance matrix, Σ_ε, using Eq. (5.136). Ignoring constant terms, the log-likelihood function, $\ell_Y\left(\vec{\theta}\right)$, can be written as

$$-\ell_Y\left(\vec{\theta}\right) = 0.5\sum_{t=1}^{m}\ln\left|\Sigma_\varepsilon\left(\vec{\theta}\right)\right| + 0.5\sum_{t=1}^{m}\vec{\varepsilon}_t^T\left(\vec{\theta}\right)\Sigma_\varepsilon^{-1}\left(\vec{\theta}\right)\vec{\varepsilon}_t\left(\vec{\theta}\right).\tag{5.143}$$

Solving this equation requires using various numerical methods. Irrespective of the approach taken, the procedure can be summarised as follows:

(1)   Obtain the initial parameter estimate, $\vec{\theta}_0$ and set $j = 1$.
(2)   Using the basic Kalman filter as given in Theorem 5.5 and the initial parameter estimate, $\vec{\theta}_0$, obtain the innovations and error covariances for $t = 1,\ldots, m$.
(3)   Perform one iteration of Newton's method with $-\ell_Y\left(\vec{\theta}_{j-1}\right)$ as the objective function to obtain a new parameter estimate, $\vec{\theta}_j$.
(4)   Augment j by 1 and repeat steps 2) and 3) until the difference between $\vec{\theta}_{j-1}$ and $\vec{\theta}_{j-1}$ or the difference between $-\ell_Y\left(\vec{\theta}\right)$ and $-\ell_Y\left(\vec{\theta}_{j-1}\right)$ is small.

Other approaches are possible, including an expectation–maximization-like method for state-space parameter estimation (Shumway and Stoffer 2011).

Theorem 5.7 (Properties of the Maximum Likelihood Estimates) *Using maximum likelihood method to obtain the state-space parameter estimates $\hat{\vec{\theta}}_j$ and assuming that the prediction errors have the usual properties, as* m $\to \infty$,

$$\hat{\vec{\theta}}_j - \vec{\theta} \sim \mathfrak{N}\left(0, m^{-0.5}\mathcal{F}^{-1}\left(\vec{\theta}\right)\right), \tag{5.144}$$

where $\mathcal{F}\left(\vec{\theta}\right)$ is the asymptotic Fisher information matrix defined as

$$\mathcal{F}\left(\vec{\theta}\right) = \lim_{m \to \infty} m^{-1} E\left(-\frac{\partial \ell_Y\left(\vec{\theta}\right)}{\partial \vec{\theta} \partial \vec{\theta}^T}\right). \tag{5.145}$$

Proof A general proof of these results based on the maximum likelihood approach is presented in Hannan and Deistler (2012).

$$Q.E.D.$$

5.7   Comprehensive Example of Time-Series Modelling

Having considered multiple different methods and approaches to modelling time series, it is now necessary to apply these methods to the problem at hand: estimating the mean summer temperature in Edmonton. The data set is described in Sect. 5.1.3 and preliminary results have already been presented (see Examples 5.11, 5.12, and 5.15).

5.7.1   Summary of Available Information

From an initial attempt to model the mean summer temperature in Edmonton as an ARMA(8,8) process, the parameter estimates and their standard deviation were determined as

$$A\left(z^{-1}\right) = 1 - 0.948(\pm 0.2)z^{-1} + 0.758(\pm 0.1)z^{-2} - 0.612(\pm 0.2)z^{-3}$$
$$- 0.045(\pm 0.155)z^{-4} - 0.451(\pm 0.2)z^{-5} + 0.508(\pm 0.2)z^{-6}$$
$$- 0.643(\pm 0.2)z^{-7} - 0.446(\pm 0.1)z^{-8}$$

$$B(z^{-1}) = 1 - 0.877(\pm 0.1)z^{-1} + 0.840(\pm 0.06)z^{-2} - 0.493(\pm 0.1)z^{-3}$$
$$- 0.137(\pm 0.09)z^{-4} - 0.598(\pm 0.08)z^{-5} + 0.700(\pm 0.1)z^{-6}$$
$$- 0.936(\pm 0.06)z^{-7} + 0.760(\pm 0.1)z^{-8}.$$

Performing model validation on this model led to the conclusion that the model obtained was not complete. First, it was noted that the 95% confidence interval for some of the middle terms in the above polynomials covered zero. This suggests that those terms are not significant and that they should be excluded from the model. Second, the model assumptions regarding the errors were all validated, suggesting that the residuals were normally distributed, white noise. Third, the predictive capability of the model was not the best. The general trend was captured, but the individual estimates were not good. This suggests that the model could be missing some component or values. Fourth, the periodogram for the undifferenced mean summer data set suggested that an integrator could be present in the data set. Similarly, the periodogram for the once-differenced data suggested two separate cycles: one around 2.5–3 years/cycle and another around 4 years/cycle.

5.7.2 Obtaining the Final Univariate Model

Based on the above discussion, different models were fit, including seasonal differencing of three and four years, differencing of one year, and model orders between one and three parameters for both the seasonal and nonseasonal components. After trying different models, the final model was determined to be

$$A_p(z^{-1}) = 1 - 0.465(\pm 0.08)z^{-1} - 0.45(\pm 0.1)z^{-2} + 0.673(\pm 0.06)z^{-3}$$
$$A_P(z^{-1}) = 1 - 0.9343(\pm 0.009)z^{-4}$$

$$B_q(z^{-1}) = 1 - 0.50(\pm 0.1)z^{-1} - 0.33(\pm 0.1)z^{-2}$$
$$B_Q(z^{-1}) = 1 - 1(\pm 0.5)z^{-4}.$$

with a seasonal difference of order one and $s = 3$. Thus, the final model can be written as

$$A_p(z^{-1})A_P(z^{-1})(1 - z^{-3})y_t = B_q(z^{-1})B_Q(z^{-1})e_t.$$

The normal probability plot of the residuals and the autocorrelation of the residuals are shown in Fig. 5.21. Both results show that the residuals are normally distributed and white. In the normal probability plot, the tails deviate a bit from what would be desirable, but given that this is real data, such behaviour is inevitable.

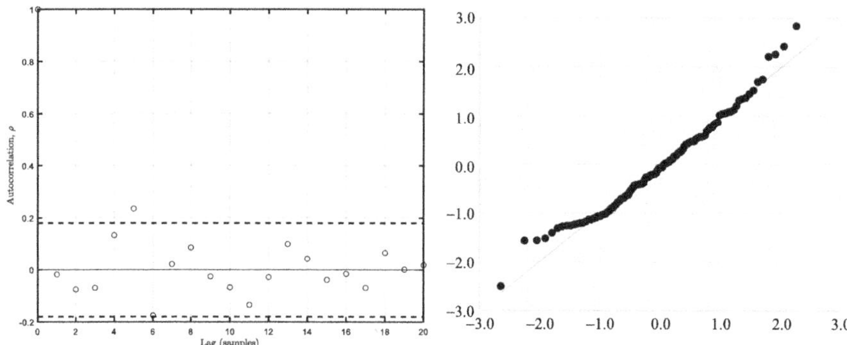

Fig. 5.21 Residual analysis for the final temperature model: (left) Autocorrelation plot of the residuals and (right) Normal probability plot of the residuals

A comparison between the predicted and measured temperatures is shown in Fig. 5.22. As before, the model gets the overall trends correct, but the individual predictions are not very good.

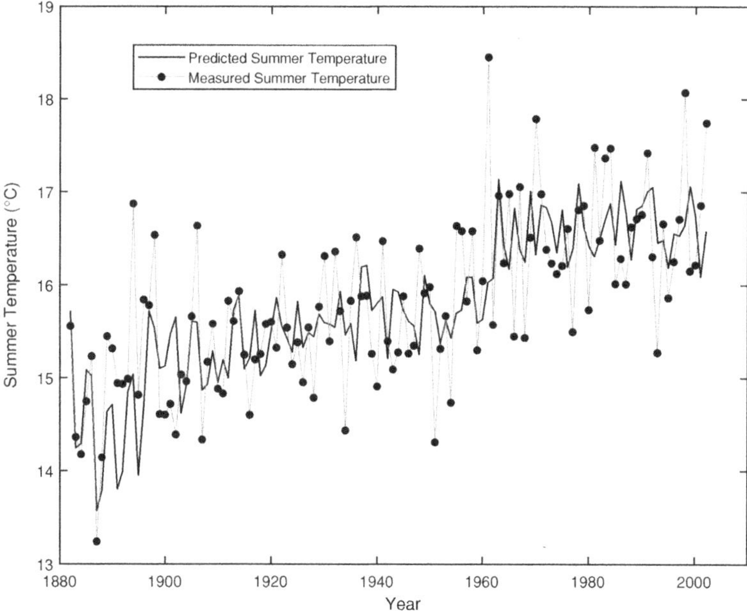

Fig. 5.22 Predicted and measured mean summer temperature using the final model

5.8 Further Reading

The following are references that provide additional information about the topic:

(1) **General Time-Series Analysis**: These sources also often contain information about transfer-function-based analysis.

a. Bloomfield, P. (2000). *Fourier Analysis of Time Series: An Introduction* (2nd ed.). New York, New York, United States of America: Wiley.

b. Box, G. E., & Jenkins, G. M. (1970). *Time Series Analysis, Forecasting, and Control.* Oakland, California, United States of America: Holden-Day.

c. Franke, J., Härdle, W. K., & Hafner, C. M. (2011). *Statistics of Financial Markets: An Introduction* (3rd ed.). Heidelberg, Germany: Springer. doi:10.1007/978-3-642-16521-4.

d. Hannan, E. J., & Deistler, M. (2012). *The statistical theory of linear systems.* Society of Industrial and Applied Mathematics.

e. Montgomery, D. C., Jennings, C. L., & Kulahci, M. (2008). *Introduction to Time Series Analysis and Forecasting.* Hoboken, New Jersey, United States of America: John Wiley & Sons, Inc.

f. Shumway, R. H., & Stoffer, D. S. (2011). *Time Series Analysis and Its Applications with R Examples* (3rd ed.). New York, New York, United States of America: Springer. doi: 10.1007978-1-4419-7865-3.

(2) **Properties of Time Series**:

a. Ashley, R. (1988). On the Relative Worth of Recent Macroeconomic Forecasts. *International Journal of Forecasting, 4,* 363–376.

b. Hassler, U. (1994). The sample autocorrelation function of I(1) processes. *Statistical Papers—Statistische Hefte, 35,* 1–16.

c. Khintchine, A. (1934). Korrelationstheorie der stationären stochastischen Prozesse (Correlation Theory of Stocastic Processes). *Mathematische Annalen, 109*(1), 604–615. doi: 10.1007/BF01449156.

d. Nelson, C. R. (1972). The Prediction Performance of the FRB-MIT-PENN Model of the U.S. Economy. *The American Economic Review, 62*(5), 902–917.

e. Shardt, Y. A. W. (2012). *Data Quality Assessment for Closed-Loop System Identification and Forecasting with Application to Soft Sensors.* Doctoral Thesis, University of Alberta, Department of Chemical and Materials Engineering, Edmonton, Alberta, Canada. doi: http://hdl.handle.net/10402/era.29018.

f. Wichern, D. W. (1973). The Behaviour of the Sample Autocorrelation Function for an Integrated Moving Average Process. *Biometrika, 60*(2), 235–239. doi: http://www.jstor.org/stble/2334535?origin=JSTOR-pdf.

(3) **Spectral Analysis**:

    a.   Priestley, M. B. (1981). *Spectral Analysis and Time Series: Vol. 1: Univariate Series and Vol 2: Multivariate Series, Prediction, and Control.* New York, New York, United States of America: Academic Press.

    b.   Stoica, P., & Moses, R. (2005). *Spectral Analysis of Signals.* Upper Saddle River, New Jersey, United States of America: Prentice Hall.

    c.   Welch, P. D. (1967). The Use of Fast Fourier Transform for the Estimation of Power Spectra: A Method Based on Time Averaging Over Short, Modified Periodograms. *IEEE Transactions on Audio and Electroacoustics,* AU-15(2), 70–73.

(4) **State-Space Analysis**:

    a.   Harvey, A. C. (1991). *Forecasting, Structural Time Series Models and the Kalman Filter.* Cambridge: Cambridge University Press.

    b.   Kalman, R. E. (1960). A new approach to linear filtering and prediction problems. *Transactions of the ASME, Journal of Basic Engineering, 82,* 35–45.

    c.   Kalman, R. E., & Bucy, R. S. (1961). New results in filtering and prediction theory. *Transactions of the ASME, Journal of Basic Engineering, 83,* 95–108.

5.9 Chapter Problems

Problems at the end of the chapter consist of three different types: (a) Basic Concepts (True/False), which seek to test the reader's comprehension of the key concepts in the chapter; (b) Short Exercises, which seek to test the reader's ability to compute the required parameters for a simple data set using simple or no technological aids. This section also includes proofs of theorems; and (c) Computational Exercises, which require not only a solid comprehension of the basic material, but also the use of appropriate software to easily manipulate the given data sets.

5.9.1 Basic Concepts

Determine if the following statements are true or false and state why this is the case.

(1) The causality of an ARMA process is determined by examining the numerator of the transfer function.

(2) An autoregressive process is always invertible.

(3)   An invertible moving-average process will give when inverted a causal autoregressive process.

(4)   A seasonal autoregressive process can be analysed by setting z^{-1} in the standard autoregressive model to be equal to z^{-s}, where s is the seasonal term.

(5)   The autocorrelation plot can be used to determine the orders of the autoregressive component.

(6)   The presence of an integrator can be detected by a slowly decaying term in the partial autocorrelation plot.

(7)   If the time series is stationary, then it should always be differenced.

(8)   If the roots of the A-polynomial of an autoregressive process are 0.5, $-0.5\pm0.75i$, then the process is causal.

(9)   If the roots of the B-polynomial of an ARMA process are -0.45, 0.15, $0.75\pm0.5i$, then the process is invertible.

(10)   If the autocorrelation function is given as $\rho(\tau) = 0.5^{\tau}$ for all $\tau \geq 0$, then it can be concluded that the process is a moving-average process.

(11)   If the autocorrelation function is given as $\rho(\tau) = 0.5^{\tau}$ for all $\tau \geq 0$, then it can be concluded that the process is causal.

(12)   If the partial autocorrelation plot has spikes at $\tau = 0, 3, 6, 9$, and 12, then we are dealing with a seasonal autoregressive process with $s = 3$ and $P = 4$.

(13)   The Yule–Walker equations are a method of moment estimator for autoregressive processes.

(14)   The method of moment estimator for moving-average processes is unbiased.

(15)   The maximum likelihood parameter estimates for an ARMA process are asymptotically normally distributed.

(16)   The maximum likelihood parameter estimates for an MA process can be obtained by solving a matrix equation without any numerical iterations.

(17)   If, when examining the autocorrelation plot of the residuals, out of 25 autocorrelations, 2 (including the zero-lag contribution) are located outside the 95% confidence bands, then it can be concluded that the residuals are Gaussian.

(18)   If a peak at $f = 0.25$ cycles/sample is observed on the periodogram, then it can be concluded that the process has a seasonal component, such that $s = 0.25$.

(19)   The Kalman filter is used to determine the parameter estimates for state-space models.

(20)   State-space parameter estimates obtained using the maximum likelihood approach are not asymptotically consistent.

5.9.2 Short Exercises

These questions should be solved using only a simple, nonprogrammable, nongraphical calculator combined with pen and paper.

Table 5.2 Autocovariance and partial autocorrelation data (for Question 24)

Lag	Autocovariance $\gamma(\tau)$	Partial autocorrelation $\rho_{X\mid X_{t+1},\dots,X_{t+\tau-1}}(\tau)$
0	5.212	–
1	3.832	0.735
2	2.919	0.045
3	2.183	0.064
4	1.645	0.022
5	1.234	0.028
6	0.923	0.013

(21) Consider an ARMA(1, 0, 1) process of the form $u_t = \dfrac{C(z^{-1})}{D(z^{-1})} = \dfrac{A-Bz^{-1}}{C-Dz^{-1}}e_t$. Derive the spectral density function for u_t in terms of the transfer-function parameters and the white-noise spectral density.

(22) For a causal AR(2) process, derive the autocorrelation and partial autocorrelation function.

(23) For an invertible MA(3) process, derive the autocorrelation function and the first four partial autocorrelation values.

(24) Given the data in Table 5.2, determine an appropriate ARIMA model for the time series. It should be noted that 1,000 data points were used to compute the samples.

(25) Given the data in Fig. 5.23, suggest an appropriate SARIMA model for this data.

5.9.3 Computational Exercises

The following problems should be solved with the help of a computer and appropriate software packages, such as MATLAB® or Excel®.

(26) Take the Edmonton temperature series and fit appropriate ARMA models to the winter, spring, fall, and annual mean temperatures. Be sure to examine the residuals and verify that the assumptions are met.

(27) Using the Kalman filter, develop a model for the Edmonton temperature series.

(28) Take the AR(2) process data in Sect. A5.2 and fit an AR(1) process to it. Analyse the residuals and fit. Comment on the results. Repeat, but using an AR(3) model. Compare the two models with the accurate AR(2) model (see Example 5.8 for the model). What happens when a model is over- or underfit?

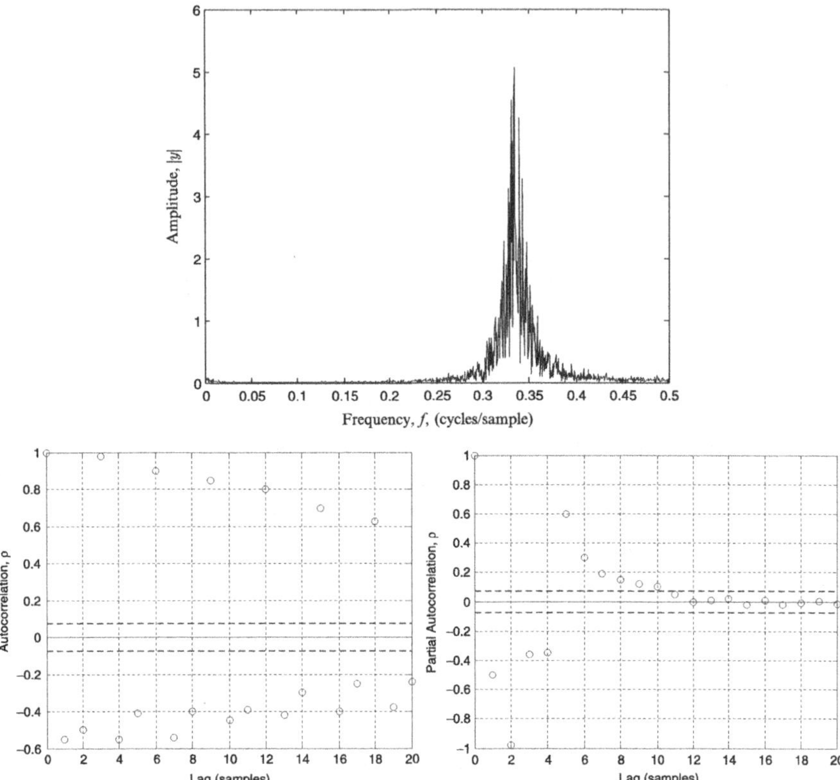

Fig. 5.23 (top) Periodogram, (bottom, left) Autocorrelation plot, and (bottom, right) partial autocorrelation plot for an unknown process

(29) Take the MA(3) process data in Sect. A5.3 and fit a MA(1) process to it. Analyse the residuals and fit. Comment on the results. Repeat, but using a MA(4) model. Compare the two models with the accurate MA(3) model (see Example 5.13 for the model). What happens when a model is over- or underfit?

(30) Take any time series of interest, analyse it, and fit an appropriate model to it.

Appendix A5: Data Sets for This Chapter

This section gives detailed information about the data set used for analysis in this chapter. All data can be downloaded as an Excel® spreadsheet or MATLAB® file from the book website.

A5.1: Edmonton Weather Data Series (1882–2002)

The raw data for the Edmonton Weather Data Series are presented in Table 5.3. This data set has been compiled using daily temperature values available from Environment Canada. The values are obtained by combining the daily temperature values from two nearby weather stations: Edmonton (C3012195) from 1880 to 1943 and Edmonton City Centre (C3012208) from 1937 to 2002. Since both locations are close to each other, the two data series were combined into a single set with the change over year being 1940: up until that year, the data were taken from the Edmonton weather station, while from January 1, 1940, the data were taken from the Edmonton City Centre weather station. It should be noted that due to missing values, most data between 1880 and 1881 have been excluded from the data series. The December 1881 values, which are complete, have been used in the computation of the mean winter temperature for 1882.

The mean temperatures were obtained by computing the mean value of the available daily high and low temperatures in the given interval. The intervals are defined as follows:

(1)   **Annual**: from January 1 to December 31 of the given year.
(2)   **Winter**: from December 1 of the previous year to February 28 (or 29, in a leap year) of the given year.
(3)   **Spring**: from March 1 to May 31 of the given year.
(4)   **Summer**: from June 1 to August 31 of the given year.
(5)   **Fall**: from September 1 to November 30 of the given year.

A5.2: AR(2) Process Data

One hundred simulations of an autoregressive process were performed and the data recorded. The values are provided in Table 5.4.

A5.3: MA(3) Process Data

One hundred simulations of a moving-average process were performed and the data were recorded. The values are provided in Table 5.5.

Table 5.3 Edmonton Weather Data Series (1882–2002)

Year	Mean temperature (°C)				
	Annual	Winter	Spring	Summer	Fall
1882	1.10	−11.12	0.79	15.56	1.53
1883	0.35	−16.05	2.60	14.36	−0.40
1884	0.93	−14.90	3.49	14.18	2.48
1885	2.69	−16.96	5.23	14.75	4.36
1886	1.48	−14.42	3.24	15.23	3.43
1887	−0.25	−19.89	4.00	13.24	2.64
1888	1.52	−16.05	−0.07	14.15	4.70
1889	5.07	−5.69	7.66	15.45	5.05
1890	2.67	−17.31	2.83	15.31	6.86
1891	3.72	−8.98	5.09	14.94	4.11
1892	2.46	−10.34	3.06	14.93	3.57
1893	1.58	−12.26	1.79	14.99	1.42
1894	2.76	−13.80	3.94	16.87	3.00
1895	2.90	−12.82	5.14	14.82	3.93
1896	1.63	−11.38	1.78	15.84	−0.66
1897	2.71	−10.23	3.01	15.78	2.79
1898	3.62	−10.00	3.51	16.54	2.84
1899	1.54	−11.09	−0.69	14.61	4.73
1900	3.31	−10.80	4.82	14.61	2.68
1901	4.01	−8.30	5.85	14.72	3.50
1902	2.80	−7.26	3.83	14.39	3.22
1903	3.24	−11.14	1.49	15.04	3.71
1904	2.32	−12.28	1.81	14.96	6.25
1905	4.24	−11.22	6.40	15.66	4.70
1906	3.85	−6.61	4.02	16.64	4.61
1907	1.84	−16.86	−0.05	14.34	6.53
1908	3.62	−7.46	3.31	15.17	4.21
1909	1.11	−14.77	1.93	15.58	2.34
1910	3.65	−12.03	6.83	14.88	3.51
1911	2.29	−14.01	4.85	14.83	2.89
1912	4.37	−9.13	4.61	15.83	4.65
1913	3.33	−11.19	3.64	15.61	4.54

(continued)

Table 5.3 (continued)

Year	Mean temperature (°C)				
	Annual	Winter	Spring	Summer	Fall
1914	3.21	−10.60	4.74	15.93	5.02
1915	4.10	−11.33	6.55	15.25	3.74
1916	1.55	−13.60	3.13	14.60	4.57
1917	1.56	−15.28	2.01	15.20	6.18
1918	3.38	−15.92	3.45	15.26	5.30
1919	1.94	−8.70	2.34	15.58	−0.16
1920	2.22	−11.07	0.15	15.60	3.95
1921	2.89	−10.32	2.93	15.33	2.83
1922	2.74	−12.90	3.37	16.33	5.44
1923	3.66	−12.95	2.53	15.54	6.79
1924	2.48	−9.21	3.37	15.15	3.86
1925	2.27	−15.50	3.82	15.38	1.47
1926	3.23	−6.58	5.38	14.95	0.60
1927	0.71	−13.16	2.19	15.54	0.63
1928	3.89	−11.40	3.67	14.79	4.07
1929	2.48	−13.01	3.49	15.77	5.45
1930	3.75	−12.69	4.47	16.31	3.35
1931	4.50	−3.02	3.75	15.39	3.62
1932	2.28	−12.15	2.42	16.36	2.77
1933	1.41	−12.67	2.80	15.72	3.86
1934	3.63	−12.28	5.57	14.44	2.88
1935	1.74	−12.51	−0.09	15.83	2.19
1936	1.56	−18.47	3.99	16.51	5.24
1937	2.22	−15.56	5.28	15.88	3.55
1938	3.72	−13.35	4.58	15.89	5.76
1939	3.46	−11.70	3.60	15.26	4.29
1940	2.12	−10.38	2.86	14.91	3.07
1941	3.21	−12.09	4.59	16.47	3.96
1942	3.01	−8.90	4.35	15.40	2.45
1943	3.42	−14.14	1.98	15.09	6.57
1944	4.04	−7.39	4.50	15.28	5.28
1945	1.93	−10.34	2.54	15.88	1.05
1946	2.65	−12.47	5.34	15.27	2.03
1947	2.65	−13.63	2.64	15.35	3.94
1948	2.22	−10.13	−0.09	16.39	4.91

(continued)

Table 5.3 (continued)

Year	Mean temperature (°C)				
	Annual	Winter	Spring	Summer	Fall
1949	2.79	−16.16	5.56	15.92	6.13
1950	0.51	−19.16	1.62	15.98	1.39
1951	0.53	−13.83	1.03	14.31	1.43
1952	3.73	−14.09	4.22	15.32	6.50
1953	4.13	−9.48	2.69	15.67	6.67
1954	2.67	−10.54	0.38	14.74	5.89
1955	1.60	−7.74	1.21	16.64	0.33
1956	2.79	−16.23	3.10	16.58	5.25
1957	3.43	−12.27	4.12	15.83	4.76
1958	4.06	−8.84	4.49	16.58	4.53
1959	3.13	−13.09	5.09	15.30	2.72
1960	3.37	−8.73	2.82	16.05	4.52
1961	3.67	−7.62	4.48	18.45	2.47
1962	3.36	−14.24	2.34	15.57	6.19
1963	4.15	−9.94	4.15	16.96	5.86
1964	2.91	−7.24	2.42	16.24	3.53
1965	2.17	−15.99	1.51	16.98	2.68
1966	1.90	−14.47	3.08	15.45	3.10
1967	3.17	−11.54	0.23	17.06	6.60
1968	3.32	−10.80	5.57	15.43	5.02
1969	2.65	−18.29	4.50	16.51	4.11
1970	2.61	−9.46	3.32	17.79	2.21
1971	3.02	−14.25	4.43	16.98	4.07
1972	2.06	−15.76	4.23	16.38	2.72
1973	3.36	−10.97	5.49	16.23	1.29
1974	3.59	−11.89	1.59	16.12	5.94
1975	3.18	−8.50	1.81	16.21	4.90
1976	5.21	−8.62	5.91	16.61	6.18
1977	4.56	−5.84	6.45	15.50	4.32
1978	3.58	−13.98	5.11	16.81	4.62
1979	3.26	−15.33	3.36	16.85	6.70
1980	3.93	−9.99	5.85	15.74	6.23
1981	6.22	−7.89	6.67	17.48	6.59
1982	2.13	−14.89	2.18	16.48	3.93
1983	3.97	−7.60	4.79	17.37	4.39

(continued)

Table 5.3 (continued)

Year	Mean temperature (°C)				
	Annual	Winter	Spring	Summer	Fall
1984	4.04	−8.18	5.14	17.47	1.01
1985	3.55	−11.80	5.94	16.02	−0.09
1986	4.76	−6.91	6.34	16.29	2.89
1987	6.19	−4.07	5.22	16.01	7.28
1988	5.31	−7.51	7.74	16.63	5.08
1989	3.75	−9.40	2.64	16.71	4.77
1990	4.10	−7.75	5.71	16.76	3.66
1991	4.79	−8.29	4.79	17.42	2.87
1992	4.67	−6.11	6.31	16.31	4.27
1993	4.40	−11.54	5.96	15.27	4.55
1994	3.59	−12.12	6.49	16.66	4.90
1995	3.27	−9.67	4.18	15.87	3.53
1996	1.42	−13.57	2.37	16.25	1.20
1997	4.45	−11.97	2.90	16.71	5.33
1998	5.02	−7.65	6.85	18.07	5.21
1999	4.84	−9.86	4.43	16.15	5.69
2000	3.65	−7.29	4.45	16.22	4.49
2001	5.29	−8.51	6.13	16.86	5.37
2002	3.85	−7.74	−0.64	17.75	4.17

Table 5.4 Sample data for the AR(2) process

Sample time	Value
1	0.5377
2	2.5866
3	1.0936
4	1.0999
5	1.3118
6	−0.0211
7	−1.119
8	−1.2135

(continued)

Table 5.4 (continued)

Sample time	Value
9	2.4391
10	6.7908
11	6.9378
12	9.3524
13	10.3499
14	9.7505
15	9.1906
16	7.7866
17	6.1818
18	6.2509
19	7.0694
20	8.1889
21	8.6013
22	6.7398
23	5.8524
24	6.4536
25	6.5978
26	7.0448
27	7.2907
28	6.3811
29	5.5821
30	3.8371
31	3.4693
32	1.7914
33	-0.2956
34	-2.119
35	-5.7631
36	-5.5705
37	-4.5919
38	-4.3984

(continued)

Table 5.4 (continued)

Sample time	Value
39	−2.4915
40	−3.0004
41	−3.057
42	−3.0211
43	−2.3818
44	−1.5112
45	−1.7896
46	−1.7799
47	−1.7619
48	−0.949
49	0.6456
50	2.4876
51	2.2962
52	2.0482
53	0.5053
54	−1.4302
55	−2.2618
56	−0.9187
57	−0.925
58	−0.4643
59	−0.4131
60	0.7712
61	0.1972
62	−0.077
63	0.3461
64	1.6237
65	3.6443
66	4.3761
67	2.8128
68	1.0076

(continued)

Table 5.4 (continued)

Sample time	Value
69	−1.0574
70	0.3664
71	0.426
72	1.1613
73	1.2204
74	2.0165
75	1.4481
76	−0.3832
77	−2.6829
78	−3.0763
79	−3.1427
80	−3.0577
81	−1.2901
82	0.0143
83	0.8628
84	2.7885
85	2.6681
86	3.0377
87	3.7538
88	3.4927
89	3.2286
90	1.6078
91	−0.5113
92	−1.4148
93	−1.0029
94	1.8889
95	2.479
96	2.7135
97	2.4769
98	0.1779
99	−1.4284
100	−3.8834

Table 5.5 Sample data for the MA(3) process

Sample time	Value
1	−1.0642
2	1.0714
3	2.3025
4	0.7850
5	−3.1322
6	−2.0663
7	0.1705
8	1.4389
9	0.2493
10	0.3608
11	0.4718
12	−0.9697
13	−2.0038
14	−1.1963
15	0.2966
16	0.3218
17	0.5350
18	−2.5619
19	−1.7343
20	1.7619
21	1.9407
22	0.4325
23	0.1520
24	0.7127
25	−0.6199
26	−1.7293
27	−0.8909
28	1.8161
29	2.0952
30	−0.247
31	−1.941
32	−0.482
33	0.3695

(continued)

Table 5.5 (continued)

Sample time	Value
34	1.1009
35	−0.1946
36	−0.7037
37	1.6789
38	−0.9078
39	0.7717
40	0.4991
41	2.3052
42	−2.9206
43	−1.9253
44	−0.9071
45	1.6793
46	−0.9468
47	−0.2606
48	−0.8765
49	0.5945
50	0.6349
51	1.5463
52	1.0334
53	0.9764
54	−0.6638
55	−1.1319
56	−1.4077
57	−1.3701
58	0.3072
59	1.5012
60	0.7052
61	0.1900
62	1.0638
63	0.4609
64	0.5934
65	1.0214
66	0.1294
67	−1.7406
68	−1.754

(continued)

Table 5.5 (continued)

Sample time	Value
69	0.8591
70	−0.3351
71	−0.6397
72	−2.4522
73	1.2611
74	1.8775
75	1.6385
76	−1.0511
77	−3.1683
78	−0.6451
79	−1.2271
80	−1.6961
81	−0.9675
82	1.3156
83	1.6567
84	1.0599
85	1.8047
86	−0.7402
87	−0.6184
88	−0.1452
89	1.2824
90	0.5870
91	0.8840
92	−0.4366
93	−0.8398
94	0.2937
95	−1.4840
96	−1.7080
97	−1.7478
98	0.7010
99	1.1535
100	2.1987

Chapter 6
Modelling Dynamic Processes Using System Identification Methods

Process system identification is a complex and involved process that can take on multiple different facets and requires understanding not only the chemical and physical aspects of the process but also the mathematical and statistical background of identification. In system identification, there are two basic approaches to this problem:

(1) **First-principle, white-box,** or *ab initio* modelling, where a description of the process is obtained from the fundamental equations (mass, energy, and force balances) and various simplifications about the overall process. Although such an approach provides a very general model that can potentially be applied over a wide range of conditions, practically, it can be very difficult to obtain a tractable and useful form for many reasons, including lack of process understanding, especially at the molecular or submolecular levels, missing information about the relevant parameters, and the complexities of the resulting equation leading to difficulties in obtaining the desired final answer.

(2) **Data-driven** or **black-box** modelling, where a description of the process is obtained solely by developing models for the available data. This approach can provide very accurate models of the system at a given set of conditions, but the model cannot generalise well to other conditions. Furthermore, developing such models can be difficult, since the selection of appropriate terms and relevant data is a nontrivial task. Unless the correlations are strong, it may be difficult to decide on an appropriate data-driven model.

Given the potential problems associated with both approaches, a third, middle way has also been considered. This approach is called **grey-box modelling**, where the initial form of the equation determined based on the first-principle model is used for data-driven modelling. This approach has the advantage that the form of the equation has some physical meaning and could provide a reasonable description of the process.

Y. A. W. Shardt, *Statistics for Chemical and Process Engineers,*
https://doi.org/10.1007/978-3-030-83190-5_6

Furthermore, another component of process system identification is that not only is the deterministic model considered, but the stochastic component is also taken into consideration to give a regression model of the form

$$y_t = f\left(\vec{u}_t, \vec{\beta}_u\right) + g\left(e_t, \vec{\beta}_e\right), \tag{6.1}$$

where u_t represents the deterministic input to the model (equivalent to x in Chap. 3). The deterministic inputs can often be treated as a time series themselves, but whose values are not necessarily driven by a white-noise process (as in Chap. 5). In order to obtain a useful solution, various assumptions are made regarding the forms of f and g and how they interact with the different systems. It should be noted that the concepts presented here are not limited in their application to solely process system identification, but can also be applied in a wide variety of different fields, including complex econometric analysis and robotics.

Although the focus of this chapter will be on linear model identification, a cursory investigation of nonlinear approaches will be presented in order to provide a complete overview of system identification.

6.1   Control and Process System Identification

In process control, the objective is to design a controller so that the overall system can track and maintain a given reference signal. Figure 6.1 shows a generalized block diagram description of the system. The signals are denoted by lines with arrows, while the blocks denote a process that converts the signals entering the block into signals leaving it. The most important signals are:

(1)   **Reference signal (r_t)**: This represents the desired or setpoint value for the process.
(2)   **Input signal (u_t)**: This represents the input into the process. This signal can also be called the manipulated variable.
(3)   **Output signal (y_t)**: This represents the measured value of the process. This signal can also be called the controlled variable.
(4)   **Disturbance signal (e_t)**: This represents the unmodelled changes in the process. The disturbance signal is often assumed to be a Gaussian, white-noise signal with zero mean and variance σ^2.

Fig. 6.1 Block diagram of the control system

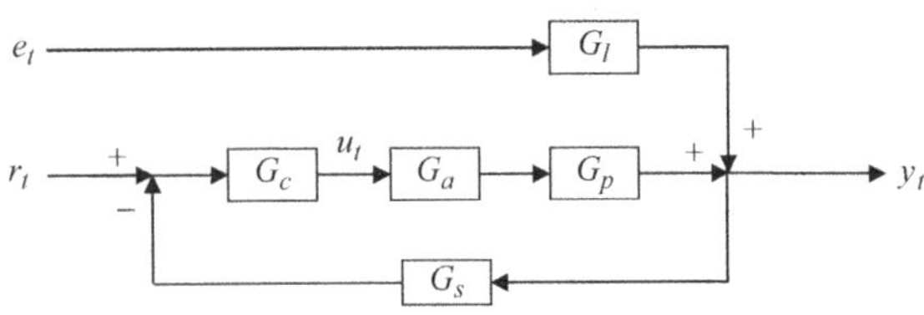

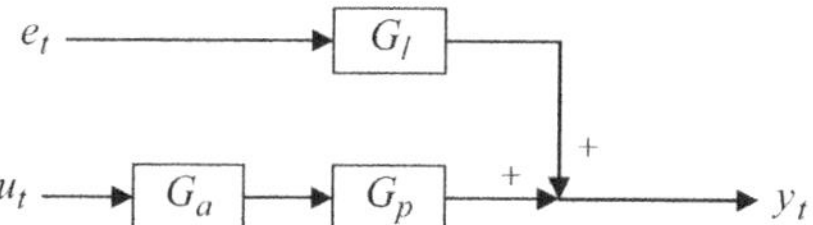

Fig. 6.2 Generic open-loop process

The most important blocks are:

(1) **Controller (G_c)**: This gives the model of the controller.
(2) **Process (G_p)**: This gives the known model of the process.
(3) **Disturbance (G_l)**: This gives the model of the unknown component.
(4) **Actuator (G_a)**: This gives the model for how the valve responds to a change in the given values. In most cases, since the response is very fast, it can be safely lumped together with the process model.
(5) **Sensor (G_s)**: This gives the model for how the sensor or measurement device works and responds to changes in the process. In most cases, since only the measured values are available, it is useful to lump this model together with the process and disturbance models. This block is useful to remind the reader that unless a variable can be measured, it cannot be used for control.

Together the process and disturbance models create the **plant model**.

The primary objective is to perform **system identification**, that is, obtain a plant model, especially that of the process, in order to design a controller. Two different situations can be considered: **open-loop system identification** and **closed-loop system identification**.

In open-loop system identification, it is assumed that the controller and reference signals are not present, that is, the control loop has not been closed. In such a case, Fig. 6.1 reduces to Fig. 6.2. The relationship between the input and output can then be written as

$$y_t = G_p u_t + G_l e_t. \tag{6.2}$$

In closed-loop system identification, the controller is fully functional and determines the value of the output based on the measured output value. Two different cases can be distinguished depending on the behaviour of the reference signal:

(1) **Routine Operating Mode**, where the reference signal does not change its value over the course of the experiment.
(2) **Externally Excited Mode**, where the reference signal does change its value in some predetermined manner over the course of the experiment.

The relationship between the reference signal and the output can be written as

$$y_t = \frac{G_c G_p}{1 + G_c G_p} r_t + \frac{G_l}{1 + G_c G_p} e_t. \tag{6.3}$$

If the reference signal is zero (or constant), then Eq. (6.3) reduces to solely the second term. This suggests that such a system can be modelled as a univariate time-series model.

Irrespective of the situation, process system identification is focused on determining the values for G_p and G_c as accurately as possible. Since most of the applications assume that the controller is digital, the system identification methods considered here will focus on the discrete-time implementation of system identification. For this reason, the models for each of the blocks will be assumed to be linear, rational functions of the backshift operator z^{-1}. Such models are most often referred to as **transfer functions**. The most general plant model is the **prediction error model**, which has the following form:

$$A(z^{-1})y_t = \frac{B(z^{-1})}{F(z^{-1})}u_{t-k} + \frac{C(z^{-1})}{D(z^{-1})}e_t, \tag{6.4}$$

where $A(z^{-1})$, $C(z^{-1})$, $D(z^{-1})$, and $F(z^{-1})$ are polynomials in z^{-1} of the form

$$1 + \sum_{i=1}^{n_a} \theta_i z^{-i}, \tag{6.5}$$

where n_a is the order of the polynomial and θ_i are the parameters, $B(z^{-1})$ is a polynomial in z^{-1} of the form

$$\sum_{i=1}^{n_b} \theta_i z^{-i}, \tag{6.6}$$

where n_b is the order of the polynomial, and k is the time delay in the system. In general, it is very rare for this system to be used directly. Instead, any of the following simplifications may be used:

1. **Box-Jenkins Model**: In this model, the $A(z^{-1})$ polynomial is ignored. Thus, this model is given as

$$y_t = \frac{B(z^{-1})}{F(z^{-1})}u_{t-k} + \frac{C(z^{-1})}{D(z^{-1})}e_t. \tag{6.7}$$

   In practice, this method is sufficient to obtain an accurate model of the system.

2. **Autoregressive Moving-Average Exogenous Model (ARMAX)**: In this model, the $D(z^{-1})$ and $F(z^{-1})$ polynomials are ignored, which gives

$$A(z^{-1})y_t = B(z^{-1})u_{t-k} + C(z^{-1})e_t. \tag{6.8}$$

This model assumes that the denominator for both the input and the error is the same.

3. **Autoregressive Exogenous Model (ARX)**: This is a simplified version of the ARMAX model, wherein it is assumed that $C(z^{-1})$ can be ignored. This gives a model of the form

$$A(z^{-1})y_t = B(z^{-1})u_{t-k} + e_t. \tag{6.9}$$

Although this model is very simple, it has the beneficial property that the estimation of its parameters can be performed using least-squares analysis. In many respects, this model is very similar to the autoregressive model previously considered for time-series analysis.

4. **Output-Error Model (OE)**: In this model, only the model for the input is fit to the data. The error terms are ignored. Thus, the model is given as

$$y_t = \frac{B(z^{-1})}{F(z^{-1})}u_{t-k} + e_t. \tag{6.10}$$

Another model that is occasionally used is the **impulse response model**, which can be written as

$$y_t = \sum_{i=0}^{\infty} h_i u_{t-i} + \sum_{j=0}^{\infty} h_j e_{t-j}, \tag{6.11}$$

where h is the impulse response coefficient that can be obtained by performing long division with the polynomials in the prediction error model. The first term of the error impulse response model, h_0, is traditionally equal to 1. This model is commonly encountered in theoretical applications.

6.1.1 Predictability of Process Models

Since the models obtained in system control are often used to predict or forecast future behaviour of a system, it is important to develop an understanding of the predictive properties of a model. This can be formalized by considering the τ-step-ahead predictor for a process, $\hat{y}_{t+\tau|t}$, which predicts the process value τ-steps ahead given all the values of the process up until the current point t and the input up until the point $t + \tau$. Let the prediction error be defined as

$$\varepsilon_{t+\tau|t} = y_{t+\tau} - \hat{y}_{t+\tau|t} \tag{6.12}$$

and the linear, τ-step-ahead predictor be defined as

$$\hat{y}_{t+\tau|t} = L_1 u_{t+\tau} + L_2 y_{t+\tau}, \tag{6.13}$$

where L is some rational function of z^{-1}.

Theorem 6.1 (τ-step-ahead linear predictor) *The τ-step-ahead linear predictor,* $\hat{y}_{t+\tau|t}$, *is*

$$\hat{y}_{t+\tau|t} = \left(\mathcal{I} - \sum_{i=0}^{\tau-1} h_i z^{-i} G_l^{-1} \right) y_{t+\tau} + \left(\sum_{i=0}^{\tau-1} h_i z^{-i} G_l^{-1} G_p \right) u_{t+\tau}, \tag{6.14}$$

where h *is the impulse coefficients of the disturbance model.*

Proof The proof of this theorem can be obtained by solving Eq. (6.12) for the case of a linear predictor given by Eq. (6.13) and the model by Eq. (6.2). Many of the steps will be similar to those used to obtain the time-series predictor in Sect. 5.4.4.

Substituting Eqs. (6.13) and (6.2) into Eq. (6.12) gives

$$\varepsilon_{t+\tau|t} = G_p u_{t+\tau} + G_l e_{t+\tau} - L_1 u_{t+\tau} - L_2 y_{t+\tau}. \tag{6.15}$$

In order to separate the available error values from those which are not, the impulse response model of the disturbance transfer function will now be split into two components: a term containing all the terms up to, but excluding, the τth impulse term (unavailable future component) and a term containing all the remaining terms (available past component). It can be noted that the last term can be rewritten as a difference between the original disturbance transfer function and the unavailable future component, that is,

$$\sum_{i=\tau}^{\infty} h_i z^{-i} = G_l - \sum_{i=0}^{\tau-1} h_i z^{-i}. \tag{6.16}$$

Thus, Eq. (6.15) can be rewritten as

$$\varepsilon_{t+\tau|t} = G_p u_{t+\tau} + \left(G_l - \sum_{i=0}^{\tau-1} h_i z^{-i} \right) e_{t+\tau} + \sum_{i=0}^{\tau-1} h_i z^{-i} e_{t+\tau} - L_1 u_{t+\tau} - L_2 y_{t+\tau}. \tag{6.17}$$

Solving Eq. (6.2) for the disturbance signal gives

$$e_t = G_l^{-1} \left(y_t - G_p u_t \right). \tag{6.18}$$

Substituting Eq. (6.18) for only the first error term in Eq. (6.17) gives

$$\varepsilon_{t+\tau|t} = G_p u_{t+\tau} + \left(G_l - \sum_{i=0}^{\tau-1} h_i z^{-i} \right) G_l^{-1} \left(y_{t+\tau} - G_p u_{t+\tau} \right)$$

$$+ \sum_{i=0}^{\tau-1} h_i z^{-i} e_{t+\tau} - L_1 u_{t+\tau} - L_2 y_{t+\tau}. \tag{6.19}$$

Re-arranging and combining like terms gives[1]

$$\varepsilon_{t+\tau|t} = \left(\sum_{i=0}^{\tau-1} h_i z^{-i} G_l^{-1} G_p - L_1 \right) u_{t+\tau} + \left(\mathcal{I} - \sum_{i=0}^{\tau-1} h_i z^{-i} G_l^{-1} - L_2 \right) y_{t+\tau}$$

$$+ \sum_{i=0}^{\tau-1} h_i z^{-i} e_{t+\tau}. \tag{6.20}$$

The last term in Eq. (6.20) cannot be simplified since it depends on future values of the error that are not yet known. Since it has been assumed that there is a time delay of at least one sample in the process transfer function, that is, $G_p\left(0, \vec{\theta}\right) = 0$, and that $G_l\left(0, \vec{\theta}\right) = \mathcal{I}$, this shows that the errors are uncorrelated either with each other or the input. Thus, in order to minimise the prediction error, both of the coefficients for $u_{t+\tau}$ and $y_{t+\tau}$ must be equal to zero, that is,

$$\sum_{i=0}^{\tau-1} h_i z^{-i} G_l^{-1} G_p - L_1 = 0 \Rightarrow L_1 = \sum_{i=0}^{\tau-1} h_i z^{-i} G_l^{-1} G_p, \tag{6.21}$$

$$\mathcal{I} - \sum_{i=0}^{\tau-1} h_i z^{-i} G_l^{-1} - L_2 = 0 \Rightarrow L_2 = \mathcal{I} - \sum_{i=0}^{\tau-1} h_i z^{-i} G_l^{-1}. \tag{6.22}$$

Thus, Eqs. (6.21) and (6.22) show that the τ-step-ahead predictor can be written as

$$\hat{y}_{t+\tau|t} = \left(\mathcal{I} - \sum_{i=0}^{\tau-1} h_i z^{-i} G_l^{-1} \right) y_{t+\tau} + \left(\sum_{i=0}^{\tau-1} h_i z^{-i} G_l^{-1} G_p \right) u_{t+\tau}. \tag{6.23}$$

$$Q.E.D.$$

Theorem 6.2 (variance of the τ-step-ahead predictor) *The variance of the τ-step-ahead predictor is*

---

[1] It should be noted that $\mathcal{I}$ represents the $n \times n$ identify matrix, where n is the size of the (square) disturbance transfer function matrix. In most cases, $\mathcal{I}$ will be equal to 1.

$$\boxed{\sigma_\tau^2 = \sigma_w^2 \sum_{i=0}^{\tau-1} h_i^2}. \tag{6.24}$$

Proof The required variance can be obtained by examining the prediction error.

Simplifying Eq. (6.20) based on the results from Theorem 6.1 gives that the prediction error can be written as

$$\varepsilon_{t+\tau|t} = \sum_{i=0}^{\tau-1} h_i z^{-i} e_{t+\tau}. \tag{6.25}$$

Since e_t is white noise, this implies that it is uncorrelated with past or future values of itself. Therefore, the variance of the estimate will be given as

$$\sigma_\tau^2 = \sigma_w^2 \sum_{i=0}^{\tau-1} h_i^2. \tag{6.26}$$

$$Q.E.D.$$

Corollary 6.1 (properties of the one-step-ahead predictor) *The **one-step-ahead predictor** is given by*

$$\boxed{\hat{y}_{t+1|t} = \left(\mathcal{I} - G_l^{-1}\right) y_t + G_l^{-1} G_p u_t} \tag{6.27}$$

and has a variance equal to

$$\boxed{\sigma_1^2 = \sigma_w^2}. \tag{6.28}$$

Proof Set $\tau = 1$ in Theorems 6.1 and 6.2 to give the above results.

$$Q.E.D.$$

The one-step-ahead predictor forms an important basis for estimating the parameters of such systems.

Corollary 6.2 (the infinite-step-ahead predictor) *The **infinite-step-ahead predictor**, or the **infinite-horizon predictor**, is given by*

$$\boxed{\hat{y}_\infty = G_p u_t} \tag{6.29}$$

and has a variance equal to

$$\boxed{\sigma_\infty^2 = \sigma_I^2}\,,\tag{6.30}$$

where σ_I^2 is the variance of the disturbance model.

Proof Take the limit as $\tau \to \infty$ of the corresponding equations in Theorems 6.1 and 6.2 to give the above results.

Q.E.D.

The infinite-horizon predictor is useful when looking at the predictive properties of a model and how generalizable the model is. Basically, the infinite-horizon predictor assumes that the errors are not known and seeks to predict the process solely on the basis of the available input information.

6.2   Framework for System Identification

The system identification framework shown in Fig. 6.3 extends the general regression framework shown in Fig. 3.1 to take into account the specific issues in process system identification. The framework consists of three steps:

(1)   **Data Collection**: During the data collection step, the required data are collected and analysed to determine if there are any obvious problems with the data set, such as missing data, faulty sensors, faulty values, or multiple operating modes. The framework presented in Fig. 6.3 assumes that a separate experiment will be designed in order to obtain the data required for system identification. In industry, the ability to perform such experiments can be limited due to various factors, including safety, economic, or reluctance on the part of the plant operators. Instead, historical data from the data historian are extracted and preprocessed to determine their usefulness for the given problem.

(2)   **Model Creation and Validation**: During this step, the data set is used to create the model and obtain parameter estimates. As well, the given model is validated to determine if it could potentially be used.

(3)   **Decision Making**: Based on the model obtained from the previous step, a decision is made whether the given model is sufficient or a better model needs to be sought. Clearly, the available time and purpose of the model will determine the amount of effort required and model accuracy. For a simple controller, a relatively crude model may be more than sufficient to obtain a good controller.

6.3   Open-Loop Process Identification

This section will examine the principles and key results for modelling an open-loop process using the general prediction error model given by Eq. (6.4). The foundation for such modelling is the prediction error method, which uses the fact that most models in system identification are used for predicting future values of the process.

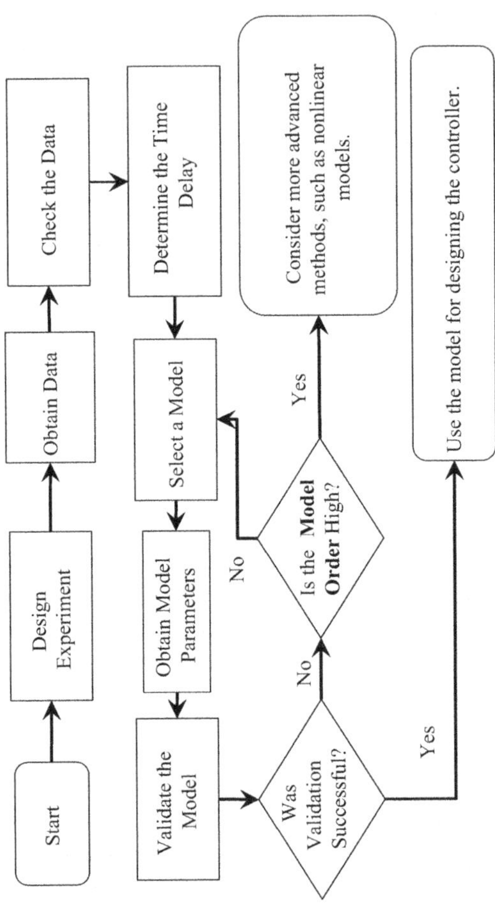

Fig. 6.3 System identification framework

6.3.1 *Parameter Estimation in Process Identification*

Although for simple models it is possible to estimate the parameters using least-squares linear regression (see, for example, Question 21 in Sect. 3.8.2), for more complex models this is not possible. Instead, more complex methods are required in order to obtain them. One very popular approach is the **prediction error method**. Parameter estimation using the prediction error method can be summarised as follows:

(1) Select an appropriate (prediction error) model and determine the corresponding one-step-ahead optimal predictors (Eq. 6.27).
(2) Using the experimental data, compute the prediction values and prediction errors as functions of the unknown parameters θ.
(3) Obtain the parameter estimates that minimise the sum of all of the prediction errors. Due to the nonlinear nature of the problem, this step is most often performed using a numerical optimisation algorithm.

In practice, this procedure is greatly simplified, since there exist appropriate computer functions that can perform the required tasks once the general form of the model (orders and time delay) is specified.

Theorem 6.3 (properties of the prediction error method) *The prediction error method produces parameter estimates that are **unbiased** if the prediction error is a white-noise signal.*

Proof This will be shown by examining the conditions under which the one-step-ahead predictor will give a white-noise signal.

From Eq. (6.12), the one-step-ahead prediction error can be written as

$$\varepsilon_{t+1|t}\left(z^{-1}, \vec{\theta}, \hat{\vec{\theta}}\right) = y_t - \hat{y}_{t+1|t} \tag{6.31}$$

where $\hat{y}_{t+1|t}$ is the one-step-ahead prediction based on the assumed model for the system, $\vec{\theta}$ is the (true) parameter vector, and $\hat{\vec{\theta}}$ is the estimated parameter vector. We know that

$$e_t = \hat{G}_l^{-1}\left(y_t - \hat{G}_p u_t\right) \tag{6.32}$$

and that the true model can be given as

$$y_t = G_p u_t + G_l e_t. \tag{6.33}$$

Substituting the above results into Corollary 6.1, the prediction error can be written as

$$\varepsilon_{t+1|t}\left(z^{-1}, \vec{\theta}, \hat{\vec{\theta}}\right) = \hat{G}_I^{-1}\left(G_p u_t + G_I e_t - \hat{G}_p u_t\right). \tag{6.34}$$

Combining the terms for input and error together gives

$$\varepsilon_{t+1|t}\left(z^{-1}, \vec{\theta}, \hat{\vec{\theta}}\right) = \hat{G}_I^{-1}\left(G_p - \hat{G}_p\right)u_t + \hat{G}_I^{-1} G_I e_t. \tag{6.35}$$

Let[2]

$$\Phi_u\left(z^{-1}, \vec{\theta}, \hat{\vec{\theta}}\right) = \hat{G}_I^{-1}\left(G_p - \hat{G}_p\right)$$
$$\Phi_e\left(z^{-1}, \vec{\theta}, \hat{\vec{\theta}}\right) = \hat{G}_I^{-1} G_I, \tag{6.36}$$

then Eq. (6.35) can be rewritten as

$$\varepsilon_{t+1|t}\left(z^{-1}, \vec{\theta}, \hat{\vec{\theta}}\right) = \Phi_u\left(z^{-1}, \vec{\theta}, \hat{\vec{\theta}}\right)u_t + \Phi_e\left(z^{-1}, \vec{\theta}, \hat{\vec{\theta}}\right)e_t. \tag{6.37}$$

Since it is desired to introduce white noise into the system, e_t will be added and subtracted from Eq. (6.37). This gives

$$\varepsilon_t\left(\vec{\theta}, \hat{\vec{\theta}}\right) = \Phi_u\left(z^{-1}, \vec{\theta}, \hat{\vec{\theta}}\right)u_t + \left(\Phi_e\left(z^{-1}, \vec{\theta}, \hat{\vec{\theta}}\right) - \mathcal{I}\right)e_t + e_t. \tag{6.38}$$

It should be noted that since this is assumed to be a sampled system, with zero-order hold, G_p will have at least one-sample time delay, that is,

$$G_p(0) = 0. \tag{6.39}$$

Furthermore, it will be assumed that

$$G_I(0) = \mathcal{I}. \tag{6.40}$$

The above two conclusions also hold for the estimated models. Thus, it can be seen that since both $\Phi_u\left(0, \vec{\theta}, \hat{\vec{\theta}}\right)$ and $\left(\Phi_e\left(0, \vec{\theta}, \hat{\vec{\theta}}\right) - \mathcal{I}\right)$ equal zero, then this implies that they both have at least a one-sample time delay. Furthermore, due to the one-sample time delay, there is no correlation between e_t and $\left(\Phi_e\left(0, \vec{\theta}, \hat{\vec{\theta}}\right) - \mathcal{I}\right)$, since the first term in Φ_e is e_{t-1}, with which white noise is by definition uncorrelated. Now, if it is further assumed that u_t and e_t are independent, then the variance of the prediction error can be written as

---

[2] The dependence of the Φ-functions on the backshift operator is made explicit in this formulation. The backshift operator has not been considered in any of the other transfer functions in order to keep the notation simple.

$$\mathrm{var}\left(\varepsilon_t\left(\vec{\theta},\hat{\vec{\theta}}\right)\right) = \mathrm{var}\left(\Phi_u\left(z^{-1},\vec{\theta},\hat{\vec{\theta}}\right)u_t\right) + \mathrm{var}\left(\left(\Phi_e\left(z^{-1},\vec{\theta},\hat{\vec{\theta}}\right) - \mathcal{I}\right)e_t\right) + \mathrm{var}(e_t).$$

$$(6.41)$$

The variance given by Eq. (6.41) is at least equal to, if not greater than, the variance of white noise, e_t. Since it is desired to minimise the covariance of the errors, this implies that by setting both Φ_u and $\Phi_e\left(z^{-1},\vec{\theta},\hat{\vec{\theta}}\right) - \mathcal{I}$ equal to zero, a minimum variance estimate can be obtained. Thus, it can be concluded that

$$\Phi_u\left(z^{-1},\vec{\theta},\hat{\vec{\theta}}\right) = \hat{G}_l^{-1}\left(G_p - \hat{G}_p\right) = 0$$

$$\Phi_e\left(z^{-1},\vec{\theta},\hat{\vec{\theta}}\right) = \hat{G}_l^{-1}G_l = \mathcal{I}. \qquad (6.42)$$

This implies that $G_p = \hat{G}_p$ and $G_l = \hat{G}_l$. Thus, this shows that the parameter estimates are unbiased.

Q.E.D.

Theorem 6.4 (asymptotic variance of the prediction error method) *The prediction error method is asymptotically a minimum variance estimator.*

Proof This will be shown by deriving the Fisher information matrix for the prediction error method.

First, define the sensitivity function, $\psi\left(t,\vec{\theta}\right)$, as

$$\psi\left(t,\vec{\theta}\right) = -\left[\frac{d\varepsilon_{t+1|t}\left(\vec{\theta}\right)}{d\vec{\theta}}\right]^T. \qquad (6.43)$$

This can be rewritten using Eq. (6.12) to give

$$\psi\left(t,\vec{\theta}\right) = \left[\frac{d\hat{y}_{t+1|t}}{d\vec{\theta}}\right]^T. \qquad (6.44)$$

It can be seen that the larger the sensitivity, and thus, the smaller the variance, the better the estimates will be. Note that the asymptotic variance for the parameters can be written as

$$\mathrm{cov}\left(\hat{\vec{\theta}}\right) = \sigma_w^2 \sum_{t=1}^{m} \psi\left(t,\hat{\vec{\theta}}\right)\psi\left(t,\hat{\vec{\theta}}\right)^T, \qquad (6.45)$$

where m is the number of data points used and σ_w^2 is the noise of the white noise. The Cramér-Rao lower bound for the parameter estimates states that

$$\text{cov}\left(\hat{\vec{\theta}}\right) \geq \mathcal{F}^{-1}, \tag{6.46}$$

where $\mathcal{F}$ is the Fisher information matrix, which is defined as

$$\mathcal{F} = \frac{1}{\sigma_w^2}\left[\sum_{t=1}^{m} \psi\left(t, \hat{\vec{\theta}}\right)\psi\left(t, \hat{\vec{\theta}}\right)^T\right]^{-1}. \tag{6.47}$$

Since Theorem 6.3 states that the prediction error method produces unbiased estimates, as $m \to \infty$, the estimated parameter values will approach the true parameter values. Thus, it can be concluded that the prediction error method **asymptotically** approaches a minimum variance estimator.

Q.E.D.

6.3.2 *Model Validation in Process Identification*

Once the model parameters have been determined, it is necessary to validate the model. As before, three different components need to be considered: (1) testing the residuals, (2) testing the adequacy of the model, and (3) taking corrective action. The general details of these components are the same as for regression analysis (see Sect. 3.3.5: Model Validation). However, some specific details are needed for model validation in process system identification.

The first system-identification-specific detail is that the goal of most such models is to predict future values. Therefore, the model validation tests are often performed on a separate set of data that was not used for model parameter estimation. This is one major difference from standard regression analysis where the same data set is used for both cases. This means that the data set is split into two parts: one part is used for model parameter estimation and the other part is used for model validation. In general, the model creation part will consist of 1/3 of the data, while the model validation part will consist of 2/3 of the data.

The second system-identification-specific detail is that testing of the residuals is commonly performed using the autocorrelation and cross-correlation tests rather than any other method. The **autocorrelation test of the residuals** seeks to determine if the residuals are white noise by plotting the autocorrelation function of the residuals for different lags (most often up to a lag of 20). If 95% of all the autocorrelations lie inside the 95% confidence interval for zero at all lags not equal to zero and there are no significant trends, then it can be concluded that the residuals are white noise. Otherwise, there is a need to take corrective action. Unfortunately, the autocorrelation plot does not provide a good indication of the source of any problems. The **cross-correlation test between the residuals and input** seeks to determine if the residuals and inputs are independent of each other. Theoretically speaking, due to the assumption of independence, the two signals should have a cross-correlation plot that is equal to zero for all lags. Practically, this can be stated as 95% of all cross-correlations should lie inside the 95% confidence interval for zero at all lags and there should not be any significant trends in the data. If this test fails, then the process model is likely to be incorrectly specified.

Testing model adequacy in process system identification is similar to the basic regression problem, for example, considering confidence intervals, comparing predicted and measured values, and using some type of index-based method. However, there are small differences. When comparing confidence intervals for the parameters, if there is a string of confidence intervals that cover zero at the end of the polynomial, then it is often the case that too large an order was selected for that polynomial. Similarly, if there are many confidence intervals that cover zero at the start of the B-polynomial, then this can be a sign that the time delay has been incorrectly specified. The time delay, n_k, should then be increased by the number of zero terms and the order of the B-polynomial decreased by the same amount. The net change will be zero, but the estimation can be made more precise.

When comparing predicted and measured values, the data set will often be different from that initially used in order to determine the adequacy of the model for forecasting new future points. Furthermore, different types of predictors can be used, for example, one-step- ahead, two-step- ahead, m-step- ahead, and infinite-horizon predictors. The infinite-horizon predictor is essentially a prediction of the process using only the deterministically available inputs. If the predictive capability of the model is good with the infinite-horizon predictor, then this implies that the model captures well most of the process behaviour in the given region. On the other hand, a poor performance with the infinite-horizon predictor can be a sign that additional information about the process may be required. This can be confirmed if low-order predictors, such as the two- or five-step-ahead predictors, give good performance. The reason for this is that the past errors can contain additional information about the process that can be useful in predicting the overall process.

It is possible to test model adequacy using various indices. These indices seek to take into consideration the trade-off between the overall model fit as measured using the variance of the residuals and the number of parameters used. The two most common indices are:

(1) **Akaike's Information Criterion (AIC)**: Akaike's information criterion seeks to find the global minimum between the variance and the number of parameters. It is defined as

$$\text{AIC} = m \log\left(m^{-1} \sum_{t=1}^{m} \varepsilon_{t+1|t}\left(\hat{\theta}\right)\varepsilon_{t+1|t}\left(\hat{\theta}\right)^{T}\right) + 2n, \qquad (6.48)$$

where n is the total number of estimated parameters in the model and m is the number of data points.

(2) **Final Prediction Error Criterion (FPE)**: The final prediction error criterion seeks to minimise the variance of the prediction errors with future data. It is defined as

$$\text{FPE} = m^{-1} \sum_{t=1}^{m} \varepsilon_{t+1|t}\left(\hat{\theta}\right) \varepsilon_{t+1|t}\left(\hat{\theta}\right)^{T}\left(\frac{1 + nm^{-1}}{1 - nm^{-1}}\right). \tag{6.49}$$

Although they can be useful for automating the model adequacy checking component, they still need to be combined with careful process knowledge in order to obtain a good final model.

In general, a model with fewer parameters is often better than a complex model with many different parameters, especially when trying to predict future process values. Analysis of complex macro-economic models has shown that these complex models can produce predictions that have variances greater than the original variables and hence are meaningless (Ashley 1988; Nelson 1972).

6.3.3 *Design of Experiments in Process Identification*

The final topic in open-loop process system identification is considering the design-of-experiments problem: under what conditions can the most information about the process be extracted from the system with minimal effort. As well, it would be useful to know the limitations on identifying the given model. Detailed information about this topic can be found in (Ljung 1999; Söderström et al. 1975). Practically, there are two topics to consider when designing a system identification experiment: theoretical constraints and practical design considerations.

Theoretically, the main concern lies with the **identifiability** of a process, that is, given a data set and model structure (order of polynomials), what are the conditions for there to be a unique solution to the parameter estimates. For open-loop experiments, the identifiability constraint for a prediction error model can be simply written as

$$n_r \geq n_a + n_b + n_c + n_d + n_f, \tag{6.50}$$

where n_r is the persistent excitation order of the input signal. **Persistency**, or the amount of information excited by a signal in the process, is defined as follows. A signal is said to have a persistent excitation order n, if the following two conditions are satisfied:

(1) The following limit exists:

$$\gamma(\tau) = \lim_{m \to \infty} \frac{1}{m} \sum_{t=1}^{m} u_{t+\tau} u_{t}^{T}. \tag{6.51}$$

(2) The matrix, $\Gamma_u(n)$, is positive definite, or, for a symmetric matrix, this is equivalent to saying that the matrix is invertible. The n-by-n matrix $\Gamma_u(n)$ is defined as

$$\Gamma_u(n) = \begin{bmatrix} \hat{\gamma}_u(0) & \hat{\gamma}_u(1) & \cdots & \hat{\gamma}_u(n-1) \\ \hat{\gamma}_u(-1) & \hat{\gamma}_u(0) & \cdots & \hat{\gamma}_u(n-2) \\ \vdots & & \ddots & \vdots \\ \hat{\gamma}_u(1-n) & \hat{\gamma}_u(2-n) & \cdots & \hat{\gamma}_u(0) \end{bmatrix}, \tag{6.52}$$

where $\hat{\gamma}$ is the estimated autocovariance of the signal.

Based on the result, it can be seen that a step input has a persistent excitation order of 1. An impulse response has a persistent excitation order of 0, while white-noise has a persistent excitation order of infinity. This would suggest that performing system identification with a white-noise signal would be ideal, as all prediction error model systems irrespective of complexity could be identified. However, practically speaking, a white-noise input signal is not useful since it contains too many random fluctuations in the values. These fluctuations would cause the actuator, such as a valve, to jump around, leading to potential mechanical stresses and equipment failure. Therefore, instead of using a white-noise signal, a **random binary signal**, which oscillates between two fixed levels (conventionally denoted as $+1$ and -1), is used. Such a signal approximates the white-noise signal and causes less mechanical stress on the system.

Designing the random binary signal requires setting the following parameters: levels, sampling time, and bandwidth. The physical values for the **levels** need to be selected carefully taking into consideration the actual system constraints, for example, overfilling a tank or leaving it empty, due to selecting too large or too small of a flow rate. Selecting a wide range between the two values can also lead to undesirable behaviour, such as exciting nonlinearities.

The **sampling time**, τ_s, represents how often the data from the system is collected. Both too fast and too slow sampling will have implications on the ability to obtain a good model. Fast sampling will force the poles of the model to approach -1, which means that the system will be difficult to identify accurately. Similarly, too slow sampling will mean that the relevant information will be lost about the process due to the Nyquist sampling theorem. The general rule for selecting the sampling time is (Zhu 2001)

$$\tau_s = 0.1\tau_{\min} \text{ to } 0.2\tau_{\min}, \tag{6.53}$$

where $\tau_{\min}$ is the smallest time constant in the process. The time constant, τ_p, of a process represents how quickly the process responds to a change in the system. The larger the time constant, the slower the response to changes.

Finally, the **bandwidth** needs to be selected. The bandwidth represents how much of the frequency domain is excited (or examined) by the given signal. Most chemical processes can be treated as low-pass filters, that is, only low frequencies are important for describing the process. Therefore, there is only a need to excite the process between $[0, f_{upper}]$, where f_{upper} is some upper bound frequency. For a first-order process, the bandwidth of a process is defined as the inverse of the time constant. For higher-order processes, the bandwidth of a process can be defined as the inverse of the smallest time constant. In order to be certain that all of the relevant bandwidth has been excited, a safety factor, k, is included. This safety factor is normally set to be equal to either 2 or 3, depending on the associated uncertainties in the initial process knowledge. If the process is known approximately, then it is possible to use a smaller k. Therefore, the bandwidth for the input signal can be defined as $\left[0, k\tau_{min}^{-1}\right]$, where, as before, τ_{min} is the smallest time constant.

From the above description, it would seem that, in order to identify the process, it is necessary to already know information about the process. In a way, this is indeed the case. However, the initial knowledge about the process need not be very precise and could be obtained using basic identification methods, such as the **step test**. The step test is a method for identifying a first-order, linear process based on making a step change in the input signal. The advantage of this approach is that it provides a quick and effective way of determining the process characteristics. The disadvantage is that the approach can only be used for simple processes that require/can be approximated with a single time constant. In practice, this is sufficient as a first approximation for most chemical engineering systems. Thus, this approach is perfect for providing the initial estimate for system identification.

The procedure for running a step test can be summarised as follows:

(1)   Once the process is at some steady state (all the process values are constant except for some minor variations), make a step change in the input signal.
(2)   Record the data until the process reaches a new steady state. The graph should look something similar to Fig. 6.4.
(3)   From the graph, compute the time delay, θ, the gain, K, and the time constant, τ.

6.3.4 Final Considerations in Open-Loop Process Identification

The above sections have provided a comprehensive view of the main issues in open-loop identification. However, there remain some final things to consider before this can be applied in practice. The most important things are time-delay estimation, drifting in the disturbance, linearity, and time invariance.

6.3.4.1 Time Delay

For both the prediction error and the linear, least-squares methods, the value of the time delay must be known beforehand, that is, neither of the methods can estimate the time delay as part of the regression problem. Estimating the time delay can be

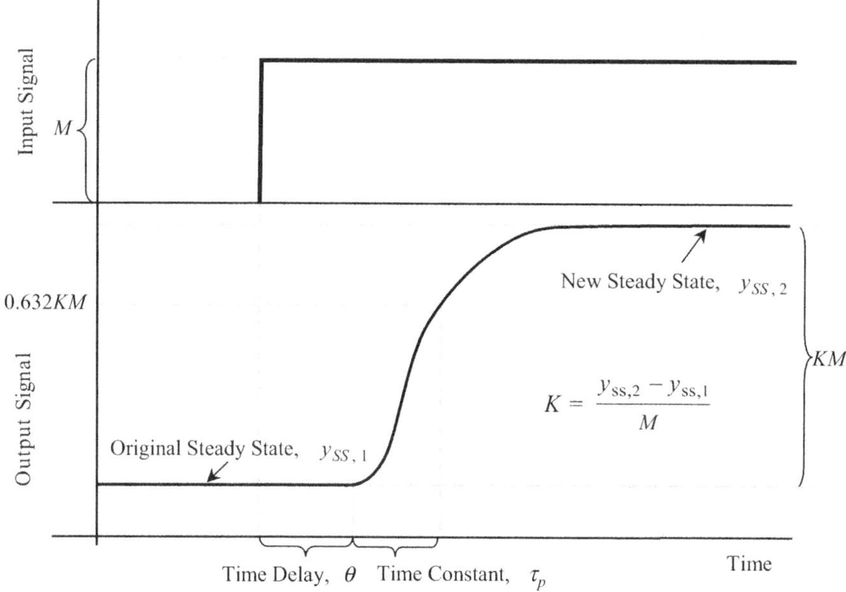

$$K = \frac{y_{ss,2} - y_{ss,1}}{M}$$

Fig. 6.4 Estimating parameters using a step test

performed using various different methods. The most common include using the values obtained from the step tests and the cross-correlation plot between the inputs and outputs.

Using step tests to estimate the time delay is a convenient and straightforward method. However, it should be noted that the time delay obtained is in continuous units that need to be converted into sampled units, using

$$n_k = \left\lfloor \frac{\theta}{\tau_s} \right\rfloor, \tag{6.54}$$

where $\lfloor \cdot \rfloor$ is the floor function that rounds down a value to the nearest integer.

Another approach to estimating the time delay is to use the cross-correlation plot between the input and the outputs. The delay will then appear as a series of zero values between a lag of zero and the time delay, n_k. A typical cross-correlation plot is shown in Fig. 6.5 (left). In this plot, the time delay would be estimated as being 4, since that is the last zero value before the significant peak. Note that using this approach requires that the input be a white-noise signal.

A related approach to estimating the time delay is to determine the impulse response coefficients, h, for the model. Similar to the cross-correlation plot, the last zero value would be assumed to be equal to the time delay. A typical impulse

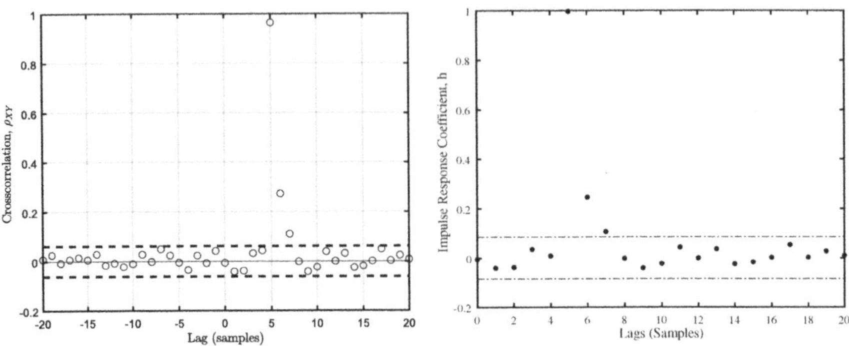

Fig. 6.5 Estimating the time delay using (left) The cross-correlation plot and (right) The impulse response method

response plot is shown in Fig. 6.5 (right). In this plot, the time delay would be estimated as being 4, since that is the last zero value before the significant peak. This method requires that the data be obtained from an open-loop experiment.

6.3.4.2   Drifting and Disturbances

When analysing a time series, the stationarity of the disturbance signal is an important characteristic to consider. If the output is not stationary, then all the data must be differenced in order to obtain a stationary model. If the data is differenced k times, then the disturbance model will be of the form

$$G_l = G_{l_d}\left(1 - z^{-1}\right)^{-k}, \tag{6.55}$$

where G_{l_d} is the disturbance model estimated using the differenced data. However, as for the univariate time-series case, it should be noted that differencing the data can lead to loss of information (excitation) in the input signal and an increase in noise. Thus, differencing the data should only be performed if no "reasonable" model can be obtained without differencing.

6.3.4.3   Linearity

When fitting a prediction error model to the data, it is assumed that the true model of the process is linear. Since very few chemical processes are truly linear, it is necessary to check the original process if a linear model is sufficient over the region of consideration of the variables. Two common tests are:

(1)   **Step-Up and Step-Down Check**: In this test, a step increase from the original conditions in the process is performed, followed by a step-down back to the

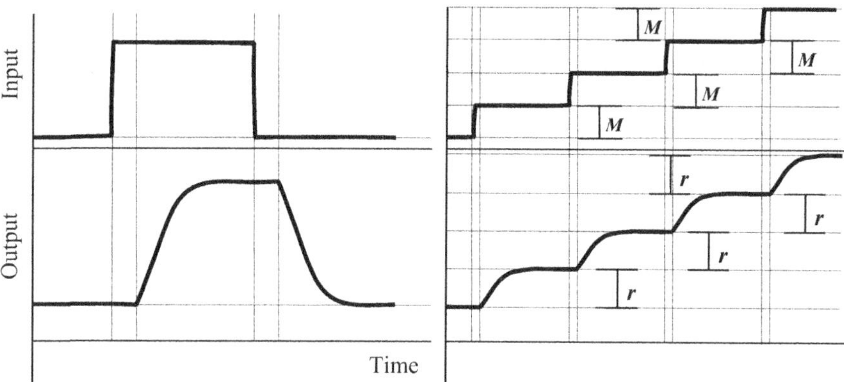

Fig. 6.6 (left) Ideal behaviour for the response for the step-up and step-down check and (right) Ideal behaviour for the response for the proportional test

original process conditions. The key process parameters, such as the time delay, gain, and process time constant, should be similar for the two responses. The ideal response is shown in Fig. 6.6 (left).

(2) **Proportional Test**: In this test, a set of step increases of magnitude M is performed. The response of the system should be the same during each step increase interval. The ideal response is shown in Fig. 6.6 (right).

6.3.4.4 Time Invariance

Finally, the prediction error model assumes that the parameter values do not change with respect to time, that is, they are time invariant. A quick and simple test of the invariance of the model is to split the data into two parts and cross-validate the models using the other data set. If both models perform successfully, then the parameters are probably time invariant, at least over the time interval considered.

6.4 Closed-Loop Process Identification

In chemical engineering, it is common to encounter cases where identifying a process using open-loop data may not be practical. Furthermore, it may be useful to extract process information from a closed-loop process without disrupting the overall process. In such cases, using and understanding closed-loop data are important. Closed-loop data comes in two flavours: routine operating data and externally excited data. In **routine operating data**, the reference signal is held constant and does not change its value during the course of the experiment. The only disturbances to the process come through the disturbance signal, e_t. On the other hand, in **externally excited data**, the value of the reference signal changes.

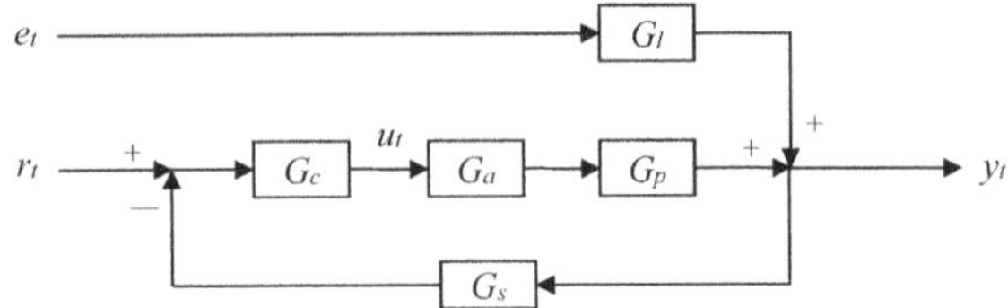

Fig. 6.7 Block diagram for a closed-loop process

The general closed-loop system is shown in Fig. 6.7, for which the transfer function can be written as

$$y_t = \frac{G_c G_p}{1 + G_c G_p} r_t + \frac{G_l}{1 + G_c G_p} e_t, \tag{6.56}$$

where G_c is the controller transfer function and r_t is the reference, or setpoint, signal. In general, closed-loop identification is more complicated than open-loop identification, since it cannot be assumed that the error and inputs are uncorrelated. For any closed-loop system, two competing equations can be fit

$$y_t = G_p u_t + G_l e_t, \tag{6.57}$$

$$y_t = r_t - G_c^{-1} u_t. \tag{6.58}$$

The primary issue with closed-loop identification is how to identify the desired process model G_p given the two competing equations.

If the data obtained are routine operating data, then the process model, G_p, can only be identified if the controller transfer function has a higher order than the process and the effect of an incorrect model on the controller transfer function is larger than on the disturbance model; or if there is significant nonlinearity in the controller and the error caused by an incorrect controller model is larger than the disturbance error.

On the other hand, if there is an external excitation, then it is easier to perform closed-loop identification. However, if the excitation is much weaker than the disturbance, the model given by Eq. (6.58) will be determined. Thus, the signal-to-noise ratio is extremely important in closed-loop identification. As well, identification depends on the model structure that has been determined for the process. If the model structure chosen for G_p and G_l is different from the structure of G_c, then the model can be identified even if the excitation is weak. Since most controllers do not have any sample time delays, if the structure chosen for G_p has at least one-sample time delay (as it should if it is a discrete system), then the closed-loop system can be easily identified even with a weak excitation.

If it is assumed that a prediction error model is being fit, then general conditions for identifiability based on the orders of the polynomials can be obtained. A process is identifiable from routine operating data if (Shardt and Huang 2011)

$$\max(n_X + n_k - n_F - n_A, n_Y - n_B) \geq n_D$$

$$+ \min(n_C + n_F + n_Y, n_A + n_F + n_Y, n_B + n_X), \tag{6.59}$$

where the controller is defined as

$$G_c = \frac{X(z^{-1})}{Y(z^{-1})}, \tag{6.60}$$

X and Y are polynomials similarly defined to the A-polynomial with order n_X and n_Y. A process is identifiable from a reference signal with persistent excitation order, n_r, if (Shardt and Huang 2014)

$$n_r \geq n_D + \min(n_C + n_F + n_Y, n_A + n_F + n_Y, n_B + n_X)$$
$$+ \min(n_F + n_A - n_X - n_k, n_B - n_Y). \tag{6.61}$$

There are 3 different approaches to determining the model structure of a closed-loop system: **indirect identification**, **direct identification**, and **joint input–output identification**.

6.4.1 Indirect Identification of a Closed-Loop Process

The first method for closed-loop identification is called **indirect identification**, where Eq. (6.56) is first fit as

$$y_t = M(z^{-1})r_t + W(z^{-1})e_t. \tag{6.62}$$

Then, given the controller transfer function, the process transfer function can be calculated as

$$G_p = \frac{1}{\frac{G_c}{M} - G_c}. \tag{6.63}$$

Since most processes have low-order dynamics and the overall transfer function, M, is likely to have a large order, plenty of cancellations must occur between G_c and M. However, many of these cancellations will not occur if the model estimates are even slightly off. Thus, there is a potential of creating a very large order model, even if it is not warranted. Furthermore, this method can only be used for identifying data obtained when the process has external excitation.

6.4.2 *Direct Identification of a Closed-Loop Process*

The second method is called **direct identification**, where the fact that the process is running in closed loop is ignored. In this type of identification, both the process and error structures must be simultaneously estimated. Thus, either a Box-Jenkins or a general prediction error model should be fit. Since this is one of the more common approaches to closed-loop system identification, it is necessary to examine the properties of this approach. It will be assumed that the prediction error method will be used.

Theorem 6.5 (properties of the prediction error method for closed-loop identification) *The prediction error method produces parameter estimates that are* **unbiased** *when the process is running in closed loop.*

Proof The proof will follow the same form as the open-loop proof.

Assume that G_p has at least one-sample time delay and the reference signal has sufficient persistent excitation. Since the direct identification method is the same as the open-loop identification method, the prediction error should be the same. Thus,

$$\varepsilon_{t+1|t}\left(z^{-1}, \vec{\theta}, \hat{\vec{\theta}}\right) = \hat{G}_l^{-1}\left(G_p u_t + G_l e_t - \hat{G}_p u_t\right). \tag{6.64}$$

From Figure 6.7, we can write the closed-loop relationship for u_t as

$$u_t = \frac{G_c}{1 + G_c G_p} r_t - \frac{G_l G_c}{1 + G_c G_p} e_t. \tag{6.65}$$

Substituting Eq. (6.65) into Eq. (6.64) and simplifying gives

$$\varepsilon_{t+1|t}\left(z^{-1}, \vec{\theta}, \hat{\vec{\theta}}\right) = \hat{G}_l^{-1}\left(G_p - \hat{G}_p\right) s G_c r_t + \frac{s G_l}{\hat{s}\hat{G}_l} e_t, \tag{6.66}$$

where s is the sensitivity function defined as

$$s = \frac{1}{1 + G_c G_p} \tag{6.67}$$

and $\hat{s}$ represents the estimated form of Eq. (6.67), that is, with G_p replaced by $\hat{G}_p$. As in the open-loop case, define

$$\Phi_r\left(z^{-1}, \vec{\theta}, \hat{\vec{\theta}}\right) = \hat{G}_l^{-1}\left(G_p - \hat{G}_p\right) s G_c$$
$$\Phi_e\left(z^{-1}, \vec{\theta}, \hat{\vec{\theta}}\right) = \hat{G}_l^{-1} G_l s \hat{s}^{-1}. \tag{6.68}$$

Thus, Eq. (6.66) can be rewritten as

$$\varepsilon_{t+1|t}\left(z^{-1},\vec{\theta},\hat{\vec{\theta}}\right) = \Phi_r\left(z^{-1},\vec{\theta},\hat{\vec{\theta}}\right)r_t + \Phi_e\left(z^{-1},\vec{\theta},\hat{\vec{\theta}}\right)e_t. \tag{6.69}$$

Similarly, a white-noise term will be introduced into Eq. (6.69) to give

$$\varepsilon_{t+1|t}\left(z^{-1},\vec{\theta},\hat{\vec{\theta}}\right) = \Phi_r\left(z^{-1},\vec{\theta},\hat{\vec{\theta}}\right)r_t + \left(\Phi_e\left(z^{-1},\vec{\theta},\hat{\vec{\theta}}\right) - \mathcal{I}\right)e_t + e_t. \tag{6.70}$$

Since it was assumed that there is at least one-sample time delay in G_p, it can be noted that $\Phi_r\left(0,\vec{\theta},\hat{\vec{\theta}}\right)$ equals zero. As well, $\Phi_e\left(0,\vec{\theta},\hat{\vec{\theta}}\right) - \mathcal{I}$ will equal zero. Thus, since both of the terms have at least a one-sample time delay, r_t and e_t are uncorrelated, and e_t and $\left(\Phi_e\left(z^{-1},\vec{\theta},\hat{\vec{\theta}}\right) - \mathcal{I}\right)e_t$ are uncorrelated, the variance can be written as

$$\mathrm{var}\left(\varepsilon_{t+1|t}\left(z^{-1},\vec{\theta},\hat{\vec{\theta}}\right)\right) = \mathrm{var}\left(\Phi_r\left(z^{-1},\vec{\theta},\hat{\vec{\theta}}\right)r_t\right) + \mathrm{var}\left(\left(\Phi_e\left(z^{-1},\vec{\theta},\hat{\vec{\theta}}\right) - \mathcal{I}\right)e_t\right) + \mathrm{var}(e_t),$$
$$\tag{6.71}$$

which must be equal to or greater than the variance of white noise. The minimum value will occur when both $\Phi_r\left(z^{-1},\vec{\theta},\hat{\vec{\theta}}\right)$ and $\left(\Phi_e\left(z^{-1},\vec{\theta},\hat{\vec{\theta}}\right) - \mathcal{I}\right)$ equal zero. In order for this to occur,

$$G_p = \hat{G}_p, \tag{6.72}$$

which implies that

$$s = \hat{s}. \tag{6.73}$$

Equation (6.73) and $\left(\Phi_e\left(z^{-1},\vec{\theta},\hat{\vec{\theta}}\right) - \mathcal{I}\right)$ equalling zero imply that

$$G_l = \hat{G}_l. \tag{6.74}$$

Thus, the parameter estimates are unbiased for the direct identification method.
   Q.E.D.

This implies that the prediction error method can be used to estimate the model parameters without taking into consideration the fact that the system is running in closed loop. Furthermore, the model of the controller is not required nor is any information about r_t needed. This implies that this approach works for both routine operating and externally excited data.

   However, when performing model validation using this approach, a few changes need to be made in the results due to correlation between the input, u_t, and the disturbance, e_t. This correlation implies that the input will be correlated with *past* values of the disturbance (and hence the residuals). Therefore, the conditions for the cross-correlation test, mentioned previously for the open-loop case, need to be changed to state "95% of all cross-correlations should lie inside the 95% confidence interval for zero for lags **greater than zero**."[3]

When using the direct approach to closed-loop identification for routine operating process data, it is important to note that due to the weak excitations present, the length of the data series is important for obtaining a good estimate of the parameters. For first-order models, about 2,000 data points are required (Shardt and Huang 2011). Furthermore, small, but consistent, changes in the overall disturbance model can render the identification of the process difficult.

6.4.3 *Joint Input–Output Identification of a Closed-Loop Process*

The third and final method for closed-loop process identification is called **the joint input–output identification method**, which uses all three signals, y_t, u_t, and r_t in order to identify a model of the system in a two-step procedure. In the first step, a model between r_t and u_t is fit to give

$$u_t = Q\!\left(z^{-1}\right)r_t + R\!\left(z^{-1}\right)e_t, \tag{6.75}$$

where

$$Q\!\left(z^{-1}\right) = \frac{G_c}{1 + G_p G_c}$$

$$R\!\left(z^{-1}\right) = -\frac{G_c G_l}{1 + G_p G_c}. \tag{6.76}$$

Then, the reference signal, r_t, is filtered using the resulting Q-polynomial to obtain an input signal, $\hat{u}_t$, uncorrelated with noise. The process model can be identified by fitting the following relationship:

$$y_t = G_p\!\left(z^{-1}\right)\hat{u}_t + P\!\left(z^{-1}\right)e_t, \tag{6.77}$$

where $P(z^{-1})$ is an arbitrary polynomial that is theoretically equal to $-R(z^{-1})$. The key advantage of this method is that there is no need to know any information about the controller. However, this approach does require that all three signals be available.

---

[3] Instead of the previous "95% of all cross-correlations should lie inside the 95% confidence interval for zero **at all lags**".

6.5 Nonlinear Process Identification

Although in many circumstances linear system identification can provide a sufficiently good model of the system for the intended purpose, it is occasionally necessary to consider nonlinear system identification.

Nonlinear system identification attempts to fit a nonlinear model to the given data. However, since there is a large number of potential nonlinear models that could be fit, nonlinear identification simplifies the available functions. Instead of choosing any arbitrary function, a **basis function**, $\kappa(x)$, is selected. The basis function can also be called the *generating function* or the *mother function*. Then, the goal becomes to fit the following model to the data:

$$y\left(\vec{\phi}\right) = \sum_{k=0}^{n} \alpha_k \kappa\left(\beta_k\left(\vec{\phi} - \gamma_k\right)\right), \tag{6.78}$$

where

$$\vec{\phi} = \left(y_{t-1}, y_{t-2}, \ldots, y_{t-n_y}, u_{t-n_k-1}, u_{t-n_k-2}, \ldots u_{t-n_k-n_u}\right), \tag{6.79}$$

where n_y is the number of past values of the output to be considered, n_u is the number of past values of the input to be considered, n_k is the time delay, n is the number of basis functions to be used, α is the coefficient, β is the dilation factor, and γ is the translation factor. Since this resembles an ARMAX model, this method is often referred to as a **nonlinear autoregressive exogenous model** (NLARX). Thus, in nonlinear system identification, the goal is to fit the three parameters, α, β, and γ, by minimising the prediction error using an appropriate nonlinear solver. In certain cases, the results obtained may not be the true global minimum, but rather a local one. Therefore, there is a need to carefully select both the function form and the solver method. A "good" nonlinear model should pass all of the standard system identification tests mentioned above for the linear case.

Common basis functions include:

(1) **Taylor Series Expansion**: $\kappa(x) = x^k$;

(2) **Fourier Transformation**: $\kappa(x) = \cos(x)$;

(3) **Piecewise Constant Function**: $\kappa(x) = \begin{cases} 1 & 0 \le x \le 1 \\ 0 & \text{otherwise} \end{cases}$;

(4) **Gaussian Function**: $\kappa(x) = \frac{1}{\sqrt{2\pi}}e^{-x^2/\sigma^2}$;

(5) **Sigmoid Function**: $\kappa(x) = \frac{1}{1+e^{-x}}$;

(6) **Wavelet Function**: $\kappa(x) = e^{-0.5x^2}$. This is only one example of many different possible basis functions using combinations of trigonometric functions and exponential functions.

It can be shown that for a sufficiently large value of n, for almost any choice of the basis function, except a polynomial basis, any reasonable nonlinear function can be approximated arbitrarily well.

6.5.1 Transformation of Nonlinear Models: Wiener–Hammerstein Models

Instead of fitting a fully nonlinear model, another approach to nonlinear system identification is to partition the nonlinearities from the linear component. A common application of this approach is the **Wiener–Hammerstein model**. A Wiener–Hammerstein model is a generalisation of the Hammerstein model, where nonlinearities are assumed only to be in the input, and the Wiener model, where nonlinearities are assumed only to be in the output, that allows nonlinearities to be present in both the input and output. The process model is assumed to be linear. Thus, the general form of the model can be written as

$$g(y_t) = \frac{B(z^{-1})}{F(z^{-1})} f(u_{t-k}) + \frac{C(z^{-1})}{D(z^{-1})} e_t, \qquad (6.80)$$

where $f(x)$ and $g(x)$ are functions of predetermined form. Identification would then proceed in the usual manner.

Wiener–Hammerstein models are useful when actuators or sensors have significant nonlinearities in their behaviour. A common application is when the valve, used as an actuator, is not behaving normally and has significant nonlinearities, such as stiction. By removing the nonlinearities from the modelling process, it becomes possible to convert the initially nonlinear problem into a linear one.

6.6 Modelling the Water Level in a Tank

Consider the task of developing models for the water level in the four tanks shown in Fig. 6.8. In this system, there are two inputs, u_1 and u_2, which represent the flow rate delivered by the two pumps. Each input is split into two and enters a bottom tank and a different top tank. Thus, input 1 enters tanks 1 and 3, while input 2 enters tanks 2 and 4. The amount of split is determined by the ratios γ_1 and γ_2. The height of water in each of the tanks can be monitored. For this experiment, the steady-state values are shown in Table 6.1.

The objective of this experiment is to determine an appropriate model for the water level in Tank 1 assuming that the splits are fixed, but the flow rate from the two pumps can vary. Design an appropriate experiment and analyse the results. Perform

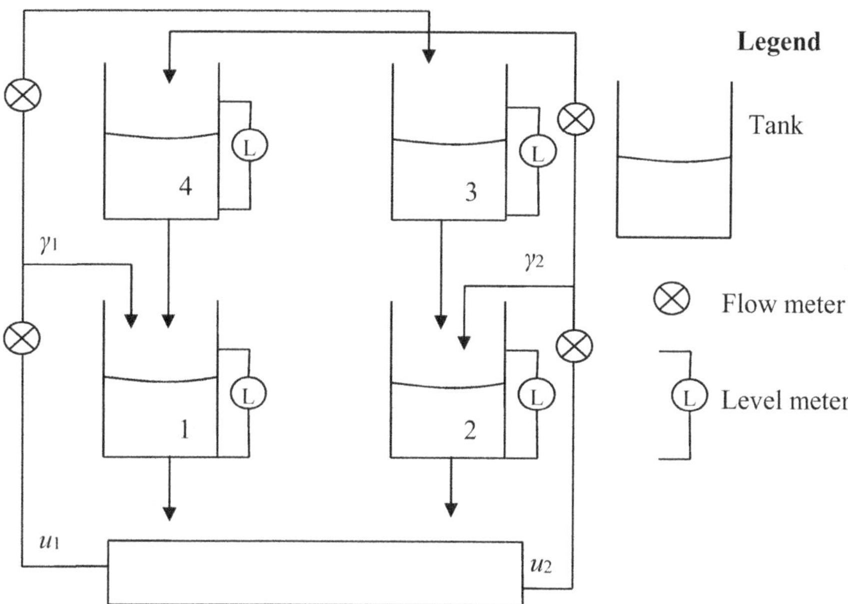

Fig. 6.8 Schematic of the four-tank system

Table 6.1 Steady-state parameter values for the system

Parameter	Left (1)	Right (2)
γ	0.2	0.3
u_{ss} (cm^3/s)	13	12
h_{ss} (cm)	18	24

both linear and nonlinear system identification and compare the resulting models. Which one would be preferred?

6.6.1 Design of Experiment

The design of the experiment can be split into two parts: preliminary identification using step tests and final identification using a random binary signal.

6.6.1.1 Preliminary Identification

In preliminary identification, the objective is to obtain a rough idea of how the system behaves under different conditions. In order to achieve this, a series of step tests will be performed on the system. Each pump will be tested separately at this point in

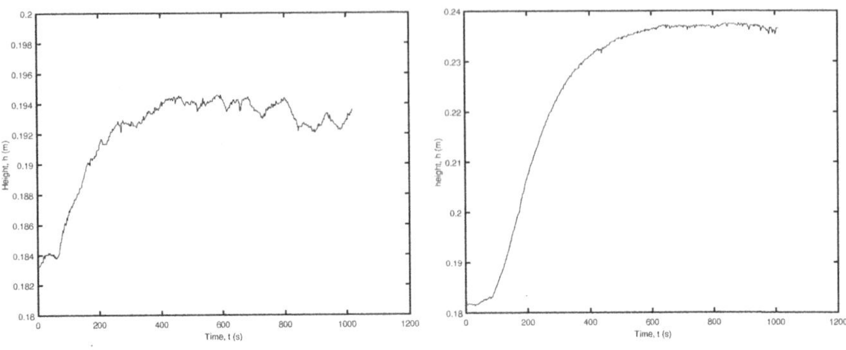

Fig. 6.9 Level in Tank 1: (left) Step change in u_1 and (right) Step change in u_2

Table 6.2 Summary of the values used to obtain the time constants, where τ_p is the time constant, h is the height, θ is the time delay, and t is the time. The subscript ss_1 refers to the initial steady-state values and ss_2 the final steady-state height. Subscripts b and c refer to specified time instants

	θ (s)	h_{ss_1} (m)	h_{ss_2} (m)	$h_c = 0.63(h_{ss_2} - h_{ss_1}) + h_{ss_1}$ (m)	$t_b = t(h_c)$ (s)	$\tau = t_b - \theta$ (s)
u_1 changed	62	0.184	0.194	0.190	162	100
u_2 changed	80	0.183	0.237	0.216	274	195

order to make the computations easier. For each pump, a step increase of $+2$ cm^3/s will be made. The resulting changes in the Tank 1 level are shown in Fig. 6.9. Table 6.2 shows the values obtained and the computation of the required time constants.

From Table 6.2, it can be seen that approximate time constants for the level in Tank 1 are 100 s for the left pump and 195 s for the right pump. This information will now be used to design an appropriate input signal.

6.6.1.2   Final Identification

Once a basic understanding of the system parameters has been obtained, the input signal can be designed. As mentioned before, the best input signal to consider is the random binary signal. In order to use this signal, three components must be selected: levels, sampling time, and bandwidth.

The levels should be selected to be symmetric about the nominal steady-state values and in most cases, less than the step changes previously made. It should be noted that selecting too large a level can lead to either the tanks overflowing or being empty during the course of the experiment. Since neither case is desirable, it is important to avoid such a situation. In this case, the level selected will be ±1.5 cm^3/s.

The sampling time is selected as some fraction of the smallest of the system time constants. Since the preliminary identification has shown that the time constants are

100 and 195 s, the base value is 100 s. The sampling time should then be between 10 and 20% of this value, or between 10 and 20 s. For the purposes of this experiment, it was assumed that the midpoint would be best, that is, the sampling time was selected to be 15 s. Note, it is very helpful when selecting the sampling time to make sure that the total experimental time can be divided by the sampling time to give no remainder.

Finally, it is necessary to select the process bandwidth. The Nyquist bandwidth is selected as the region $\left[0, \frac{k\tau_s}{\tau_p \pi}\right]$ with k equal to 2 and $\tau_p = 100$ s, the smallest time constant. The final signal was generated using the idinput command in MATLAB$^{\circledR}$

```
>>u=idinput(2*60*60/Ts,'rbs',[0,2*Ts/Tp/pi],[-1.5, 1.5])
```

It was assumed that the experiment was going to be run for 2 h $(= 2 \times 60 \times 60$ s). The values where then sent to the distributed control system (DCS) controlling the four-tank system. The data was recorded every second and the values were collected in MATLAB$^{\circledR}$.

6.6.2 Raw Data

The data obtained for both Tank 1 and Tank 2 levels is shown in Fig. 6.10. A total of 2 h of data were collected. It can be seen that there are no obvious issues with the data collected, such as missing values or abnormal values. In order to use the data, it is necessary to downsample the 1-s data to the desired sampling rate of 15 s. This can be accomplished by taking every 15th data point from the original data set for the new downsampled data vector. The downsampled data are presented in Sect. A6.1: Water Level in Tanks 1 and 2 Data.

For the purposes of modelling, the data set will be split into two parts: training and validation. The training set will consist of the first 2/3, while the validation set will consist of the remaining 1/3. Although this split is different from the suggested division, it will be used in order to obtain better initial parameter estimates. All modelling will be performed on Tank 1. The modelling of Tank 2 will be left as an exercise.

6.6.3 Linear Model Creation and Validation

The first step will be to create a linear model of the system and validate it. This procedure will be split into three steps: time-delay estimation, model creation, and model validation. The last two steps are iterative, in that if the model validation fails, a new model structure may be created and then fit. This procedure is repeated until a sufficiently good model is obtained. For the purpose of this section, the initial model

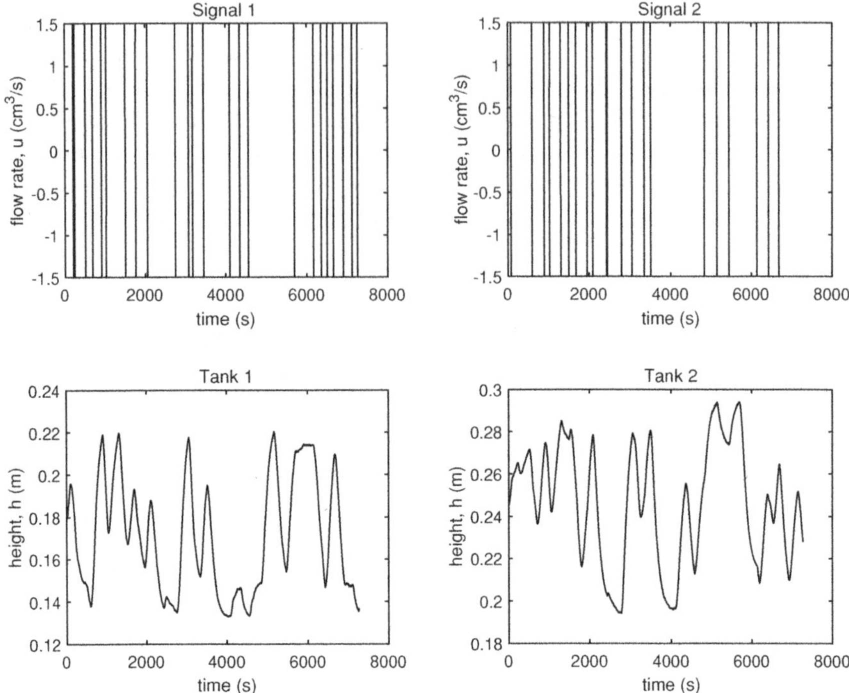

Fig. 6.10 The signals and heights as a function of time

and the final model will be presented, as well as any intermediate steps that present any special challenges.

6.6.3.1 Time-Delay Estimation

The normalized signal values, as well as the actual heights of the tanks, are shown in Fig. 6.10. In general, the data look quite good. Figure 6.11 shows the impulse response calculations between each of the inputs and outputs. This is the only approach that will work given the fact that the input was not white noise. The results suggest a time delay of zero for u_1 and 1 for u_2.

6.6.3.2 Initial Model

Unless there is additional information about the process, it is useful to always start with the simplest model and work one's way up. A good recommended initial guess is a first-order Box-Jenkins model and then, based on the fit, to advance to more complex models until the fit becomes good or the model order is too large.

For the initial, first-order Box-Jenkins model, the parameter estimates and their standard deviation are

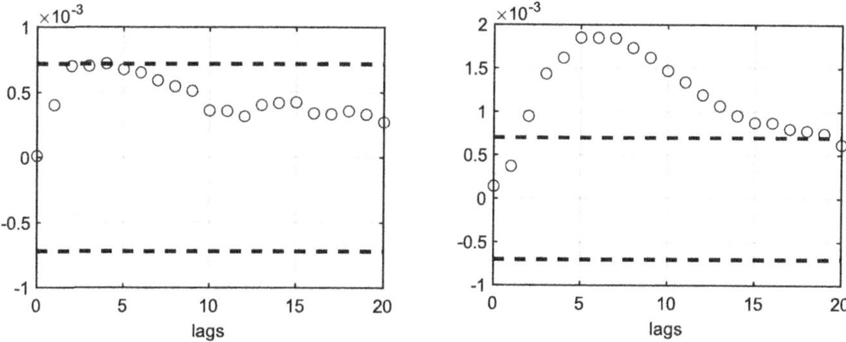

Fig. 6.11 Impulse responses for Tank 1 level (left) For u_1 and (right) For u_2

$$h_1 = \left[\frac{(5.7\times10^{-4}\pm7\times10^{-5})z^{-1}}{1-(0.91\pm0.03)z^{-1}} \quad \frac{(1.76\times10^{-3}\pm7\times10^{-5})z^{-2}}{1-(0.956\pm0.006)z^{-1}} \right] \begin{bmatrix} u_1 \\ u_2 \end{bmatrix}$$

$$+ \frac{1+(0.38\pm0.05)z^{-1}}{1-(0.99\pm0.01)z^{-1}} e_t. \tag{6.81}$$

The auto- and cross-correlation plots are shown in Fig. 6.12. A comparison between the predicted and actual levels is shown in Fig. 6.13. Both figures use the validation data set for fitting the model. From Fig. 6.12, it is clear that the residuals are not uncorrelated with each other or the inputs. Therefore, the initial model needs to be improved. Since there is a suggestion that the process model is incorrectly specified, it will be first changed. The best approach is to increase the order of the numerator and denominator (of the B- and F-polynomials) until either the cross-correlation plot shows the desired behaviour or the confidence intervals for the parameters cover zero. If the second case is reached, then this could be a suggestion that a linear model is insufficient/inappropriate for the given data set. Furthermore, the fit between the predicted and measured levels is not great (55.4%).

6.6.3.3 Final Model

After performing a series of iterations in increasing the model orders, the final model for the system can be written as

$$h_1 = \frac{\left(4.5\times10^{-4}\pm4\times10^{-5}\right)z^{-1} + \left(3.4\times10^{-4}\pm4\times10^{-5}\right)z^{-2}}{1-(0.85\pm0.01)z^{-1}} u_1$$

$$+ \frac{\left(7.8\times10^{-4}\pm2\times10^{-5}\right)z^{-2}}{1-(1.664\pm0.007)z^{-1}+(0.695\pm0.007)z^{-2}} u_2$$

$$+ \frac{1+(0.3\pm0.2)z^{-1}-(0.23\pm0.06)z^{-2}}{1-(1.3\pm0.2)z^{-1}+(0.4\pm0.2)z^{-2}} e_t. \tag{6.82}$$

The auto- and cross-correlation plots are shown in Fig. 6.14. A comparison between the predicted and actual levels is shown in Fig. 6.15. Both figures use the validation

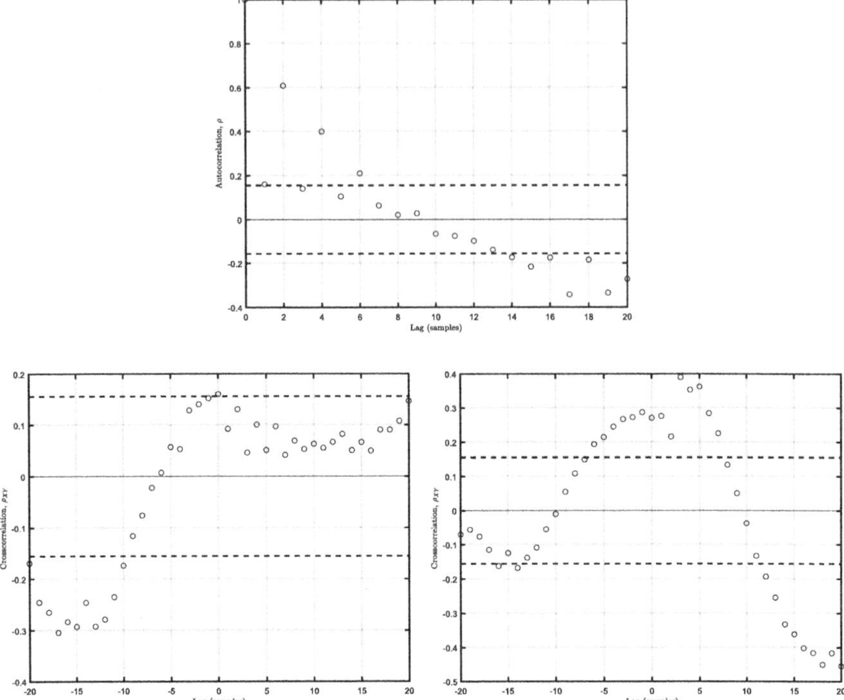

Fig. 6.12 (top) Autocorrelation plot for the residuals and (bottom) Cross-correlation plots between the inputs (left) u_1 and (right) u_2 and the residuals for the initial linear model

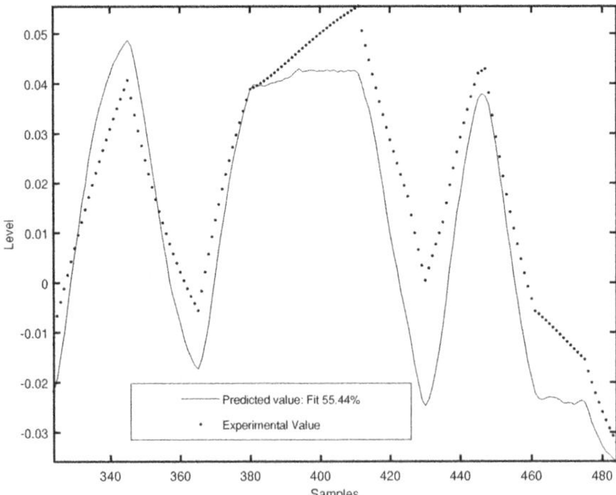

Fig. 6.13 Predicted and experimental tank levels for the initial linear model

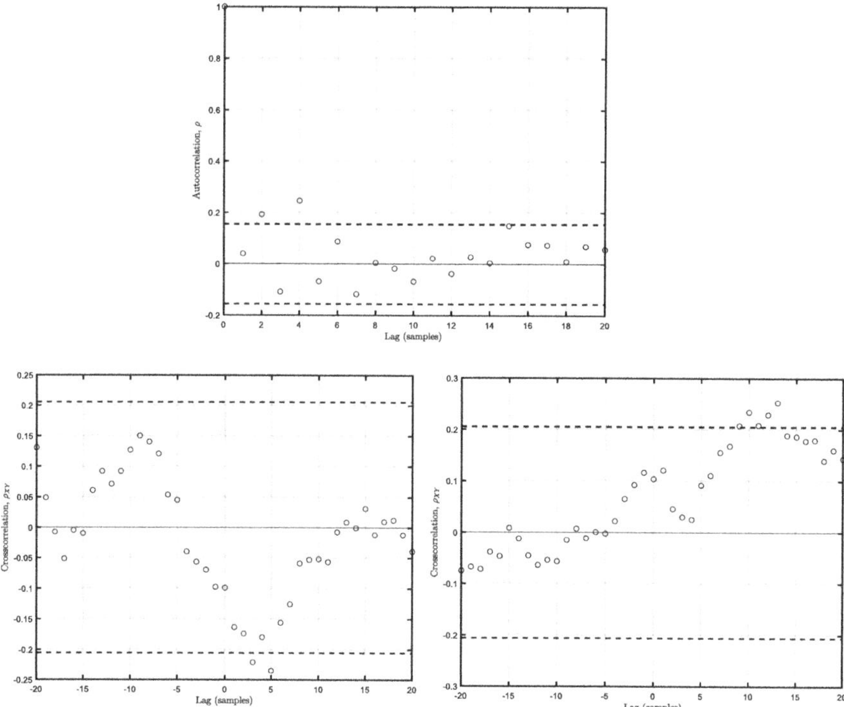

Fig. 6.14 (top) Autocorrelation plot for the residuals and (bottom) Cross-correlation plots between the inputs (left) u_1 and (right) u_2 and the residuals for the final linear model

Fig. 6.15 Predicted and experimental tank levels for the final linear model

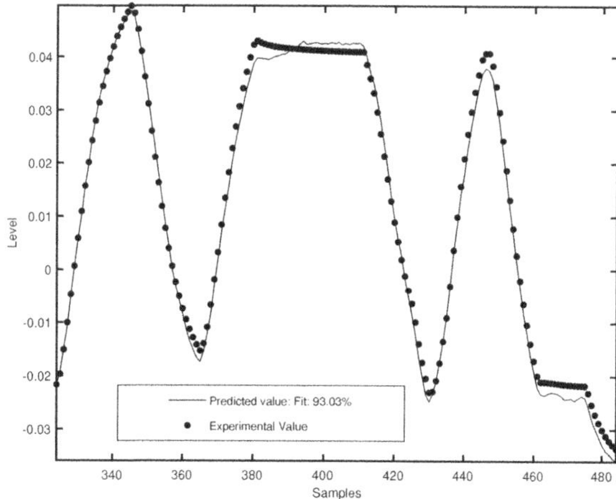

data set for fitting the model. The amount of deviation has now been significantly decreased from the initial model. Although some of the correlations values are still above the 95% confidence intervals, the values are closer to what should be expected. The fit for the data is an excellent 93.15%, which is not significantly improved by changing the model structure further. Therefore, this will be considered the final model.

6.6.4 Nonlinear Model Creation and Validation

Finally, a nonlinear model will be fit to the data to see if any improvement can be obtained. Using MATLAB®, the `wavenet` basis function will be used. This implies that the model will be given as

$$h_1 = (\vec{x} - \overline{x})\mathcal{P}\mathcal{L} + \sum_{k=1}^{n_w} \alpha_k g(\beta_k(\vec{x} - \overline{x})\mathcal{Q} - \beta_k \gamma_k), \tag{6.83}$$

where $\vec{x}$ is the vector of regressors; $\bar{x}$ is the mean value of the regressors; $\mathcal{P}$ and $\mathcal{Q}$ are projection matrices resulting from a principal component analysis of the estimation data; $\mathcal{L}$ is another projection matrix; α, β, and γ are the unknown parameters; $g(\vec{x})$ is the wavelet basis function

$$g(\vec{x}) = \left(n_r - \vec{x}\vec{x}^T\right)e^{-0.5\vec{x}\vec{x}^T}; \tag{6.84}$$

n_w is the number of wavelets to be used, and n_r is the number of regressors present in the model. Notice the complexity of the model and the number of parameters being estimated.

A wavelet model for the data consisted of the following regressors: y_{t-1}, y_{t-2}, y_{t-3}, and y_{t-4}, plus for each of the inputs the terms between $n_k + 1$ and $n_k + 5$. The total time delay was assumed to be 1 for the process between the u_1 and h_1 and 2 between the u_2 and h_1. Note that the time delay used here must include the one-sample time delay introduced by sampling a system, that is, the total time delay equals $n_k + 1$. The estimated model parameters will not be included here as they are quite complex and provide no real insight into the results. The number of wavelets was set at 6. The auto- and cross-correlation plots are shown in Fig. 6.16. A comparison between the predicted and actual levels is shown in Fig. 6.17. It can be seen that the fit has improved to 96.71%. However, this has come at the cost of a much more complex model whose physical understanding and computational requirements are much greater. This is often the problem encountered in system identification: the trade-off between the

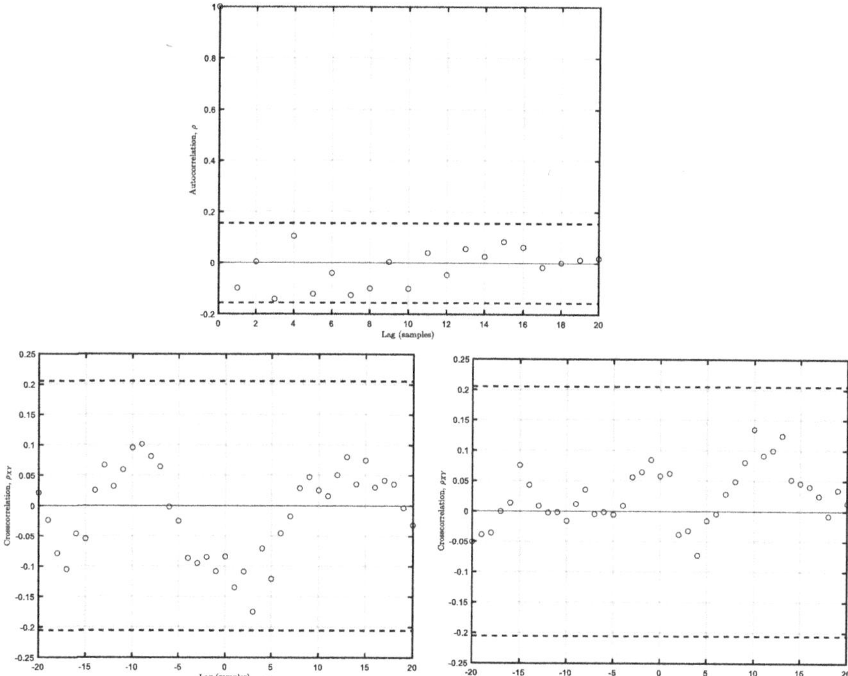

Fig. 6.16 (top) Autocorrelation plot for the residuals and (bottom) Cross-correlation plots between the inputs (left) u_1 and (right) u_2 and the residuals for the nonlinear model

model complexity and fit. Furthermore, it can be noted that the nonlinear approach requires the regressors to be estimated using some other method. This may not always be so easy to do, especially if the number of potential regressors is very large.

6.6.5 Final Comments

This brief example has shown some of the issues and concepts involved in the development of models for real systems. The procedure involved becomes more complicated as the complexity of the system increases, but the basic steps remain the same. As well, it can be seen that obtaining a good fit can require trying multiple different models and comparing the results. Although, in general, the fit will improve with increased model complexity, it does not always mean that such a model is better or more appropriate for the given application. There is always the need to compare the model obtained against the purpose for which the model will be used.

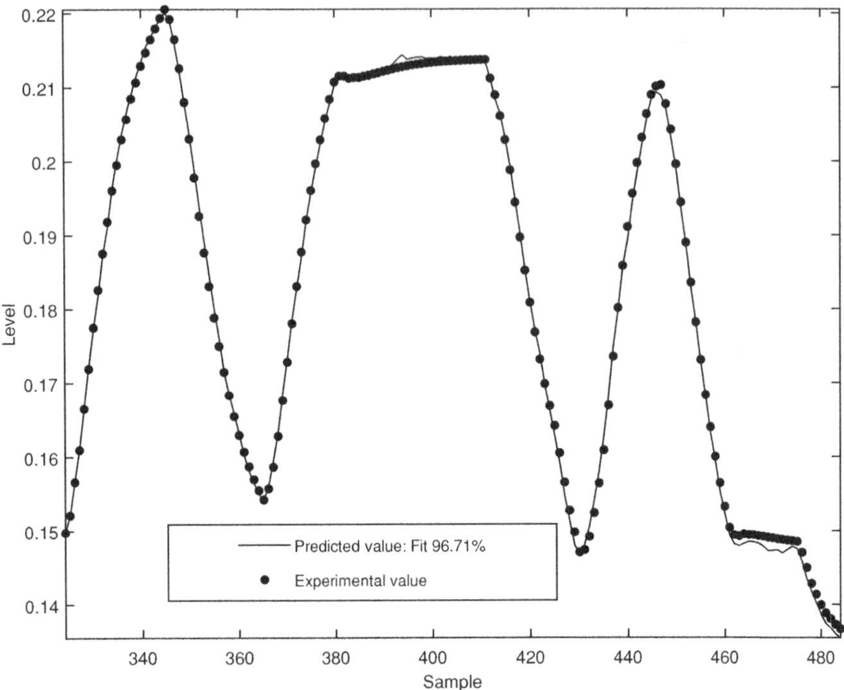

Fig. 6.17 Predicted and experimental tank level for the nonlinear model

6.7 Further Reading

The following are references that provide additional information about the topic:

(1) **General System Identification**:

    a.   Huang, B., & Kadali, R. (2008). *Dynamic Modeling, Predictive Control, and Performance Monitoring*. Springer-Verlag.

    b.   Ljung, L. (1999). *System Identification Theory for the User*. Upper Saddle River, New Jersey, United States of America: Prentice Hall, Inc.

    c.   Zhu, Y. (2001). *Multivariable system identification for process control*. Oxford, United Kingdom: Elsevier Science Ltd.

(2) **Properties of System Identification:**

    a.   Ashley, R. (1988). On the Relative Worth of Recent Macroeconomic Forecasts. *International Journal of Forecasting*, 4, 363–376.

    b.   Nelson, C. R. (1972). The Prediction Performance of the FRB-MIT-PENN Model of the U.S. Economy. *The American Economic Review*, 62(5), 902–917.

    c.   Shardt, Y. A. W., & Huang, B. (2011). Closed-loop identification with routine operating data: Effect of time delay and sampling time. *Journal of Process Control, 21,* 997–1010.

    d.   Shardt, Y. A. W., & Huang, B. (2014). Minimal Required Excitation for Closed-Loop Identification: Implications for PID Control Loops. *ADCONIP Conference Proceedings,* (pp. 296–301). Hiroshima, Japan.

    e.   Söderström, T., Gustavsson, I., & Ljung, L. (1975). Identifiability conditions for linear systems operating in closed loop. *International Journal of Control, 21*(2), 243–255.

6.8 Chapter Problems

Problems at the end of the chapter consist of three different types: (a) Basic Concepts (True/False), which seek to test the reader's comprehension of the key concepts in the chapter; (b) Short Exercises, which seek to test the reader's ability to compute the required parameters for a simple data set using simple or no technological aids. This section also includes proofs of theorems; and (c) Computational Exercises, which require not only a solid comprehension of the basic material, but also the use of appropriate software to easily manipulate the given data sets.

6.8.1 Basic Concepts

Determine if the following statements are true or false and state why this is the case.

(1)   Data-driven models can be used for arbitrary conditions and operating points.

(2)   Grey-box modelling combines the advantages of first-principle and data-driven models.

(3)   The controller, process, and disturbance models together create the plant model.

(4)   In the Box-Jenkins model, the A-polynomial has a fixed order of 3.

(5)   All prediction error models can be fit using standard, linear regression.

(6)   Only the one-step-ahead predictor has a variance equal to the white-noise variance.

(7)   The prediction error method provides consistent parameter estimates.

(8)   Many nonzero autocorrelation and cross-correlation values imply that the fit of the model is poor.

(9)   If the cross-correlation plot shows many nonzero correlations, then the likely problem is a misspecified process model.

(10)   In open-loop system identification, a first-order Box-Jenkins model (that is, all polynomials have order 1) requires a signal to have at least a persistent excitation of 4.

(11)   White noise has a persistent excitation order of exactly 1,000.

(12)   A random binary signal does not approximate a white-noise signal.

(13)   To specify a random binary signal, the physical values for the levels, the sampling time, and bandwidth are required.

(14)   A step test can provide information about high-order processes.

(15)   The time delay can be estimated using the autocorrelation plot.

(16)   All chemical processes are linear and time invariant.

(17)   In closed-loop identification, it is not necessary to accurately specify both the process and disturbance models.

(18)   Indirect identification of closed-loop processes requires that only the input and output signals be available.

(19)   A polynomial basis function can fit any nonlinear function arbitrarily well.

(20)   The Wiener transformation removes nonlinearities from the process output.

6.8.2 Short Exercises

These questions should be solved using only a simple, nonprogrammable, nongraphical calculator combined with pen and paper.

(21)   What is the three-step-ahead predictor and its variance? If $G_p = 5/(1 - 0.5z^{-1})$ and $G_l = 1/(1 - 0.25z^{-1})$, what is the three-step-ahead predictor?

(22)   Show that, for open-loop identification irrespective of the true plant model, an output-error model will provide an unbiased estimate of the process parameters. Provide a useful implication of this result.

(23)   Show that, for closed-loop identification irrespective of the true plant model, an output-error model will provide a biased estimate of the process parameters.

(24)   What is the time delay for the figures provided in Fig. 6.18? Assume open-loop conditions.

(25)   Comment on the validation figures shown in Fig. 6.19. Is the model adequate?

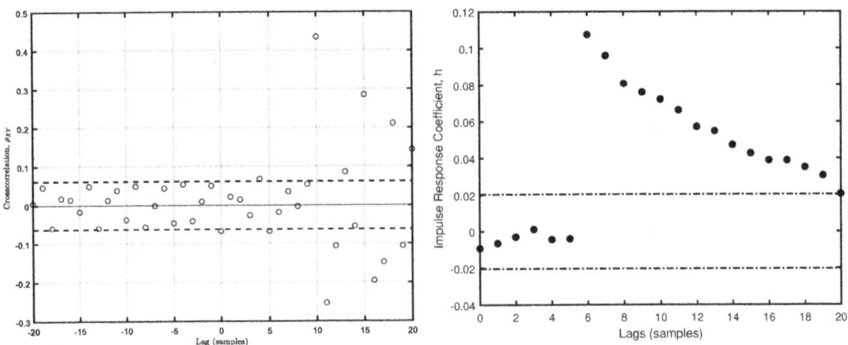

Fig. 6.18 Estimating time delay: (left) Cross-correlation plot and (right) Impulse response coefficients

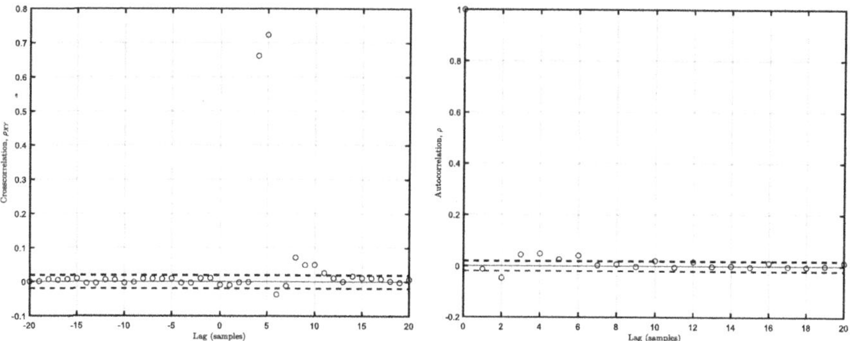

Fig. 6.19 Model validation for the open-loop case: (left) Cross-correlation between the input and the residuals and (right) Autocorrelation of the residuals

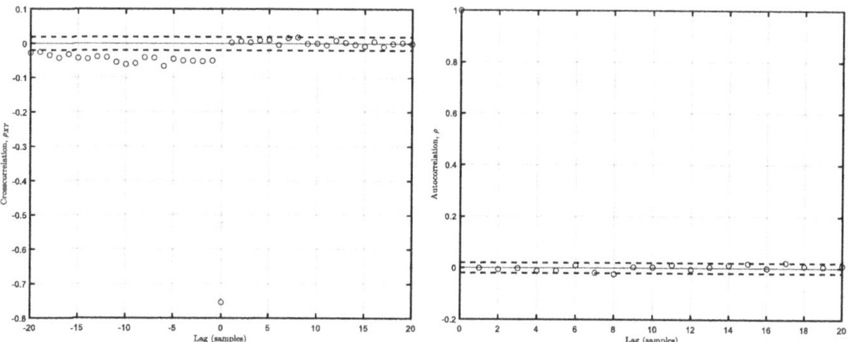

Fig. 6.20 Model validation for the closed-loop case: (left) Cross-correlation between the input and the residuals and (right) Autocorrelation of the residuals

Assume open-loop conditions.

(26)   Comment on the validation figures shown in Fig. 6.20. Is the model adequate? Assume closed-loop conditions.

6.8.3   Computational Exercises

The following problems should be solved with the help of a computer and appropriate software packages, such as MATLAB® *or* Excel®.

(27)   Take the Edmonton temperature series from Sect. A5.1: Edmonton Weather Data Series (1882–2002) and model the winter temperature as a function of the other available temperatures. Validate the model.

(28)   Model the height in Tank 2 using the data provided in Sect. A6.1: Water Level in Tanks 1 and 2 Data.

(29) Take a process that you are familiar with and design an experiment to identify
      the process. If possible perform the experiment and obtain an appropriate
      model of the system. Make sure to clearly explain the design of the experiment,
      model creation, and model validation used. Consider both linear and nonlinear
      modelling.

Appendix A6: Data Sets for This Chapter

This appendix gives detailed information about the data set used for analysis in this
chapter. All data can be downloaded as an Excel® spreadsheet or MATLAB® file
from the book website.

A6.1: Water Level in Tanks 1 and 2 Data

The water level in both Tanks 1 and 2, as well as the corresponding pump flow rates,
is presented in Table 6.3, in 15-s intervals.

Table 6.3 Water tank data set

Time (s)	Level (m)		Pump flow rate (cm^3/s)	
	Tank 1 (h_1)	Tank 2 (h_2)	Left pump (u_1)	Right pump (u_2)
1	0.1792	0.2456	14.5	13.5
16	0.1801	0.2462	14.5	13.5
31	0.1814	0.2480	14.5	13.5
46	0.1855	0.2507	14.5	13.5
61	0.1881	0.2538	14.5	13.5
76	0.1913	0.2565	14.5	10.5
91	0.1944	0.2579	14.5	10.5
106	0.1955	0.2588	14.5	10.5
121	0.1952	0.2593	14.5	10.5
136	0.1939	0.2603	14.5	10.5
151	0.1917	0.2611	14.5	10.5
166	0.1887	0.2617	14.5	10.5
181	0.1855	0.2626	14.5	10.5
196	0.1815	0.2639	14.5	10.5
211	0.1776	0.2649	11.5	10.5
226	0.1720	0.2654	11.5	10.5

(continued)

Table 6.3 (continued)

Time (s)	Level (m)		Pump flow rate (cm^3/s)	
	Tank 1 (h_1)	Tank 2 (h_2)	Left pump (u_1)	Right pump (u_2)
241	0.1672	0.2647	14.5	10.5
256	0.1635	0.2629	14.5	10.5
271	0.1611	0.2615	14.5	10.5
286	0.1592	0.2607	14.5	10.5
301	0.1583	0.2605	14.5	10.5
316	0.1566	0.2605	14.5	10.5
331	0.1552	0.2609	14.5	10.5
346	0.1536	0.2615	14.5	10.5
361	0.1524	0.2625	14.5	10.5
376	0.1511	0.2638	14.5	10.5
391	0.1505	0.2647	14.5	10.5
406	0.1500	0.2654	14.5	10.5
421	0.1494	0.2666	14.5	10.5
436	0.1489	0.2674	14.5	10.5
451	0.1495	0.2682	14.5	10.5
466	0.1489	0.2692	14.5	10.5
481	0.1486	0.2701	14.5	10.5
496	0.1480	0.2703	14.5	10.5
511	0.1477	0.2714	11.5	10.5
526	0.1457	0.2714	11.5	10.5
541	0.1438	0.2700	11.5	10.5
556	0.1419	0.2677	11.5	10.5
571	0.1404	0.2641	11.5	10.5
586	0.1393	0.2601	11.5	10.5
601	0.1378	0.2553	11.5	13.5
616	0.1385	0.2524	11.5	13.5
631	0.1415	0.2499	11.5	13.5
646	0.1451	0.2476	11.5	13.5
661	0.1501	0.2450	11.5	13.5
676	0.1550	0.2422	11.5	13.5
691	0.1612	0.2395	14.5	13.5
706	0.1671	0.2372	14.5	13.5
721	0.1748	0.2365	14.5	13.5
736	0.1814	0.2373	14.5	13.5
751	0.1873	0.2394	14.5	13.5
766	0.1923	0.2421	14.5	13.5

(continued)

Table 6.3 (continued)

Time (s)	Level (m)		Pump flow rate (cm^3/s)	
	Tank 1 (h_1)	Tank 2 (h_2)	Left pump (u_1)	Right pump (u_2)
781	0.1966	0.2457	14.5	13.5
796	0.2001	0.2492	14.5	13.5
811	0.2039	0.2535	14.5	13.5
826	0.2069	0.2570	14.5	13.5
841	0.2095	0.2611	14.5	13.5
856	0.2120	0.2647	14.5	13.5
871	0.2141	0.2681	14.5	13.5
886	0.2164	0.2712	14.5	13.5
901	0.2179	0.2736	14.5	10.5
916	0.2189	0.2747	11.5	10.5
931	0.2162	0.2740	11.5	10.5
946	0.2115	0.2725	11.5	10.5
961	0.2060	0.2692	11.5	10.5
976	0.2003	0.2657	11.5	10.5
991	0.1936	0.2607	11.5	10.5
1,006	0.1880	0.2562	11.5	10.5
1,021	0.1816	0.2504	11.5	10.5
1,036	0.1765	0.2461	14.5	13.5
1,051	0.1732	0.2427	14.5	13.5
1,066	0.1730	0.2422	14.5	13.5
1,081	0.1742	0.2430	14.5	13.5
1,096	0.1770	0.2449	14.5	13.5
1,111	0.1812	0.2480	14.5	13.5
1,126	0.1848	0.2507	14.5	13.5
1,141	0.1892	0.2541	14.5	13.5
1,156	0.1930	0.2572	14.5	13.5
1,171	0.1973	0.2604	14.5	13.5
1,186	0.1999	0.2636	14.5	13.5
1,201	0.2027	0.2672	14.5	13.5
1,216	0.2053	0.2701	14.5	13.5
1,231	0.2075	0.2731	14.5	13.5
1,246	0.2105	0.2759	14.5	13.5
1,261	0.2126	0.2790	14.5	13.5
1,276	0.2147	0.2808	14.5	13.5
1,291	0.2161	0.2832	14.5	13.5
1,306	0.2183	0.2847	14.5	10.5

(continued)

Table 6.3 (continued)

Time (s)	Level (m)		Pump flow rate (cm^3/s)	
	Tank 1 (h_1)	Tank 2 (h_2)	Left pump (u_1)	Right pump (u_2)
1,321	0.2196	0.2844	14.5	10.5
1,336	0.2190	0.2834	14.5	10.5
1,351	0.2168	0.2820	14.5	10.5
1,366	0.2126	0.2807	14.5	10.5
1,381	0.2080	0.2800	14.5	10.5
1,396	0.2027	0.2794	14.5	10.5
1,411	0.1979	0.2789	14.5	10.5
1,426	0.1924	0.2784	14.5	10.5
1,441	0.1879	0.2778	14.5	10.5
1,456	0.1833	0.2770	14.5	10.5
1,471	0.1793	0.2770	14.5	10.5
1,486	0.1746	0.2769	14.5	10.5
1,501	0.1715	0.2768	14.5	13.5
1,516	0.1691	0.2784	11.5	13.5
1,531	0.1678	0.2803	11.5	13.5
1,546	0.1671	0.2810	11.5	13.5
1,561	0.1682	0.2799	11.5	13.5
1,576	0.1705	0.2778	11.5	13.5
1,591	0.1728	0.2744	11.5	13.5
1,606	0.1759	0.2710	11.5	13.5
1,621	0.1789	0.2668	11.5	13.5
1,636	0.1821	0.2627	11.5	13.5
1,651	0.1851	0.2583	11.5	13.5
1,666	0.1884	0.2543	11.5	13.5
1,681	0.1909	0.2499	11.5	10.5
1,696	0.1929	0.2447	11.5	10.5
1,711	0.1929	0.2381	11.5	10.5
1,726	0.1907	0.2332	11.5	10.5
1,741	0.1880	0.2278	11.5	10.5
1,756	0.1843	0.2240	11.5	10.5
1,771	0.1807	0.2199	14.5	10.5
1,786	0.1775	0.2173	14.5	10.5
1,801	0.1759	0.2161	14.5	10.5
1,816	0.1739	0.2170	14.5	10.5
1,831	0.1719	0.2191	14.5	10.5
1,846	0.1691	0.2217	14.5	10.5

(continued)

Table 6.3 (continued)

Time (s)	Level (m)		Pump flow rate (cm^3/s)	
	Tank 1 (h_1)	Tank 2 (h_2)	Left pump (u_1)	Right pump (u_2)
1,861	0.1671	0.2249	14.5	10.5
1,876	0.1641	0.2284	14.5	10.5
1,891	0.1625	0.2321	14.5	10.5
1,906	0.1605	0.2356	14.5	10.5
1,921	0.1592	0.2390	14.5	10.5
1,936	0.1578	0.2426	14.5	10.5
1,951	0.1563	0.2458	14.5	13.5
1,966	0.1568	0.2507	14.5	13.5
1,981	0.1588	0.2556	14.5	13.5
1,996	0.1623	0.2608	14.5	13.5
2,011	0.1662	0.2649	14.5	13.5
2,026	0.1709	0.2687	14.5	13.5
2,041	0.1752	0.2721	14.5	13.5
2,056	0.1805	0.2751	11.5	13.5
2,071	0.1834	0.2775	11.5	13.5
2,086	0.1857	0.2784	11.5	10.5
2,101	0.1878	0.2759	11.5	10.5
2,116	0.1878	0.2709	11.5	10.5
2,131	0.1866	0.2660	11.5	10.5
2,146	0.1836	0.2593	11.5	10.5
2,161	0.1805	0.2543	11.5	10.5
2,176	0.1772	0.2483	11.5	10.5
2,191	0.1738	0.2435	11.5	10.5
2,206	0.1697	0.2379	11.5	10.5
2,221	0.1663	0.2333	11.5	10.5
2,236	0.1627	0.2286	11.5	10.5
2,251	0.1588	0.2251	11.5	10.5
2,266	0.1554	0.2214	11.5	10.5
2,281	0.1533	0.2186	11.5	10.5
2,296	0.1503	0.2155	11.5	10.5
2,311	0.1479	0.2134	11.5	10.5
2,326	0.1457	0.2108	11.5	10.5
2,341	0.1440	0.2091	11.5	10.5
2,356	0.1422	0.2075	11.5	10.5
2,371	0.1407	0.2062	11.5	10.5
2,386	0.1397	0.2047	11.5	10.5

(continued)

Table 6.3 (continued)

Time (s)	Level (m)		Pump flow rate (cm^3/s)	
	Tank 1 (h_1)	Tank 2 (h_2)	Left pump (u_1)	Right pump (u_2)
2,401	0.1387	0.2036	11.5	10.5
2,416	0.1377	0.2027	11.5	10.5
2,431	0.1371	0.2025	11.5	13.5
2,446	0.1378	0.2033	11.5	10.5
2,461	0.1398	0.2036	11.5	10.5
2,476	0.1415	0.2025	11.5	10.5
2,491	0.1421	0.2019	11.5	10.5
2,506	0.1419	0.2009	11.5	10.5
2,521	0.1418	0.2004	11.5	10.5
2,536	0.1412	0.1999	11.5	10.5
2,551	0.1403	0.1993	11.5	10.5
2,566	0.1397	0.1990	11.5	10.5
2,581	0.1394	0.1983	11.5	10.5
2,596	0.1388	0.1975	11.5	10.5
2,611	0.1388	0.1972	11.5	10.5
2,626	0.1384	0.1967	11.5	10.5
2,641	0.1381	0.1966	11.5	10.5
2,656	0.1379	0.1958	11.5	10.5
2,671	0.1376	0.1953	11.5	10.5
2,686	0.1366	0.1951	11.5	10.5
2,701	0.1358	0.1947	11.5	10.5
2,716	0.1361	0.1946	11.5	10.5
2,731	0.1355	0.1946	11.5	10.5
2,746	0.1353	0.1946	11.5	10.5
2,761	0.1352	0.1945	14.5	10.5
2,776	0.1361	0.1949	14.5	10.5
2,791	0.1378	0.1971	14.5	13.5
2,806	0.1416	0.2022	14.5	13.5
2,821	0.1457	0.2086	14.5	13.5
2,836	0.1517	0.2156	14.5	13.5
2,851	0.1569	0.2223	14.5	13.5
2,866	0.1635	0.2297	14.5	13.5
2,881	0.1692	0.2355	14.5	13.5
2,896	0.1755	0.2417	14.5	13.5
2,911	0.1808	0.2471	14.5	13.5
2,926	0.1866	0.2524	14.5	13.5

(continued)

Table 6.3 (continued)

Time (s)	Level (m)		Pump flow rate (cm^3/s)	
	Tank 1 (h_1)	Tank 2 (h_2)	Left pump (u_1)	Right pump (u_2)
2,941	0.1906	0.2568	14.5	13.5
2,956	0.1960	0.2615	14.5	13.5
2,971	0.2002	0.2648	14.5	13.5
2,986	0.2044	0.2687	14.5	13.5
3,001	0.2074	0.2715	14.5	13.5
3,016	0.2104	0.2746	14.5	13.5
3,031	0.2127	0.2766	14.5	13.5
3,046	0.2155	0.2790	14.5	10.5
3,061	0.2171	0.2790	14.5	10.5
3,076	0.2173	0.2780	14.5	10.5
3,091	0.2145	0.2771	11.5	10.5
3,106	0.2099	0.2759	11.5	10.5
3,121	0.2028	0.2734	11.5	10.5
3,136	0.1968	0.2703	11.5	10.5
3,151	0.1898	0.2663	11.5	10.5
3,166	0.1841	0.2610	11.5	10.5
3,181	0.1775	0.2563	11.5	10.5
3,196	0.1722	0.2506	14.5	10.5
3,211	0.1682	0.2456	14.5	10.5
3,226	0.1660	0.2423	14.5	10.5
3,241	0.1638	0.2402	14.5	10.5
3,256	0.1612	0.2400	14.5	10.5
3,271	0.1587	0.2406	14.5	10.5
3,286	0.1566	0.2421	14.5	10.5
3,301	0.1548	0.2438	14.5	10.5
3,316	0.1533	0.2458	14.5	10.5
3,331	0.1529	0.2481	14.5	10.5
3,346	0.1518	0.2502	14.5	13.5
3,361	0.1522	0.2539	14.5	13.5
3,376	0.1548	0.2588	14.5	13.5
3,391	0.1585	0.2630	14.5	13.5
3,406	0.1624	0.2674	14.5	13.5
3,421	0.1682	0.2707	14.5	13.5
3,436	0.1733	0.2737	14.5	13.5
3,451	0.1781	0.2761	14.5	13.5
3,466	0.1834	0.2788	11.5	13.5

(continued)

Table 6.3 (continued)

Time (s)	Level (m)		Pump flow rate (cm^3/s)	
	Tank 1 (h_1)	Tank 2 (h_2)	Left pump (u_1)	Right pump (u_2)
3,481	0.1871	0.2801	11.5	13.5
3,496	0.1900	0.2801	11.5	13.5
3,511	0.1927	0.2787	11.5	10.5
3,526	0.1946	0.2738	11.5	10.5
3,541	0.1941	0.2676	11.5	10.5
3,556	0.1920	0.2609	11.5	10.5
3,571	0.1886	0.2548	11.5	10.5
3,586	0.1848	0.2486	11.5	10.5
3,601	0.1802	0.2430	11.5	10.5
3,616	0.1758	0.2373	11.5	10.5
3,631	0.1712	0.2326	11.5	10.5
3,646	0.1672	0.2276	11.5	10.5
3,661	0.1629	0.2240	11.5	10.5
3,676	0.1592	0.2201	11.5	10.5
3,691	0.1555	0.2171	11.5	10.5
3,706	0.1525	0.2143	11.5	10.5
3,721	0.1496	0.2122	11.5	10.5
3,736	0.1472	0.2095	11.5	10.5
3,751	0.1446	0.2078	11.5	10.5
3,766	0.1428	0.2060	11.5	10.5
3,781	0.1412	0.2053	11.5	10.5
3,796	0.1400	0.2034	11.5	10.5
3,811	0.1389	0.2018	11.5	10.5
3,826	0.1381	0.2015	11.5	10.5
3,841	0.1371	0.2009	11.5	10.5
3,856	0.1365	0.2002	11.5	10.5
3,871	0.1357	0.1994	11.5	10.5
3,886	0.1353	0.1991	11.5	10.5
3,901	0.1348	0.1985	11.5	10.5
3,916	0.1343	0.1984	11.5	10.5
3,931	0.1339	0.1981	11.5	10.5
3,946	0.1337	0.1972	11.5	10.5
3,961	0.1337	0.1969	11.5	10.5
3,976	0.1337	0.1965	11.5	10.5
3,991	0.1336	0.1965	11.5	10.5
4,006	0.1332	0.1964	11.5	10.5

(continued)

Table 6.3 (continued)

Time (s)	Level (m)		Pump flow rate (cm^3/s)	
	Tank 1 (h_1)	Tank 2 (h_2)	Left pump (u_1)	Right pump (u_2)
4,021	0.1331	0.1966	11.5	10.5
4,036	0.1332	0.1962	11.5	10.5
4,051	0.1332	0.1959	11.5	10.5
4,066	0.1332	0.1966	11.5	10.5
4,081	0.1332	0.1967	11.5	10.5
4,096	0.1335	0.1967	14.5	10.5
4,111	0.1355	0.1967	14.5	10.5
4,126	0.1376	0.1977	14.5	10.5
4,141	0.1401	0.2009	14.5	10.5
4,156	0.1412	0.2044	14.5	10.5
4,171	0.1417	0.2088	14.5	10.5
4,186	0.1425	0.2141	14.5	10.5
4,201	0.1429	0.2187	14.5	10.5
4,216	0.1438	0.2232	14.5	10.5
4,231	0.1446	0.2280	14.5	10.5
4,246	0.1455	0.2328	14.5	10.5
4,261	0.1459	0.2361	14.5	10.5
4,276	0.1459	0.2405	14.5	10.5
4,291	0.1459	0.2437	14.5	10.5
4,306	0.1459	0.2471	14.5	10.5
4,321	0.1465	0.2498	14.5	10.5
4,336	0.1463	0.2527	11.5	10.5
4,351	0.1452	0.2548	11.5	10.5
4,366	0.1431	0.2550	11.5	10.5
4,381	0.1409	0.2544	11.5	10.5
4,396	0.1395	0.2523	11.5	10.5
4,411	0.1377	0.2498	11.5	10.5
4,426	0.1374	0.2457	11.5	10.5
4,441	0.1364	0.2423	11.5	10.5
4,456	0.1356	0.2376	11.5	10.5
4,471	0.1352	0.2341	11.5	10.5
4,486	0.1347	0.2296	11.5	10.5
4,501	0.1341	0.2260	11.5	10.5
4,516	0.1338	0.2221	11.5	10.5
4,531	0.1337	0.2190	11.5	10.5
4,546	0.1333	0.2163	14.5	10.5

(continued)

Table 6.3 (continued)

Time (s)	Level (m)		Pump flow rate (cm^3/s)	
	Tank 1 (h_1)	Tank 2 (h_2)	Left pump (u_1)	Right pump (u_2)
4,561	0.1345	0.2141	14.5	10.5
4,576	0.1358	0.2132	14.5	10.5
4,591	0.1384	0.2144	14.5	10.5
4,606	0.1399	0.2166	14.5	10.5
4,621	0.1411	0.2192	14.5	10.5
4,636	0.1416	0.2227	14.5	10.5
4,651	0.1432	0.2264	14.5	10.5
4,666	0.1444	0.2304	14.5	10.5
4,681	0.1452	0.2338	14.5	10.5
4,696	0.1459	0.2371	14.5	10.5
4,711	0.1467	0.2402	14.5	10.5
4,726	0.1470	0.2437	14.5	10.5
4,741	0.1473	0.2463	14.5	10.5
4,756	0.1479	0.2490	14.5	10.5
4,771	0.1481	0.2518	14.5	10.5
4,786	0.1483	0.2536	14.5	10.5
4,801	0.1486	0.2554	14.5	10.5
4,816	0.1488	0.2571	14.5	10.5
4,831	0.1489	0.2593	14.5	13.5
4,846	0.1498	0.2622	14.5	13.5
4,861	0.1521	0.2663	14.5	13.5
4,876	0.1566	0.2699	14.5	13.5
4,891	0.1610	0.2735	14.5	13.5
4,906	0.1665	0.2762	14.5	13.5
4,921	0.1721	0.2786	14.5	13.5
4,936	0.1773	0.2807	14.5	13.5
4,951	0.1817	0.2831	14.5	13.5
4,966	0.1873	0.2848	14.5	13.5
4,981	0.1910	0.2863	14.5	13.5
4,996	0.1962	0.2877	14.5	13.5
5,011	0.2000	0.2887	14.5	13.5
5,026	0.2036	0.2892	14.5	13.5
5,041	0.2065	0.2897	14.5	13.5
5,056	0.2089	0.2907	14.5	13.5
5,071	0.2114	0.2909	14.5	13.5
5,086	0.2134	0.2922	14.5	13.5

(continued)

Table 6.3 (continued)

Time (s)	Level (m)		Pump flow rate (cm^3/s)	
	Tank 1 (h_1)	Tank 2 (h_2)	Left pump (u_1)	Right pump (u_2)
5,101	0.2155	0.2920	14.5	13.5
5,116	0.2167	0.2928	14.5	13.5
5,131	0.2181	0.2931	14.5	13.5
5,146	0.2192	0.2936	14.5	10.5
5,161	0.2201	0.2920	14.5	10.5
5,176	0.2189	0.2895	14.5	10.5
5,191	0.2158	0.2878	14.5	10.5
5,206	0.2121	0.2857	14.5	10.5
5,221	0.2080	0.2841	14.5	10.5
5,236	0.2037	0.2823	14.5	10.5
5,251	0.1989	0.2815	14.5	10.5
5,266	0.1943	0.2803	14.5	10.5
5,281	0.1891	0.2792	14.5	10.5
5,296	0.1848	0.2786	14.5	10.5
5,311	0.1802	0.2778	14.5	10.5
5,326	0.1762	0.2766	14.5	10.5
5,341	0.1716	0.2764	14.5	10.5
5,356	0.1685	0.2751	14.5	10.5
5,371	0.1658	0.2749	14.5	10.5
5,386	0.1632	0.2747	14.5	10.5
5,401	0.1603	0.2747	14.5	10.5
5,416	0.1581	0.2744	14.5	10.5
5,431	0.1565	0.2747	14.5	10.5
5,446	0.1549	0.2743	14.5	13.5
5,461	0.1542	0.2760	14.5	13.5
5,476	0.1560	0.2784	14.5	13.5
5,491	0.1594	0.2807	14.5	13.5
5,506	0.1632	0.2827	14.5	13.5
5,521	0.1685	0.2844	14.5	13.5
5,536	0.1733	0.2860	14.5	13.5
5,551	0.1785	0.2872	14.5	13.5
5,566	0.1828	0.2884	14.5	13.5
5,581	0.1870	0.2892	14.5	13.5
5,596	0.1919	0.2906	14.5	13.5
5,611	0.1957	0.2908	14.5	13.5
5,626	0.1991	0.2917	14.5	13.5

(continued)

Table 6.3 (continued)

Time (s)	Level (m)		Pump flow rate (cm^3/s)	
	Tank 1 (h_1)	Tank 2 (h_2)	Left pump (u_1)	Right pump (u_2)
5,641	0.2023	0.2924	14.5	13.5
5,656	0.2054	0.2931	14.5	13.5
5,671	0.2078	0.2936	14.5	13.5
5,686	0.2102	0.2937	11.5	13.5
5,701	0.2112	0.2937	11.5	13.5
5,716	0.2111	0.2922	11.5	13.5
5,731	0.2111	0.2900	11.5	13.5
5,746	0.2109	0.2856	11.5	13.5
5,761	0.2111	0.2815	11.5	13.5
5,776	0.2115	0.2762	11.5	13.5
5,791	0.2116	0.2712	11.5	13.5
5,806	0.2119	0.2655	11.5	13.5
5,821	0.2120	0.2607	11.5	13.5
5,836	0.2125	0.2553	11.5	13.5
5,851	0.2126	0.2510	11.5	13.5
5,866	0.2131	0.2464	11.5	13.5
5,881	0.2138	0.2428	11.5	13.5
5,896	0.2144	0.2389	11.5	13.5
5,911	0.2138	0.2357	11.5	13.5
5,926	0.2139	0.2327	11.5	13.5
5,941	0.2141	0.2303	11.5	13.5
5,956	0.2141	0.2281	11.5	13.5
5,971	0.2139	0.2265	11.5	13.5
5,986	0.2139	0.2247	11.5	13.5
6,001	0.2139	0.2233	11.5	13.5
6,016	0.2141	0.2220	11.5	13.5
6,031	0.2139	0.2210	11.5	13.5
6,046	0.2142	0.2199	11.5	13.5
6,061	0.2137	0.2191	11.5	13.5
6,076	0.2140	0.2184	11.5	13.5
6,091	0.2137	0.2178	11.5	13.5
6,106	0.2140	0.2171	11.5	13.5
6,121	0.2139	0.2167	11.5	13.5
6,136	0.2140	0.2161	11.5	10.5
6,151	0.2134	0.2140	11.5	10.5
6,166	0.2113	0.2111	14.5	10.5

(continued)

Table 6.3 (continued)

Time (s)	Level (m)		Pump flow rate (cm^3/s)	
	Tank 1 (h_1)	Tank 2 (h_2)	Left pump (u_1)	Right pump (u_2)
6,181	0.2085	0.2091	14.5	10.5
6,196	0.2065	0.2090	14.5	10.5
6,211	0.2031	0.2105	14.5	10.5
6,226	0.1994	0.2128	14.5	10.5
6,241	0.1949	0.2161	14.5	10.5
6,256	0.1904	0.2202	14.5	10.5
6,271	0.1853	0.2238	14.5	10.5
6,286	0.1810	0.2285	14.5	10.5
6,301	0.1772	0.2323	14.5	10.5
6,316	0.1741	0.2364	14.5	10.5
6,331	0.1698	0.2397	14.5	10.5
6,346	0.1663	0.2434	14.5	10.5
6,361	0.1629	0.2469	11.5	10.5
6,376	0.1592	0.2494	11.5	10.5
6,391	0.1548	0.2504	11.5	10.5
6,406	0.1514	0.2496	11.5	10.5
6,421	0.1480	0.2480	11.5	13.5
6,436	0.1467	0.2472	11.5	13.5
6,451	0.1481	0.2465	11.5	13.5
6,466	0.1506	0.2450	11.5	13.5
6,481	0.1542	0.2432	11.5	13.5
6,496	0.1580	0.2408	11.5	13.5
6,511	0.1628	0.2390	14.5	13.5
6,526	0.1683	0.2373	14.5	13.5
6,541	0.1748	0.2371	14.5	13.5
6,556	0.1797	0.2383	14.5	13.5
6,571	0.1856	0.2405	14.5	13.5
6,586	0.1894	0.2436	14.5	13.5
6,601	0.1944	0.2467	14.5	13.5
6,616	0.1985	0.2507	14.5	13.5
6,631	0.2021	0.2540	14.5	13.5
6,646	0.2056	0.2581	14.5	13.5
6,661	0.2083	0.2612	11.5	13.5
6,676	0.2093	0.2644	11.5	10.5
6,691	0.2089	0.2642	11.5	10.5
6,706	0.2072	0.2617	11.5	10.5

(continued)

Table 6.3 (continued)

Time (s)	Level (m)		Pump flow rate (cm³/s)	
	Tank 1 (h_1)	Tank 2 (h_2)	Left pump (u_1)	Right pump (u_2)
6,721	0.2032	0.2573	11.5	10.5
6,736	0.1991	0.2525	11.5	10.5
6,751	0.1936	0.2485	11.5	10.5
6,766	0.1888	0.2440	11.5	10.5
6,781	0.1829	0.2391	11.5	10.5
6,796	0.1782	0.2352	11.5	10.5
6,811	0.1728	0.2303	11.5	10.5
6,826	0.1682	0.2271	11.5	10.5
6,841	0.1632	0.2228	11.5	10.5
6,856	0.1595	0.2198	11.5	10.5
6,871	0.1553	0.2167	11.5	10.5
6,886	0.1524	0.2140	11.5	10.5
6,901	0.1493	0.2115	14.5	10.5
6,916	0.1481	0.2100	14.5	10.5
6,931	0.1479	0.2100	14.5	10.5
6,946	0.1482	0.2113	14.5	10.5
6,961	0.1485	0.2141	14.5	10.5
6,976	0.1484	0.2169	14.5	10.5
6,991	0.1483	0.2213	14.5	10.5
7,006	0.1478	0.2246	14.5	10.5
7,021	0.1472	0.2289	14.5	10.5
7,036	0.1473	0.2320	14.5	10.5
7,051	0.1473	0.2360	14.5	10.5
7,066	0.1469	0.2396	14.5	10.5
7,081	0.1474	0.2428	14.5	10.5
7,096	0.1478	0.2460	14.5	10.5
7,111	0.1475	0.2486	11.5	10.5
7,126	0.1458	0.2511	11.5	10.5
7,141	0.1434	0.2514	11.5	10.5
7,156	0.1417	0.2508	11.5	10.5
7,171	0.1402	0.2486	11.5	10.5
7,186	0.1386	0.2457	11.5	10.5
7,201	0.1374	0.2428	11.5	10.5
7,216	0.1368	0.2390	11.5	10.5
7,231	0.1360	0.2358	11.5	10.5
7,246	0.1355	0.2314	14.5	10.5
7,261	0.1367	0.2285	14.5	10.5

Chapter 7
Using MATLAB® for Statistical Analysis

MATLAB® is a mathematical programme developed by the company called The MathWorks, Inc. Examples in this chapter have been tested on MATLAB® versions between 2011a and 2019b. It is expected that most of the commands presented will work with some earlier versions, as well as most later versions. It will be assumed that the reader has a basic understanding of MATLAB®, can write MATLAB® statements, understands basic MATLAB® commands, and can plot a simple MATLAB® graph. This chapter will examine in detail additional features, such as the different toolboxes and formatting features. In order to clearly distinguish between the code required for the MATLAB® function and text, all MATLAB® commands and variables are shown in bold `Courier New`.

7.1 Basic Statistical Functions

The functions given in Table 7.1 are available with all standard MATLAB® installations and do not require purchasing any additional toolboxes or licences.

7.2 Basic Functions for Creating Graphs

A list of functions for creating different types of graphs is listed in Table 7.2. Functions with an asterisk after them require installation of the Statistics and Machine Learning Toolbox in MATLAB®.

© The Author(s), under exclusive license to Springer Nature Switzerland AG 2022, corrected publication 2023
Y. A. W. Shardt, *Statistics for Chemical and Process Engineers*,
https://doi.org/10.1007/978-3-030-83190-5_7

Table 7.1 Basic statistics functions

Function	Description
mean(x)	For vectors, this function determines the mean of **x**. For matrices, this function determines the mean for each column and returns a row vector containing these values
median(x)	For vectors, this function determines the median value of **x**. For matrices, this function determines the median value of each column and returns a row vector containing these values
std(x)	For vectors, this function determines the observational (sample) standard deviation of **x**. For matrices, this function determines the standard deviation of each column and returns a row vector containing these values
std(x,1)	For vectors, this function determines the sample space (population) standard deviation of **x**. For matrices, this function determines the sample space standard deviation of each column and returns a row vector containing these values

7.3   The Statistics and Machine Learning Toolbox

This section lists those statistical functions that require the Statistics and Machine Learning Toolbox in MATLAB® to be installed.

7.3.1   Probability Distributions

Detailed information regarding the definitions of the different probability density functions and the meaning of the required variables can be found in Sect. 2.4. The functions are described in Table 7.4.

7.3.2   Advanced Statistical Functions

The functions listed in Table 7.5 are useful for computing more advanced statistical properties.

7.3.3   Advanced Probability Functions

The functions listed in Table 7.6 are useful for advanced more advanced probability problems.

Table 7.2 Basic plotting functions (functions followed by an asterisk (*) require the Statistics and Machine Learning Toolbox)

Function	Description
`bar(x)`	Creates the vertical bar graph for the data in **x**
`barh(x)`	Creates the horizontal bar graph for the data in **x**. When using the set method to set the labels, **xticklabel** should be replaced by **yticklabel**
`boxplot(x,nameArray)` *	Creates a box-and-whisker plot for the data in **x**. The chart labels are provided in the list array **nameArray**. Multiple box-and-whisker plots can be combined by entering multiple columns in the matrix **x**. A separate box-and-whisker plot will be made for each column
`colorbar`	Sets the colour bar in 3-D plots
`colormap(NAME)`	Sets the colour map to the given **NAME**
`hist(x)`	Creates the histogram for the data in **x**
`imagesc(data)`	Creates an image plot so that the information in **data** is centred and displays properly. Useful for creating cross-correlation plots. In order to create the classical correlation plot, it is necessary to use the following two additional comments: 1. **set(gca, 'xtick', 1:n),** where **gca** is a handle to the current figure and **n** is the number of data points in the plot. This command centres the bins so that the labels are properly set in the next step. The handle **gca** can be replaced by the actual handle to the figure 2. **set (gca,'xticklabel',L})** where **gca** is a handle to the current figure and **L** is an array that gives the names of the individual points It should be noted that the above comments are only for the x-axis. This must be repeated for the y-axis by replacing the **x** by **y**, so that **xtick** is replaced by **ytick**, for example, **set(gca,'yticklabel',L})**
`legend(nameArray)`	Adds a legend to the graph. The **nameArray** contains an array ordered so that the first entry corresponds with the name for the first drawn line. If more names are provided, then only the first n is used. Latex commands can be used
`loglog(x,y, 'format')`	Plots given the vector x on the x-axis and the vector **y** on the y-axis following the formatting rules presented in `format` string (see Table 7.3 for some common examples). Both axes will be logarithmic
`normplot(x)` *	Plots a normal probability graph for the data given by **x**. The axes are flipped from what is recommended in this textbook, that is, the x-axis has the data and the y-axis has the expected scores

(continued)

Table 7.2 (continued)

Function	Description
`pie(x,nameArray)`	Creates a pie chart for the data in **x**. The pie chart labels are provided in the list array **nameArray**. The data should be provided as percentages that total 100%
`plot(x,y, 'format')`	Plots given the vector **x** on the **x**-axis and the vector **y** on the y-axis following the formatting rules presented in format string (see Table 7.3 for some common examples). The vector **x** can be omitted, in which case a time-series plot will be drawn, that is, the **x**-axis will increment by one after each data point
`Plot3(x,y,z)`	Plots a three-dimensional line plot with **x** on the x-axis, **y** on the y-axis and **z** on the z-axis. This is the three-dimensional analogue to the **plot** function
`plotmatrix(array)`	Creates an $n \times n$ plot using columns of **array**. It is assumed that the rows of **array** represent sample values and the columns different variables. The plot that is displayed contains on the diagonal, the (i, i)-entry, a histogram of the data in the ith column. The off-diagonal entries, that is, the (i, j)-entries, represent the correlation between the ith and jth column of **array**
`plotmatrix(x,y)` `plotmatrix(y)`	Plots the columns of data matrix **x** against the columns of the data matrix **y** to show the relationships between the different columns. Providing a single entry is the same as `plotmatrix(y,y)` except that the diagonals are replaced by histograms
`polar(th,r, 'format')`	Plots a polar graph using the angle vector th and the radius vector **r** following the formatting rules presented in format string (see Table 7.3 for some common examples). The angle vector can be omitted, in which case it is assumed that the angle increases by exactly 1 rad ($57.296°$) for each data point
`rose(x,bins)`	Plots a polar histogram using the data vector **x** and the requested number of **bins**. The number of bins is optional
`scatter(x,y,g)`	Creates a scatter plot with **x** on the x-axis and **y** on the y-axis. Cell array **g** is a grouping variable by which one can plot multiple groups on a single scatter plot. The values in **g** then become the default legend labels. For example, with two runs, **g** would be written as **{'Run 1'; 'Run 2'; 'Run 1'; 'Run 1'; 'Run 2'; 'Run 2'}**. This would assign the first entry in **x** and **y** to Run 1, the second to Run 2 and so on

(continued)

Table 7.2 (continued)

Function	Description
`semilogx(x,y, 'format')` `semilogy(x,y, 'format')`	Plots given the vector x on the *x*-axis and the vector **y** on the *y*-axis following the formatting rules presented in the **format** string (see Table 7.3 for some common examples). Either the *x*- or *y*-axis will be logarithmic
`set(gca, 'xticklabel', listarray)`	Creates for the vertical bar graph referenced by the handle **gca** the *x*-axis labels given in the **listarray**. The current graph is referenced by the handle **gca**, while the handle for a specific graph can be obtained by setting **h** = **plot(…)** (or any similar method to obtain a figure)
`surf(x,y,z)`	Creates a three-dimensional surface plot with **x** on the *x*-axis, **y** on the *y*-axis and **z** on the *z*-axis
`title('name')`	Draws the title for the graph. Latex commands can be used
`xlabel('name')` `ylabel('name')` `zlabel ('name')`	Draws the label for the *x*- (*y*- or *z*-) axis. Latex commands can be used
`ylabel(colorbar,'My colorbar')`	Sets the label for the colour bar

7.3.4 Linear Regression Analysis

There are two main functions for performing linear regression in MATLAB®: `regress` and `nlinfit`. The first works for linear regression, while the second works for nonlinear regression. Weighted, linear least-squares can be performed using `lscov`. Detailed information about the different commands and their requirements is given in Table 7.7

7.3.5 Design of Experiments

The functions listed in Table 7.8 are useful when performing design of experiments or their analysis.

7.4 The System Identification Toolbox

The System Identification Toolbox in MATLAB® is a very useful toolbox when fitting models for system identification using the prediction error model. It provides a convenient and concise way of storing, accessing, and manipulating different data sets

Table 7.3 Useful formatting options

Name	Description
b	Blue
g	Green
r	Red
w	White
c	Cyan
y	Yellow
k	Black
m	Magenta
. (period)	Dot
+	Cross, +
*	Star,*
s	Square, □
d	Diamond, ◇
v	▽
^	△
<	◁
>	▷
p	☆
h	Hexagram ✶
–	Solid line
:	Dotted line
–.	Dash-dotted line
– –	Dashed line

and their associated models. Although most time-series analysis can be performed using the System Identification Toolbox, at times it is easier to use the econometric toolbox described below. In order to fully appreciate and use the System Identification Toolbox, it is first useful to examine in detail the special data objects that store and hold the information: the **iddata** and the **idpoly** objects.

The **iddata** object, which will be denoted by a generic **z**, stores the data that is used in determining the models. It consists of two main fields:

(1)   The inputs to the system are stored in **z.u**, which is a matrix. Each of the columns contains a different input.

Table 7.4 Probability distribution functions

Function	Description
chi2inv(p,df)	Calculates the inverse χ^2-distribution, given a probability, **p**, and the number of degrees of freedom, **df**
chi2pdf(x,df)	Calculates the probability density function for the χ^2-distribution at **x** given the number of degrees of freedom, **df**
finv(p,ndf,ddf)	Calculates the inverse F-distribution at a probability, **p**, given the number of degrees of freedom for the numerator, **ndf**, and the number of degrees of freedom for the denominator, **ddf**
fpdf(x,ndf,ddf)	Calculates the probability density function for the F-distribution at **x** given the number of degrees of freedom for the numerator, **ndf**, and the number of degrees of freedom for the denominator, **ddf**
normpdf(x,m,s)	Calculates the probability density function for the normal distribution for a given **x**, given a mean of **m** and a standard deviation of **s**. If **m** and **s** are not given, then the default values of $m = 0$ and $s = 1$ will be used
norminv(p)	Calculates the Z-score for the probability **p**, which must be between 0 and 1
tinv(p,df)	Calculates the inverse t-distribution at a probability, **p**, given the number of degrees of freedom, **df**
tpdf(x,df)	Calculates the probability density function for the t-distribution at **x** given the number of degrees of freedom, **df**

Table 7.5 Advanced statistical functions

Function	Description
mad(y,1)	Computes the median absolute difference for the data vector **y**
zscore(u)	Normalizes the data matrix **u** by columns, that is, it computes for each column the mean and standard deviation to normalize the values in that column

Table 7.6 Advanced probability functions

Function	Description
quantile(X,p)	Creates the **p**-tiles for the given data set **X**
rand(x,y)	Creates an **x**-by-**y** matrix of pseudorandom numbers generated from a uniform distribution on $]0, 1[$
randn(x,y)	Creates an **x**-by-**y** matrix of pseudorandom numbers generated from a normal distribution with mean equal to zero and standard deviation equal to 1
randperm(N)	Creates a vector containing a random permutation of the numbers from 1 to **N**

(2) The outputs from the system are stored in **z.y**, which is a matrix. Each of the columns contains a different output. Thus, the second output of the system would be accessed as **z.y(:,2)**, regardless of how the variables may be named on the screen.

Table 7.7 Linear regression functions

Function	Description
`[coeff,Cint,res,resint,stats] = regress(y,A,alpha)`	Computes a multivariate linear regression model of the form $\mathbf{y} = \mathbf{A}\beta$ and returns the following parameters: a. **coeff**, which is the vector containing the estimated coefficients using least squares, $\hat{\beta}$ b. **Cint**, which is a vector that contains the $100(1 - \textbf{alpha})\%$ confidence intervals for the coefficients in **coeff** c. **Res**, which is a vector containing the residuals d. **resint**, which is a vector that contains the confidence interval for the residuals e. **stats**, which is a vector that contains the following entries (in order): R^2, F-statistics, p-value, and $\hat{\sigma}^2$ The variable **alpha** is optional. The default value of **alpha** is 0.05
`[beta,r,J,covb] = nlinfit(x,y,'FUN',beta0)`	Performs nonlinear regression using the Gauss–Newton estimation method. The x-data is given as **x**, while the y-data is given as **y**. The function, **FUN**, that is to be fit must be written as an m-file. It will take three arguments: the coefficient values, x, and y (in this order). The function should be written to allow for matrix evaluation. The initial guess is specified in **beta0**. The vector **beta** contains the estimated values of the coefficients, the vector **r** contains the residuals, and **covb** is the estimated covariance matrix for the problem. **J** is the Jacobian matrix evaluated with the best estimate for the parameters
`Ci = nlparci(beta,r,'covar', covb)`	Calculates the $100(1 - \textbf{alpha})\%$ confidence intervals for the coefficients, **beta**, given the residuals, **r**, and the covariance matrix, **covb**
`[y,delta] = nlpredci('Fun',x, beta,r,'covar',covarb)`	Calculates the $100(1 - \textbf{alpha})\%$ mean confidence intervals for the nonlinear function, **Fun**, given the values of x, **x**, the estimated coefficients, **beta**, the residuals, **r**, and the covariance matrix, **covarb**. The function returns the predicted y-values, **y**, and the half-width lengths, **delta**. This implies that the mean confidence interval will be given as **y**±**delta**

(continued)

Table 7.7 (continued)

Function	Description
`[y,delta] = nplpredci('Fun',x,beta,r, 'covar',covarb,'predopt', 'observation')`	Calculates the $100(1 - $ **alpha**$)\%$ predicted confidence intervals for the nonlinear function, **Fun**, given the values of x, **x**, the estimated coefficients, **beta**, the residuals, **r**, and the covariance matrix, **covarb**. The function returns the predicted y-values, **y**, and the half-width lengths, **delta**. This implies that the prediction confidence interval will be given as **y$\pm$delta**
`[coeff,stdevc,mse,s] = lscov(A,y,W)`	Computes a weighted, multivariate linear regression model of the form **y** $=$ **A**β given the weighting matrix . The weighting vector **W** is equal to the diagonal terms of the weighting matrix $\mathcal{W}$, that is, **W** $=$ diag$(\mathcal{W})$. The parameter estimates are returned as **coeff**, the standard deviation for the parameter estimates is returned as **stdevc**, the variance of the model is returned as **mse**, and **s** is the covariance matrix for the parameters

Table 7.8 Design of experiment functions

Function	Description
`A = ff2n(n)`	Determines the 2^n factorial experimental regression matrix using codes 0 and 1
`[A, conf] = fracfact('a b c ab bc')`	Determines a fractional factorial experimental regression matrix, **A**, using the stated generating strings. The matrix **conf** contains the confounding pattern for the given design
`[setting,A] = cordexch(nf,nr,'q')`	Determines a D-optimal regression matrix, **A**, given the number of runs, **nr**, the number of factors, **nf**

Additional fields include:

- **z.Tstart**, which stores the value of the starting time for the object.
- **z.Ts**, which stores the value of the sampling time.

The **iddata** object can be treated as a vector to access all the relevant data between two end points. For example, to take the data located from the 1st to 100th point in the object, the command would be **z(1:100)**.

The **idpoly** object, which will be denoted by a generic **m**, stores information about the model that has been fit to the data. It consists of five main fields that consist of the coefficients, a_i, ordered in descending powers of z^{-1}, given by

$$1 + \sum_{i=1}^{n} a_i z^{-i},$$

where n is the order of the system. Each of the fields is the same length. The fields are given the names: A, B, C, D, and F, and represent the coefficients of the following model:

$$A\left(z^{-1}\right)y_k = \frac{B\left(z^{-1}\right)}{F\left(z^{-1}\right)}u_{k-d} + \frac{C\left(z^{-1}\right)}{D\left(z^{-1}\right)}e_t,$$

where d is the discrete time delay, which for a zero-order hold is one more than the continuous time delay. The coefficients of $A(z^{-1})$ would be accessed using **m.A**. Note that the definition of the B-polynomial in MATLAB® is different from the definition used in the other chapters. This difference implies that the time delay, d, must be increased by one from the values obtained in the other chapters, that is, $d = n_k + 1$. This is because MATLAB® requires the time delay due to sampling to be explicitly noted in the definition of the function.

For multi-input systems, where there are multiple inputs, the model representation is converted into a matrix form, so that each row represents a different input and the columns represent the parameter specifications. The orders are then stored as the augmented column matrix with each column representing the orders of a different row, for example, the B-order would be specified as **[2, 3]** for a two-input system.

The most important functions from the System Identification Toolbox are given in Table 7.9 for creating the data object, Table 7.10 for creating the model, Table 7.11 for validating the model, and Table 7.12 for designing the system identification experiment.

7.5   The Econometrics Toolbox

The Econometrics Toolbox contains some useful tools for analysing and preprocessing time-series data. It is especially useful for fitting seasonal models. Unfortunately, not all the validation functions can be as easily obtained with this toolbox. Table 7.13 contains the required functions for creating an econometric model, Table 7.14 contains the functions for creating various types of correlation plots, Table 7.15 contains the functions for estimating the model parameters of econometric functions, and Table 7.16 contains useful functions for model validation.

7.6   The Signal Processing Toolbox

The Signal Processing Toolbox contains complementary functions that can be used to create cross- and autocorrelation plots without using the econometric toolbox. It can also be used to effectively create periodograms. Table 7.17 contains a summary of the useful functions.

Table 7.9 System Identification Toolbox: Functions for creating the data object

Function	Description
`z = iddata(yk,uk,Ts)`	Creates an **iddata** object, **z**, based on the input, **uk**, and output, **yk**, with a sample time of **Ts**. For time-series data, **uk** can be left blank by replacing it by **[]**
`zd = detrend(z,0)`	Removes a trend from the given **iddata** object, **z**, and returns the **iddata** object, **zd**. The option **0** removes the mean value from the data allowing the data to vary about the mean of zero. This should be performed on all data before carrying out any further analysis
`idplot(z)`	This allows an **iddata** object, **z**, to be plotted. Two basic figures are produced: the output as a function of time and the input as a function of time. The programme will pause between each set of inputs and outputs. To continue, return to the main MATLAB® window and press any key. Note: typing **plot(z)** will also work
`model = idpoly(A,B,C,D,F,S,Ts)`	Creates an **idpoly** object, **model**, that describes the model based on the generalized prediction error model. The elements are entered in descending powers of z^{-1}, starting with the constant term, even if it is absent. **S** is the variance of the error term and **Ts** is the sampling time. The last two terms are optional

7.7 MATLAB® Recipes

This section provides useful MATLAB® code for various functions that are not provided by default in MATLAB®. This code can be re-used, but full attribution both to the author and this book must be made.

7.7.1 Periodogram

Requirements: None

Goal: Given a data series **yt** return the corresponding periodogram on the region [0, 0.5].

File Name: `periodogram2.m`

Table 7.10 System Identification Toolbox: Functions for creating a model

Function	Description
`cra(z)`	Determines the impulse response coefficients between the input and output, as defined in the **iddata** object, **z**. The output is the value of the impulse responses. A graph is plotted showing the individual coefficients, as well as the confidence intervals. This function can be used to estimate the time delay
`mARa = ar(z,[na])`	Finds for the data in the **iddata** object, **z**, an autoregressive model with order **na**. This model is stored as an **idpoly** object, **mARa**
`mARMAXacd = armax(z,[na,nc,d])`	Finds for the data in the **iddata** object, **z**, an autoregressive moving average with exogenous input model with orders **na** and **nc** with a delay of **d**. This model is stored as an **idpoly** object, **mARMAXacd**
`mARXabd = arx(z,[na,nb,d])`	Finds for the data in the **iddata** object, **z**, an autoregressive exogenous model with orders **na** and **nb** with a delay of **d**. This model is stored as an **idpoly** object, **mARXabd**
`mBJbcdfd = bj(z,[nb,nc,nd,nf,d])`	Finds for the data in the **iddata** object, **z**, a Box–Jenkins model with orders **nb**, **nc**, **nd**, and **nf** with a delay of **d**. This model is stored as an **idpoly** object, **mBJbcdfd**
`mNL = nlarx(z,nn,basis)`	Finds for the data in the **iddata** object, **z**, a nonlinear ARX model with ARX orders **nn** and the basis function given by **basis**. When defining the basis function, it is important to include the number of functions to be used; for a wavelet basis function this can be done as follows **wavenet('num',nfun)**, where **nfun** is the number of basis functions to be used. The ARX orders are entered in the same manner as the standard ARX model. This model is stored as an **idpoly** object, **mNL**
`mOEabd = oe(z,[nb,nf,d])`	Finds for the data in the **iddata** object, **z**, an output-error model with orders **nb** and **nf** with a delay of **d**. This model is stored as an **idpoly** object, **mOEbfd**
`nk = delayest(z,na,nb,minnk,maxnk)`	Estimates the time delay for the **iddata** object, z, by searching all feasible ARMAX models and selecting the one with the lowest error. The estimated time delay is returned as **nk**. If a large order model is to be searched, then **na** is the order of the **A** polynomial and **nb** is the order of the **B**-polynomial, **minnk** is the minimum time delay, and **maxnk** is the maximum time delay. The last four values need not be specified

Table 7.11 System Identification Toolbox: Functions for validating a model

Function	Description
`compare(z,mabd)`	Compares the original data stored in the **iddata** object, **z**, with the model stored in **mabd**, to produce a plot showing the fit between the original data and the predicted data. This function call uses the infinite-horizon predictor for the modelled data
`compare(z,mabd,n)`	Compares the original data stored in the **iddata** object, **z**, with the model stored in **mabd** and an **n**th-step-ahead predictor to produce a plot showing the fit between the original data and the predicted data. This function call uses the **n**th -step-ahead predictor to forecast the values
`present(mabd)`	Displays the **idpoly** object, **mabd**, with the estimated parameters and their standard deviation
`resid(mabd,z)`	Determines and plots the residuals between the **idpoly** model, **mabd**, and the original **iddata** object, **z**. The plot shows the correlation between the residuals as well as the cross-correlation between the residuals and the inputs. **N.B.:** The bounds are the 99% confidence intervals
`residual = resid(mabd,z)`	Same as **resid(mabd,z)**, also returns the residuals as a vector

Table 7.12 System Identification Toolbox: Functions for designing a system identification experiment

Function	Description
`u = idinput(N,Type,band,Range)`	Creates an input, **u**, with number of values, **N**, with the following characteristics: a. **Type**, which describes what kind of input is desired. Permitted types include: **'RGS'**, which gives random, Gaussian signal and **'RBS'**, which gives a random, binary signal b. **band**, which is a 1-by-2 row vector that describes the region of the passband in terms of the Nyquist frequency. It must, thus, lie within [0, 1] c. **Range**, which describes the region over which the data ranges

Table 7.13 Econometrics Toolbox: Functions for creating the data object

Function	Description
`modStruc = arima('Constant',0,'D',d, 'Seasonality',S,'SMALags',nSMA,'MALags', nMA,'ARlags',nAR,'SARLags',nSAR)`	Creates the initial model structure for fitting a seasonal (or more advanced model) in MATLAB®. The model structure is given as **modStruc**. The degree of normal differencing is given as **d**, and the seasonal differencing order is **S**. The seasonal moving-average orders are explicitly stated in the vector **nSMA**, the moving-average order in the vector **nMA**, the seasonal autoregressive orders as **nSAR**, and the autoregressive orders as **nAR**. The order vector given as $[1, 2, 3, 5]$ would specify that the coefficients at the points z^{-1}, z^{-2}, z^{-3}, and z^{-5} are to be estimated. For the seasonal terms, it is necessary to clearly state the values including any seasonal component, so that $[4\ 8]$ would correspond to the seasonal vector z^{-4} and $z^{-2\times4}$

Table 7.14 Econometrics Toolbox: Functions for creating various correlation plots

Function	Description
`autocorr(data)`	Displays the autocorrelation graph for the data set **data**, including appropriate confidence intervals
`crosscorr(data1,data2)`	Displays the cross-correlation graph between two data sets **data1** and **data2** including appropriate confidence intervals
`parcorr(data)`	Displays the partial autocorrelation graph for the data set **data**, including appropriate confidence intervals

Table 7.15 Econometrics Toolbox: Functions for estimating model parameters

Function	Description
`model = estimate(modStruc,data)`	Estimates a **model** given the initial model structure **modStruc** and the data vector **data**
`residual = infer(model,data)`	Obtains the residuals **residual** given the estimated **model** and the data vector **data**

Table 7.16 Econometrics Toolbox: Functions for validating the model

Function	Description
`[h,p,stat,c] = lbqtest(residuals, 'lags',lags)`	Performs the Ljung–Box–Pierce test on the data set **residuals**. The optional vector **lags** contains the lag values at which the statistic is to be evaluated. It can be either a vector or a scalar. The default number of lags is 20. Let l be equal to the number of lags tested. The Boolean value of the hypothesis test is returned as the $l \times 1$ column vector **h**, the corresponding p-values for the test statistic as the $l \times 1$ column vector **p**, the actual test statistic values as the $l \times 1$ column vector **stat**, and the corresponding critical values as the $l \times 1$ column vector **c**

Table 7.17 Signal Processing Toolbox: Functions for analysing signals

Function	Description
`[q] = xcorr(data1,data2)`	Obtains the cross-correlation between two data series **data1** and **data2**, each of length m, and returns it as the $(2m - 1) \times 1$ vector **q**. Autocorrelation can be obtained by setting **data2** to be equal to **data1**
`periodogram(yt)`	Displays the periodogram for the signal **yt**

```
function periodogram2(yt)
%Custom-built function that creates the periodogram for a
given signal%
%Inputs:
%    yt: signal for which the periodogram is desired.
%Copyright 2014: Yuri Shardt
%Distributed as part of the book Statistics for Chemical and
Process
%Engineers: A Modern Approach, published by Springer Verlag.

%Checking the data
q=size(yt);
N=length(yt);
if (N<5)
    error('Please make sure that the size of yt is at least 5
samples.');
end
if (N==q(1) && q(2)~=1) || (N==q(2) && q(1)~=1)
    error('Please make sure that yt is either a row or column
vector.');
```

```
end
y=detrend(yt,0);
%Perform the Fast Fourier Transform
X1=abs(fft(y,N));
X=fftshift(X1);
if (mod(N,2)==0)
    F=[-(N)/2:(N)/2-1]/N;
    count=N/2+1;
else
    F=[-(N-1)/2:(N-1)/2]/N;
    count=(N-1)/2+1;
end
%Creating the plot
plot(F(count:end),X(count:end)/N*2,'-k')
xlabel(['Frequency, $f$,
(cycles/sample)'],'interpreter','latex'),ylabel('Amplitude,
$|y|$','interpreter','latex')
end
```

7.7.2   *Autocorrelation Plot*

Requirements: Signal Processing Toolbox (`xcorr`).

Goal: Given a data series **yt** return the corresponding autocorrelation plot for twenty lags.

File Name: `autocorrelation.m`

```
function autocorrelation(yt)
%Custom-built function that creates the autocorrelation plot
for a given
%signal. A lag of 20 is the maximum lag considered.
%Inputs:
%   yt: signal for which the autocorrelation plot is desired.
%%Copyright 2014: Yuri Shardt
%Distributed as part of the book Statistics for Chemical and
Process
%Engineers: A Modern Approach, published by Springer Verlag.
```

```
%Checking the data
q=size(yt);
N=length(yt);
lag=20;
if (N<lag)
    error('Please make sure that the size of yt is at least 20
samples.');
end
if (N==q(1) && q(2)~=1) || (N==q(2) && q(1)~=1)
    error('Please make sure that yt is either a row or column
vector.');
y=detrend(yt,0);
%Obtain the autocorrelation values
q=xcorr(y,y);
count=N;
%Creating the plot
plot([0:lag],q(count:count+lag)/max(q),'ok')
grid on
hold on;plot([0,lag],tinv(0.975,N)*[1,1]/sqrt(N),'--
k','linewidth',2)
plot([0,lag],-tinv(0.975,N)*[1,1]/sqrt(N),'--k','linewidth',2)
plot([0,lag],[0,0],'-k')
xlabel(['Lag
(samples)'],'interpreter','latex'),ylabel('Autocorrelation,
$\rho$','interpreter','latex')
end
```

7.7.3 Correlation Plot

Requirements: none.

Goal: Given a matrix of `correlations` **correlation** create the two-dimensional correlation plot.

File Name: `corrplot1.m`

```
function corrplot1(correlation,tags,title1,axis1)
%Custom-built function that creates the correlation plot given
a correlation matrix
%Inputs:
%    correlation: correlation matrix
%    tags: alphanumeric explanation of the columns in the data
set (should not be left blank)
%    title1: a title for the figure (can be left blank)
%    axis1: the label for the axis (can be left blank)
%Copyright 2014: Yuri Shardt
%Distributed as part of the book Statistics for Chemical and
Process
%Engineers: A Modern Approach, published by Springer Verlag.


[a,b]=size(correlation);
imagesc(abs(correlation));colorbar;colormap(jet)
set(gca, 'XTick', 1:a); % center x-axis ticks on bins
set(gca, 'YTick', 1:a); % center y-axis ticks on bins
set(gca, 'XTickLabel', tags); % set x-axis labels
set(gca, 'YTickLabel', tags); % set y-axis labels
title(title1); % set title
xlabel(axis1);
ylabel(axis1);
```

7.7.4 Cross-Correlation Plot

Requirements: Signal Processing Toolbox (`xcorr`).

Goal: Given two data series **yt** and **zt** return the corresponding cross-correlation plot for ±20 lags.

File Name: `crosscorrelation.m`

```
function crosscorrelation(xt,yt)
%Custom-built function that creates the cross-correlation plot
for 2 signals.
%A maximal lag of +-20 is assumed.
%Inputs:
```

```matlab
%   x: signal 1
%   y: signal 2
%Copyright 2014: Yuri Shardt
%Distributed as part of the book Statistics for Chemical and
Process
%Engineers: A Modern Approach, published by Springer Verlag.

%Checking the data
q1=size(xt);
N1=length(xt);
q2=size(yt);
N2=length(yt);
lag=20;
if (N1~=N2)
    error('Please make sure that both signals have the same
length');
if (N1<2*lag || N2<2*lag)
    error('Please make sure that the size of yt and zt are at
least 40 samples.');
end
if (N1==q1(1) && q1(2)~=1) || (N1==q1(2) && q1(1)~=1) ||
(N2==q2(1) && q2(2)~=1) || (N2==q2(2) && q2(1)~=1)
    error('Please make sure that both yt and zt is either a
row or column vector.');
end
x=detrend(xt,0);
y=detrend(yt,0);
%Obtain the cross-correlation values
q=xcorr(y,x);
count=N1;
N=N1;
%Creating the plot
plot([-lag:lag],q(count-lag:count+lag)/std(x)/std(y)/N1,'ok')
```

```
grid on
hold on;plot([-lag,lag],tinv(0.975,N)*[1,1]/sqrt(N),'--
k','linewidth',2)
plot([-lag,lag],-tinv(0.975,N)*[1,1]/sqrt(N),'--
k','linewidth',2)
plot([-lag,lag],[0,0],'-k')
xlabel(['Lag
(samples)'],'interpreter','latex'),ylabel('Cross-correlation,
$\rho_{YZ}$','interpreter','latex')
end
```

7.8 MATLAB® Examples

This section presents three examples that show how to implement various forms of regression analysis in MATLAB®. The topics considered are linear regression, nonlinear regression, and system identification. All examples are based on real data obtained from experiments. Appropriate MATLAB® code, as well as the final results, is provided so that the readers can modify these examples to fit their particular needs.

7.8.1 Linear Regression Example in MATLAB®

This example examines the problem of fitting a theoretical equation to experimental data in order to obtain the values of the different constants in the system. Detailed information about the problem can be found in (Prickett et al. 2011, 2010; Elliott et al. 2007; Jochem and Körber 1987). Data used for this example come from Prickett et al. (2011).

7.8.1.1 Problem Statement for the Linear Regression Example

Consider the problem of obtaining the values of the parameters in a theoretical equation that describes the osmotic pressure of the sodium chloride (NaCl) salt and hydroxyethyl starch (HES, chemical formula $(C_6H_{10}O_5)_m(C_2H_5O)_n$). Based on the virial equation of state, it is assumed that the following equation can be used to describe the osmolality (Π) of such a mixture

$$\Pi = B_3 m_3^2 + B_3 k_{diss} m_2 m_3 + C_3 m_3^3 + k_c, \tag{7.1}$$

Table 7.18 Fitting the virial equation (MATLAB® example)

m_2 (millimol/kg solv)	m_3 (millimol/kg solv)	k_c (milliosm/kg solv)	Π (milliosm/kg solv)
0	0.0000	0	0
600	0.0390	1,052	1,314
1,268	0.0823	2,326	2,267
2,013	0.1307	3,879	3,712
2,852	0.1852	5,792	5,496
3,803	0.2469	8,170	8,035
4,889	0.3175	11,161	11,513

where B_3 and C_3 are the virial parameters to be determined, m_2 is the molality of NaCl in millimol/kg of solvent, m_3 is the molality of HES in millimol/kg of solvent, k_{diss} is the disassociation constant that is equal to 1.678, and k_c is a known constant that depends on the system being analysed. An experiment was run where the ratio of the mass of HES to the mass of NaCl was fixed to 0.5. The results obtained are shown in Table 7.18.

7.8.1.2 Solution for the Linear Regression Example

Before linear regression can be applied, the above equation must be re-arranged so that all known constant information is on the left-hand side and all the unknown variables are on the right-hand side. Thus, the equation would be rewritten as

$$\Pi - k_c = B_3\left(m_3^2 + k_{diss}m_2 m_3\right) + C_3 m_3^3. \tag{7.2}$$

The required variables would be defined as

$$y = \Pi - k_c$$
$$\vec{x} = \left\langle m_3^2 + k_{diss}m_2 m_3, m_3^3 \right\rangle \tag{7.3}$$
$$\vec{\beta} = \langle B_3, C_3 \rangle^T.$$

In order to obtain the parameter estimates and analyse the results, the following MATLAB® script will be used.

```matlab
%Script for solving linear regression problems in MATLAB
%Copyright 2015 Yuri Shardt
%To be used in conjunction with Chapter 7 of the Springer
book, Statistics
%for Chemical and Process Engineers: A Modern Approach.

%Entering the raw data
m2=[0 600 1268 2013 2852 3803 4889]';
m3=[0 0.039 0.0823 0.1307 0.182 0.2469 0.3175]';
kc=[0 1052 2326 3879 5792 8170 11161]';
pi=[0 1314 2267 3712 5496 8035 11513]';
kdiss=1.678;
%Creating the required data matrices for solving the problem.
y=pi-kc;
A=[m3.^2+kdiss*m2.*m3 m3.^3]; %Note the use of the dot
operator
%Add code here for Part 2.
%Solve the problem to obtain the parameter estimates and
%information
[param,CI,residual,sr,info]=regress(y,A);
%display the results
fprintf(['B_3: %f±%f\n'],param(1),(CI(1,2)-CI(1,1))/2);
fprintf(['C_3: %f±%f\n'],param(2),(CI(2,2)-CI(2,1))/2);
%display the statistics
fprintf(['R^2 = %f\n'],info(1));
%examine the residuals
normplot(residual);
figure;plot(y,residual,'ok');xlabel('Measured
value');ylabel('Residual');
figure;plot(A(:,1),residual,'ok');xlabel('First
Regressor');ylabel('Residual');
figure;plot(A(:,2),residual,'ok');xlabel('Second
Regressor');ylabel('Residual');
figure;plot(A*param,residual,'ok');xlabel('Predicted
value');ylabel('Residual');
figure;plot(residual,'ok');xlabel('Sample');
ylabel('Residual');
```

The output from MATLAB® is.

```
B₃: -0.8206±0.624
C₃: 77,469±55,563
R² = 0.731103
```

The figures are shown in Fig. 7.1. It is clear from examining these figures that the second data point seems to be quite the outlier with an error that is much larger than any of the other data points. Other than this single outlier, the data set looks quite good. Even though the data sample is small, it would be worthwhile to remove this point and see how the regression changes. The previous MATLAB® code is changed by adding, after defining the regression matrices, the lines

```
y=y([1, 3:length(y)]);
A=A([1, 3:size(A,1)],:);
```

The output from MATLAB® becomes

```
B₃: -0.8535±0.168
C₃: 80,344±14,965
R² = 0.982121
```

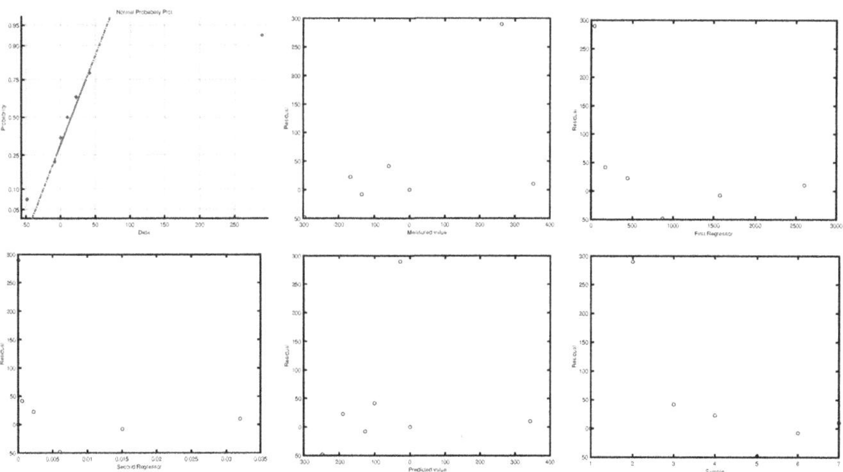

Fig. 7.1 Linear regression example: MATLAB® plots of the (top) Normal probability plot of the residuals, Residuals as a function of y, and Residuals as a function of the first regressor, x_1; and (bottom) Residuals as a function of x_2, Residuals as a function of $\hat{y}$, and A time-series plot of the residuals

It can be seen that the confidence intervals have decreased markedly and the R^2 is now almost 1. This strongly suggests that the removed data point was an outlier. Given the small sample size, the residual analysis graphs do not give any additional information. Practically speaking, the background regarding the outlier would need to be investigated in order to confirm that it is indeed an outlier. If after examining that there were no data collection or input errors, then the presence of the outlier could suggest that the model was not appropriate for the data set. It is always important to provide detailed reasons for why a given point was removed as an outlier, especially if there is access to the original data.

7.8.2   *Nonlinear Regression Example in MATLAB®*

This example examines the problem of fitting a theoretical equation to experimental data in order to obtain the values of the different constants in the system. Unlike the previous case, nonlinear regression must be performed in order to obtain a result. Detailed information about the problem can be found in (Ross-Rodriguez, 2009). Data provided courtesy of Dr. Lisa Ross-Rodriguez.

7.8.2.1   Problem Statement for the Nonlinear Regression Example

Consider the problem of obtaining a relationship for the ratio between the equilibrium and isotonic cell volumes given the osmotic pressure. The theoretical relationship can be written as

$$\frac{V}{V_0} = \left(1 - b^*\right)\frac{-1 + \sqrt{1 + 4B\Pi_0}}{-1 + \sqrt{1 + 4B\Pi}} + b^*, \qquad (7.4)$$

where both B and b^* are the parameters to be determined and Π_0 is a known osmotic value. The experimental data is provided in Table 7.19. For this data set, Π_0 has a value of 0.293.

	V/Vo	Π
Table 7.19 Equilibrium cell volume data (MATLAB® example)	1.000 34	0.292 78
	0.804 65	0.571 72
	0.753 58	0.855 14
	0.715 48	1.135 95
	0.685 88	1.433 49
	0.666 00	1.729 08
	0.659 13	2.028 15
	0.640 04	2.326 60
	0.626 61	2.667 04

7.8.2.2 Problem Solution for Nonlinear Regression Example

In order to solve the problem in MATLAB®, the function for which the parameter estimates are being obtained needs to be written as a MATLAB® function. It is very important that the following points be considered when writing the function:

(1) First, it must be able to deal with vector entries, that is, the dot operators should be used with times ($*$) and divide ($/$) to give ($.*$) and ($./$).

(2) Second, the header of the function must be correctly specified. The order of the inputs is parameter values, regressor values, and measured values. Each entry is assumed to be a matrix of appropriate size. The output is a single vector containing the results. Therefore, the header will be of the form

```
y = functionName(parameters,A,y);
```

Based on these constraints, the following MATLAB® function was written. It should be saved in the same location as the script that will be used to run the nonlinear regression.

```
function [y1]=volume(beta,x,y)
%Function to compute the predicted cell volumes given
%   beta:   the parameter coefficients
%   x:      the corresponding regressors
%   y:      the corresponding measured values
%Copyright 2015 Yuri Shardt
%Written as part of the Springer book Statistics for Chemical
and Process
%Engineers: A Modern Approach


y1=(1-beta(1))*(-1+sqrt(1+4*beta(2)*0.293))./(-
1+sqrt(1+4*beta(2)*x))+beta(1);
```

The following script was used to solve the nonlinear regression problem. The initial guess for the parameter estimates needs to be made carefully, as it can impact the ability of the system to give an answer. If possible, using the estimate obtained using the linearized model is a good idea.

```
%Script for solving nonlinear regression problems in MATLAB
%Copyright 2015 Yuri Shardt
%To be used in conjunction with Chapter 7 of the Springer
book. Statistics

%Entering the raw data
VVo=[1.00034 0.80465 0.75358 0.71548 0.68588 0.66600 0.65913
0.64004 0.62661];
pi=[0.29278 0.57172 0.85514 1.13595 1.43349 1.72908 2.02815
2.32660 2.66704];
%Solve the problem to obtain the parameter estimates and
associated
%information
[param,residual,J,covb] = nlinfit(pi,VVo,'volume',[0.2,0.56]);
CI=nlparci(param,residual,'covar',covb);
%display the results
fprintf(['b*: %f±%f\n'],param(1),(CI(1,2)-CI(1,1))/2);
fprintf(['B: %f±%f\n'],param(2),(CI(2,2)-CI(2,1))/2);
%examine the residuals
normplot(residual);
figure;plot(VVo,residual,'ok');xlabel('Measured
value');ylabel('Residual');
figure;plot(pi,residual,'ok');xlabel('First
Regressor');ylabel('Residual');
figure;plot(residual,'ok');
xlabel('Sample');ylabel('Residual');
```

Note that the parameter estimates may be slightly different from those obtained here
due to differences in the way the optimizing engine works. The MATLAB® output
is

```
b*: 0.5245±0.0436
B: 2.408±3.616
```

From here, it is easy to note that the B parameter is not significant and its value
could be zero. This suggests that potentially not enough data have been collected to
make an appropriate estimate. The residual plots are shown in Fig. 7.2. This figure

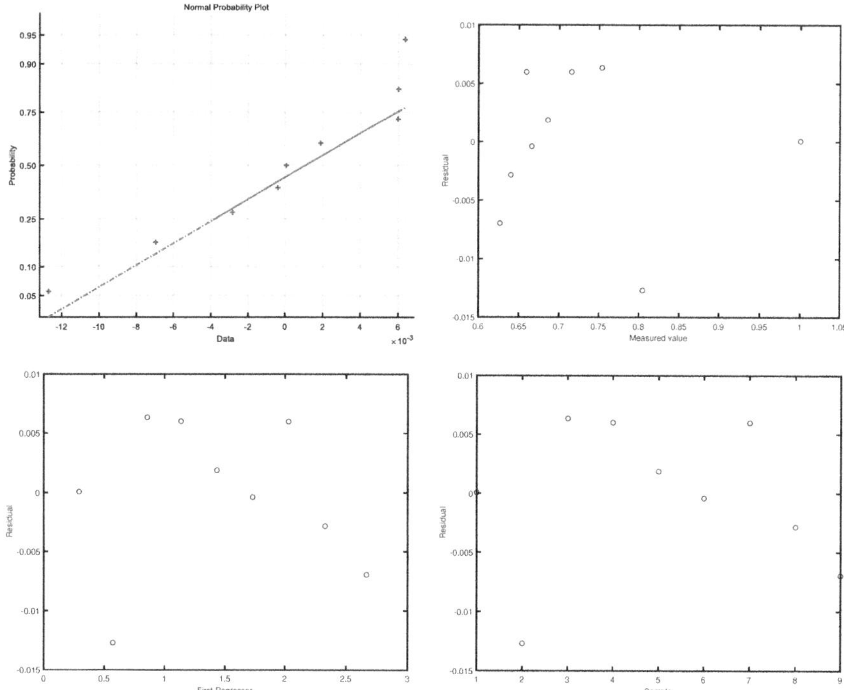

Fig. 7.2 Nonlinear regression example: MATLAB® plots of the (top) Normal probability plot of the residuals and Residuals as a function of Π; and (bottom) Residuals as a function of $\hat{y}$ and A time-series plot of the residuals

seems to show that there is some trend to the residuals. However, given the rather small sample, it is hard to discern exactly what this trend may be. Since it has been assumed that the given equation holds, in order to obtain a better understanding of the data, additional experiments should be provided.

7.8.3 System Identification Example in MATLAB®

The final example will consider the problem of system identification using the same data as used in Sect. 6.6: Modelling the Water Level in a Tank. For this reason, only the code required to model the level in Tank 1 will be presented. After making the relevant changes in the figure formatting the given figures will be obtained. This code requires the use of the System Identification Toolbox. The function can be called as follows: `systemidentification ([1 2], 1 1, [1 2], [1 2])` where, since there are two inputs, the values of n_b, n_f, and n_k are entered as vectors where each entry represents the individual cases.

```
function z=experiment(nb,nc,nd,nf,nk)
%Function to obtain system identification models of the data
assuming a
%Box-Jenkins model with parameter orders nb, nc, nd, and nf
with a time
%delay of nk.
%Copyright 2015 Yuri Shardt
load SystemIdentificationData;
newy1=Lower_Left_Level;
qnew=size(newy1);
%Plot the raw data
subplot(2,2,1), plot(U1)
xlabel('time (s)')
ylabel('flow rate, u (cm^3/s)')
title('Signal 1')
subplot(2,2,2), plot(U2)
xlabel('time (s)')
ylabel('flow rate, u (cm^3/s)')
title('Signal 2')
subplot(2,2,3), plot(newy1)
xlabel('time (s)')
ylabel('height, h (m)')
title('Tank 1')
%Create the data to store the object
z1=iddata([newy1],[U1,U2],1);
z1=detrend(z1,0);
%Obtain the parameter estimates
z=processbj(z1,nb,nc,nd,nf,nk,qnew);
end
function modelBJ=processbj(z1,nb,nc,nd,nf,nk,q)
%Partition the data set
split=ceil(2*q(1)/3);
```

```matlab
%Obtain the parameter estimates
modelBJ=bj(z1(1:split),'nb',nb,'nc',nc,'nd',nd,'nf',nf,'nk',nk
);
%Display the results
present(modelBJ)
%Plot the required residual analysis figures
figure
compare(modelBJ,z1(split+1:end))
figure
resid(modelBJ,z1(split+1:end)) %Note that the programme will
pause here in order to for the first graph to be examined
before displaying the next one.
r=resid(modelBJ,z1(split+1:end));
figure
normplot(r.OutputData);
end
```

7.9 Further Reading

The following are references that provide additional information about the topic:

(1) **General MATLAB® Help**:

    a.   Sizemore, J., & Mueller, J. P. (2015). *MATLAB® for Dummies.* Hoboken, New Jersey, United States of America: John Wiley & Sons, Inc.

    b.   Hunt, B. R., Lipsman, R. L., & Rosenberg, J. (2014). *A guide to MATLAB®: for beginners and experienced users: Updated for MATLAB® 8 and Simulink 8* (3rd ed.). Cambridge, England, United Kingdom: Cambridge University Press.

(2) **Linear Regression Data set**:

    a.   Elliott, J. A., Prickett, R. C., Elmoazzen, H. Y., Porter, K. R., & McGann, L. E. (2007). A Multisolute Osmotic Virial Equation for Solutions of Interest in Biology. *Journal of Physical Chemistry B, 111*, 1775–1785.

    b.   Prickett, R. C., Elliott, J. A., & McGann, L. E. (2010). Application of the osmotic virial equation in cyrobiology. *Cryobiology, 2010*, 30–42.

    c.   Prickett, R. C., Elliott, J. A., & McGann, L. E. (2011). Application of the Multisolute Osmotic Virial Equation to Solutions Containing Electrolytes. *The Journal of Physical Chemistry B, 115*, 14,531–14,543.

d. Jochem, M., & Körber, C. (1987). Extended Phase Diagrams for the Ternary Solutions $H_2O - NaCl - Glycerol$ and $H_2O - NaCl - $ Hydroxyethylstarch (HES) Determined by DSC. *Cryobiology, 24,* 513–536.

(3) **Nonlinear Regression Data set**:

a. Ross-Rodriguez, L. U. (2009). *Cellular Osmotic Properties and Cellular Responses to Cooling.* Edmonton, Alberta, Canada: University of Alberta.

Chapter 8
Using Excel® to Do Statistical Analysis

Microsoft Excel® is a spreadsheet programme developed by Microsoft®, which comes bundled with Microsoft Office®. The most recent version of Microsoft Office is Office 2021. Not only can Excel® perform most basic spreadsheet commands, it also contains a programming language called Visual Basic that can be used to create powerful and useful macros. Most, if not all, of the methods presented in the previous chapters can be easily implemented in Excel®. It will be assumed that the reader has a basic understanding of Excel®, can write simple formulae (equations), understands what a column and row are, and can create basic graphs. Basic background information about Excel® can be found from such sources as the *Excel® for Dummies* series (Harvey 2019). This chapter will examine in detail additional features, such as array functions, using Solver, and writing basic Excel® macros. In order to clearly distinguish between the code required for Excel® functions and text, all Excel® commands and variables are shown in bold **`Courier New`**. Locations on the ribbon, menu paths, and keystrokes are shown in plain `Courier New`.

8.1 Ranges and Arrays in Excel®

Ranges and **arrays** are how Excel® refers to groups of columns and rows. The difference lies in how they are used. A **range** is a group of rows and columns in an Excel® spreadsheet, while an **array** is a group of rows and columns used in an Excel® function or code. In Excel®, an array most closely approximates a matrix.

One useful property of Excel® is the ability to write an equation in one cell and then drag it to other cells. When dragging such an equation, any references to a range will be changed as the cells are dragged, for example, the cell **A4** will become **A5** if the formula is dragged down one row and **B4** if it is dragged right by one column. This is called **relative** referencing, since it depends on the location. Although relative referencing is useful, it is not always desirable. Excel® allows a cell reference to be made **absolute**, that is, it will not change its value as the cell is dragged. This is done by placing a dollar sign **$** before the element that one wishes to freeze. There are three options (illustrated using cell **A2**):

(1) **Absolute Rows and Columns (A2)**: In this case, the reference will always be to this cell.
(2) **Absolute Row, but Relative Column (A$2)**: In this case, the row will stay the same, but the column can change.
(3) **Absolute Column, but Relative Row ($A2)**: In this case, the column will stay the same, but the row can change.

Another useful feature in Excel® is the ability to name a range. This means that rather than having to drag and select a very large range of cells, it can be conveniently referred to using the corresponding range name, for example, instead of **average(B4:B1000)** we could write **average(y)**, where **y** has been defined to equal **B4:B1000**. Naming a range is shown in Fig. 8.1 and consists of two steps: select the range and enter a name in the location shown. In order to make your life easier, make sure that the name selected is unique to the workbook and not the worksheet.

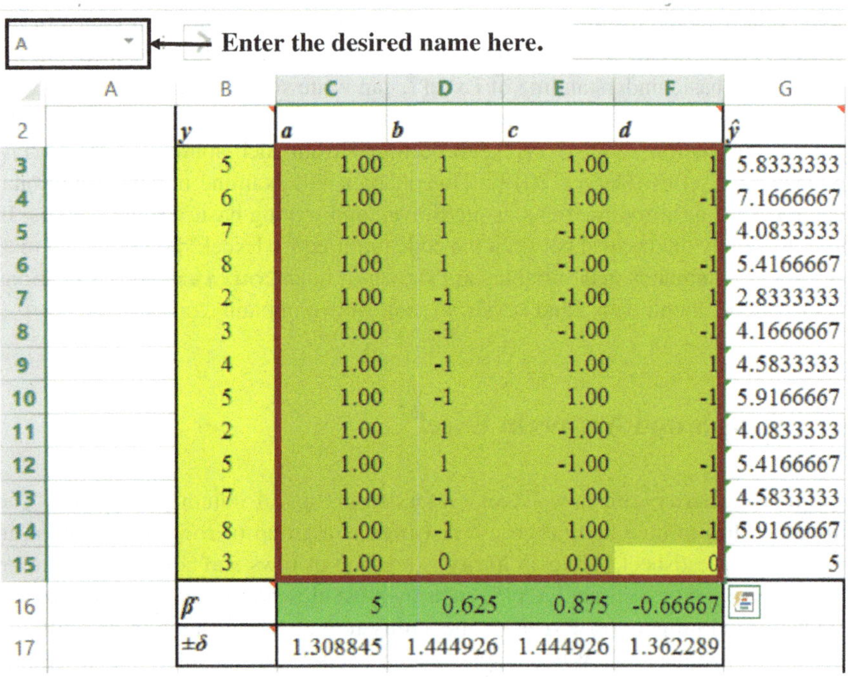

Enter the desired name here.

A

y	a	b	c	d	ŷ
5	1.00	1	1.00	1	5.8333333
6	1.00	1	1.00	-1	7.1666667
7	1.00	1	-1.00	1	4.0833333
8	1.00	1	-1.00	-1	5.4166667
2	1.00	-1	-1.00	1	2.8333333
3	1.00	-1	-1.00	-1	4.1666667
4	1.00	-1	1.00	1	4.5833333
5	1.00	-1	1.00	-1	5.9166667
2	1.00	1	-1.00	1	4.0833333
5	1.00	1	-1.00	-1	5.4166667
7	1.00	-1	1.00	1	4.5833333
8	1.00	-1	1.00	-1	5.9166667
3	1.00	0	0.00	0	5
β		5	0.625	0.875	-0.66667	
±δ		1.308845	1.444926	1.444926	1.362289	

Fig. 8.1 Naming a range (Excel® 2019)

Table 8.1 Excel® array functions

Function	Description
mdeterm(array)	Determines the determinant of an **array**. The result will be a single scalar value
minverse(array)	Determines the inverse of the $n \times n$ **array**. The result will be the same size as the initial array
mmult(array1, array2)	Multiples two arrays **array1** and **array2** together. If **array1** has size $m \times n$, then **array2** must have size $n \times p$. The result will have size $m \times p$
transpose(array)	Transposes an **array**, that is, the rows and column are exchanged. If the array was originally $n \times m$, then the output will be $m \times n$

8.2 Useful Excel® Functions

This section will examine various Excel® functions that can be used for solving statistical problems.

8.2.1 Array Functions in Excel®

Array functions are Excel®'s equivalent to matrices. Arrays are defined as a range of cells that are treated together. When using array functions, the following steps must be followed:

(1) Select the output range.
(2) Enter the array formula into one of the cells in the selected array.
(3) Once the formula has been entered press Ctrl + Shift + Enter to register the formula as an array formula. Normally, one would simply press Enter.

A summary of the most common array functions is given in Table 8.1.

8.2.2 Statistical Functions in Excel®

Table 8.2 lists some common statistical functions in Excel®. A detailed explanation of the functions and differences can be found in Sect. 2.4: Common Statistical Distributions.

Table 8.2 Excel® statistical functions

Function	Description
average(range)	Determines the mean value of the numbers in **range**
count(range)	Counts the number of nonempty cells in **range**
f.inv (p,df1,df2)	Returns the critical value of the F-test for the given left probability **p**, the degrees of freedom in the numerator **df1**, and the degrees of freedom in the denominator **df2**
norm.inv(p,mean,stdev)	Returns the critical value for the normal distribution with mean **mean** and standard deviation **stdev** given the left probability **p**
norm.s.inv(p)	Returns the critical value for the standard normal distribution ($\mu = 0$ and $\sigma = 1$) given the left probability **p**
rank(value,range,order)	Returns the order rank of **value** given the **range**. The option **order** determines in which manner the list is ordered: **0** implies descending order and **1** implies ascending order
stdev(range)	Determines the sample standard deviation of the numbers in **range**
sum(range)	Determines the sum of the numbers in **range**
t.inv(p,df)	Returns the critical value of Student's t-test for the given left probability **p** and degrees of freedom **df**

8.3 Excel® Macros and Security

Macros are Excel®'s version of functions, or user-written code, that Excel® can execute. The programming language used by Excel® is called Visual Basic (VB).

In Excel® 2007 or newer, code can be inserted by going to the View Ribbon, selecting the Macro icon, and then View Macro. In the window that appears, enter the name of the function that you desire to create (or edit) and press Create (Edit). If a new function is being created then, in the new window that opens, replace **Sub** with **Public Function**. This will allow the new code to be directly accessed from the spreadsheet by typing = **=FunctionName(Required Parameters)**. Below, some sample code has been provided that implements the Michaelis–Menten equation.

```
Public Function MichaelisMenten(Concentration, vmax, KM) As
Double
'This function will contain a single line of code that imple-
ments the Michaelis-Menten equation
MichaelisMenten = vmax * Concentration / (KM + Concentration)
End Function
```

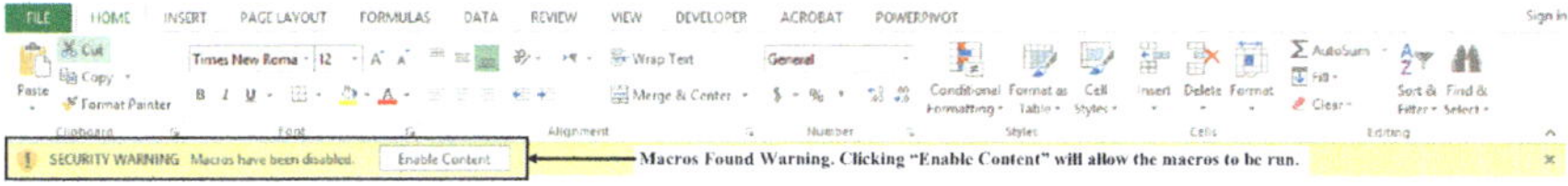

Fig. 8.2 Security warning when macros are present (Excel® 2019)

8.3.1 Security in Excel®

Unfortunately, when a macro is designed, Excel® has the tendency to be paranoid and think that it is always a nasty virus. Thus, the appropriate parameters should be set for security. The procedure in dealing with security in Excel® depends on the version of Excel® installed. The following sections explain the procedures for Excel® 2016 (and newer).

In Excel® 2016 or newer, to set the security, go to the `File` menu, and select `Options`. Select `Trust Center` in the window that appears. After this, select `Trust Center Settings….` In the new window, go to `Macro Settings` and select the appropriate level of security you desire. A good choice is to select the option `Disable All Macros with Notification` because the macros will be disabled, but you will be notified of their existence. Press `OK` on all the open Windows to save the changes. A file with a macro must be saved as an `.xlsm` file. When opening the file with a macro and the above-suggested settings, a warning similar to that shown in Fig. 8.2 should appear. Clicking on `Enable Content` will permit the macros to be used.

8.4 The Excel® Solver Add-In

Solver is an Excel® add-in that allows the user to iteratively solve systems of equations. Unfortunately, it is not installed by default on most computers.

8.4.1 Installing the Solver Add-In

In Excel® 2016 or newer, in order to install the Solver add-in, go to the `File` menu and select `Options` at the bottom of the Menu that appears. In the new window that appears, select `Add-ins`. Finally, click the `Go…` button. The last two steps are shown in Fig. 8.3. A window similar to Fig. 8.4 should appear.

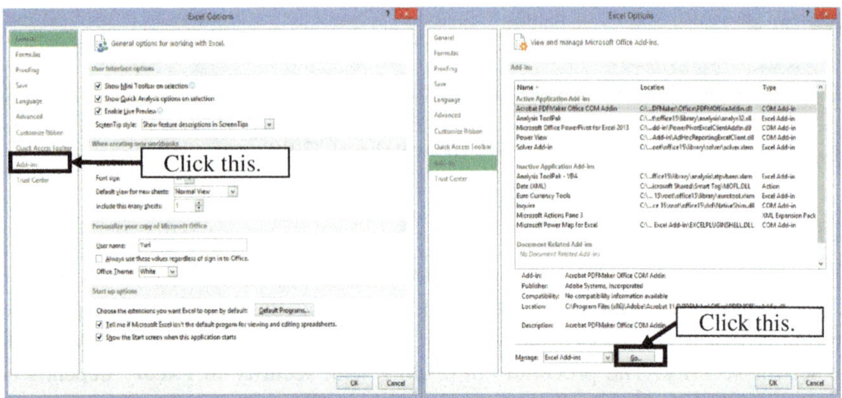

Fig. 8.3 Navigating to the Solver installation menu (Excel® 2019)

Fig. 8.4 Installing Solver

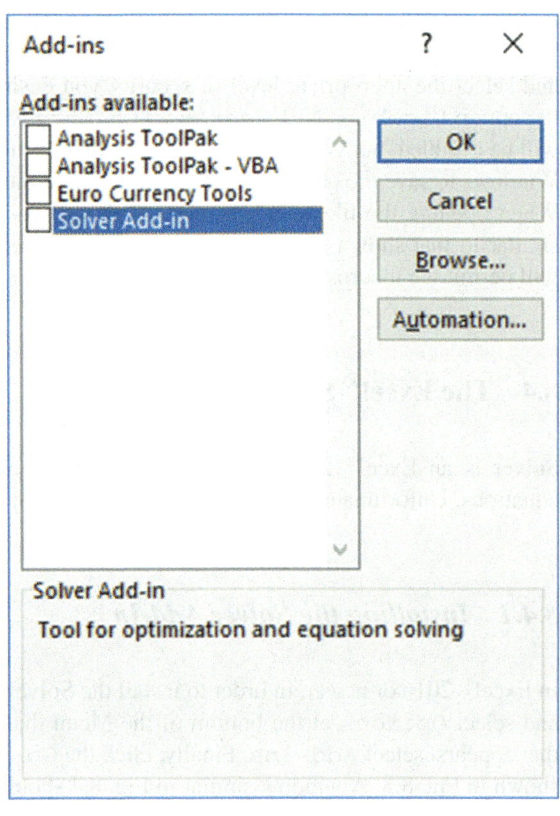

8.4.2 Using the Solver Add-In

In order to start Solver, in Excel® 2016 or newer, locate the `Data ribbon` and go to the extreme right-hand side in the area marked `Analysis`. Solver should be there as shown in Fig. 8.5.

Figure 8.6 shows the `Main Solver Window` that appears in Excel® 2016 or newer. It is a must that the option `Make Unconstrained Variables Non-Negative` be **unchecked**, as it can lead to wrong results otherwise. The following sections are important for use in regression analysis:

(1) **Objective Function Value**: This is the value of the objective function that is to be optimized.
(2) **Type of Optimisation**: What type of optimization is desired: maximization (`Max`), minimization (`Min`), or force the solver to obtain a particular value (`Value of`). For regression, the minimization option should be used.

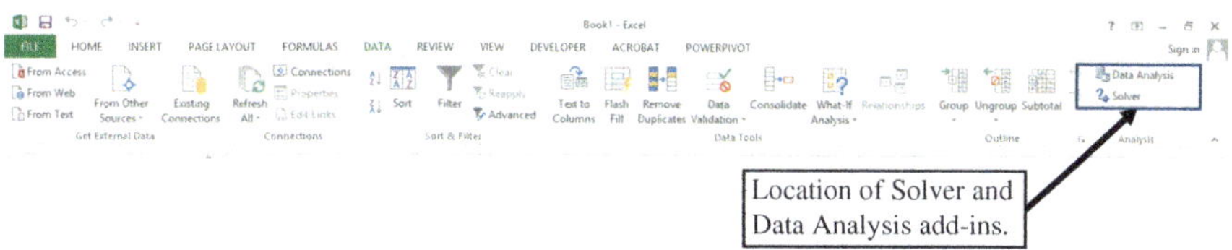

Fig. 8.5 Location of the Solver and Data Analysis add-ins (Excel® 2019)

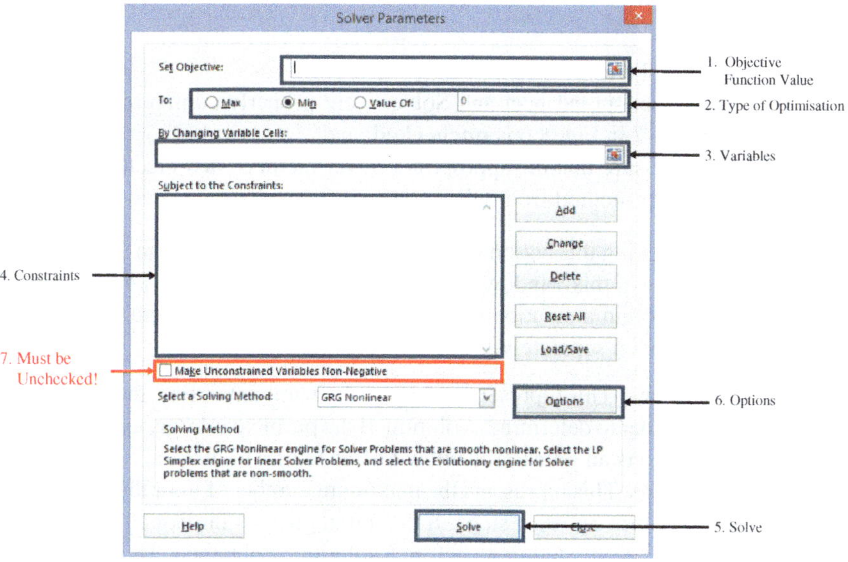

Fig. 8.6 Main Solver window (Excel® 2016 or newer)

Fig. 8.7 Add Constraint
window

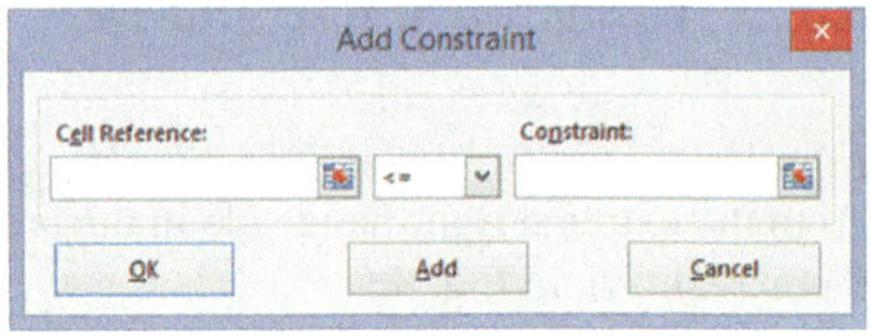

(3)   **Variables**: This is the range of the cells (variables) that the computer can vary
to determine the solution. For regression, this would represent the cells where
the parameter values have been entered.

(4)   **Constraints**: This box lists the constraints for the problem. In order to add
a constraint, click on the Add button. The window shown in Fig. 8.7 should
appear. Once the desired form of the constraint has been selected, click Add to
add the constraint to the list of constraints. Selecting a constraint from the box
and clicking Change will cause the same window to appear and the properties
of the constraint can be changed. Finally, selecting a constraint and clicking
Delete will remove the constraint.

(5)   **Solve**: Clicking this button will start the solver. The solution of the problem
may take some time. Solver will either state that a solution was found (Fig. 8.8
(left)) or that no solution was found (Fig. 8.8 (right)). In general, if a solution
is found, select Keep Solver Solution and press OK; otherwise select
Restore Original Values and press OK. If the Solver fails to find a
solution, an error message will be included. It can give a suggestion as to how
to fix the problem. Three common things to check (in order of preference) are
that:

a.   The number of iterations was not exceeded;

b.   The Excel spreadsheet and Solver were properly configured, especially
that Box 7 in Fig. 8.6 is **unchecked**; and

c.   To make sure that the appropriate solver method was selected. Changing
the solver method from GRG nonlinear to evolutionary can be useful.

(6)   **Options**: Clicking this button will bring up the window shown in Fig. 8.9. Each
of the choices in this window can speed up or slow down the amount of time
required to obtain a solution or even if a solution can be found. Each option
will be discussed separately:

a.   **Max Time**: This represents the maximum amount of time that Solver will
run in order to determine a solution. If the problem is large, then increasing
this option can potentially allow Solver to find a solution.

b.   **Iterations**: This represents the maximum number of iterations that Solver
will perform before it stops. If the initial guess is far from the solution, it
may take many iterations before a solution is obtained. Thus, increasing
the number of iterations can be a good idea.

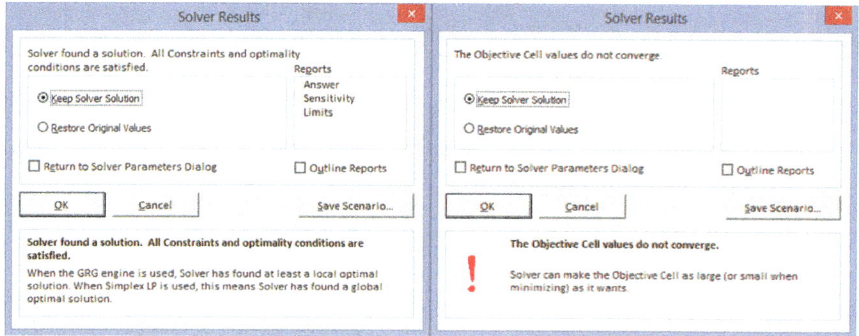

Fig. 8.8 (left) Solver found a solution and (right) Solver failed to find a solution (one possible result)

c. **Precision**: This represents the largest possible difference between the calculated value of the constraints and the specified value of the constraints. The smaller the number the longer it will take to find a solution.

d. **Tolerance**: This is similar to precision but is used for integer constraints. It represents the percentage by which the calculated values differ from the specified values.

e. **Convergence**: This is similar to precision but is used to compute the maximum allowable difference between two iterations of the parameters (or cells that can change). Since for most purposes, a relative value would be better, this entry should be changed whenever the parameters are expected to either be all very large numbers or very small numbers.

f. **Use Automatic Scaling**: This should always be selected as it minimises the effect the magnitude of the different variables can have on the solution. It is especially important if one of the variables ranges from 100 to 1,000, but the other variable ranges from 0.01 to 1.

The options in the other tabs are mostly irrelevant and should be left at their default values unless the problem at hand requires special treatment. However, the correct approach to take requires consulting an appropriate source on numerical methods.

8.5   The Excel® Data Analysis Add-In

The Data Analysis add-in in Excel® is another very useful Excel® add-in that can improve the ability to perform certain statistical tasks. It is installed using the same procedure as installing the Solver add-in (see Sect. 8.4.1: Installing the Solver). In order to start the Data Analysis Add-In, locate the `Data ribbon` and go to the extreme right-hand side in the area marked `Analysis`. The Data Analysis Add-In should be there as shown in Fig. 8.5.

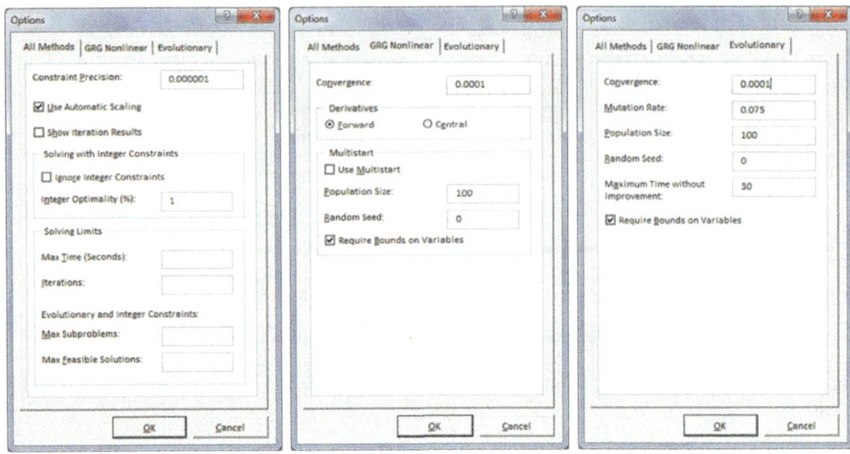

Fig. 8.9 Solver Option Window (Excel® 2016 or newer)

The Data Analysis window is shown in Fig. 8.10. Although there are many different options, the main problem with the data analysis add-in is that the results are static and that any changes made in the original data set require the given programme to be rerun. As well, the display of information is not always the best. Nevertheless, for the purposes of this book, the only useful option is the **Fourier Analysis** option, which will compute, given a data set, the appropriate Fourier coefficients, which can then be used to create a periodogram for the data set. An Excel® template file has been created to simplify the process (see Sect. 8.6.3: Periodogram Template).

Selecting the Fourier Analysis option will give the window shown in Fig. 8.11. There are only two key areas to consider. First, the input range must have a length of 2^n, where $n \in \mathbb{N}$, that is, the length must be an integer power of 2. If the particular list is less than the desired value, then add extra zeros to the end of the list to make it an integer power of 2. The output range should have the same size and orientation as the input range, that is, if the input range is a column, then the output range should also be one; similarly for a row. Clicking OK will give the required Fourier coefficients.

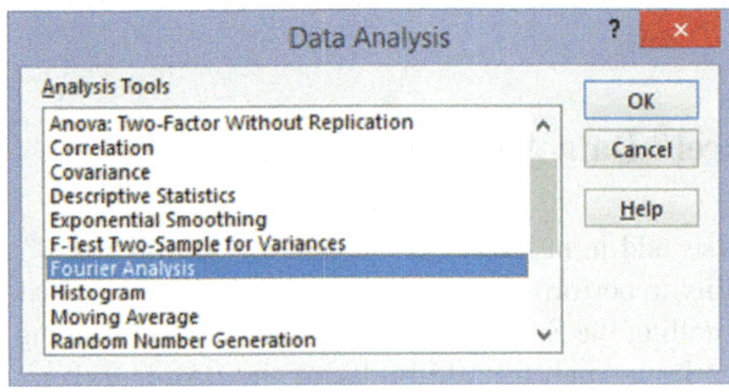

Fig. 8.10 Data Analysis window (Excel® 2016 or newer)

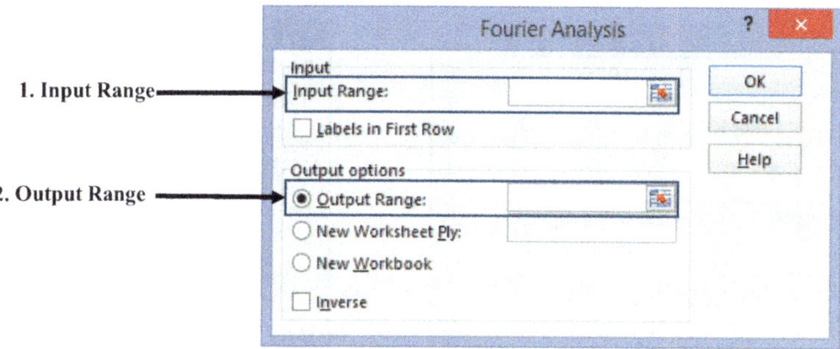

Fig. 8.11 Fourier Analysis window (Excel® 2016 or newer)

8.6 Excel® Templates

This section describes the Excel® templates available from the book website. All templates have been tested on Excel® 2019, 2016 and 2013.

The following are some useful reminders when using the templates:

(1) **Adding new rows**: New rows should be entered inside the thick bordered area. This will automatically update all formulae to include the new row. This can be accomplished by right clicking on the appropriate row name and selecting `Insert`. A row will be inserted above the selected row. See Fig. 8.12 for an example.

(2) **Adding new columns**: New columns should be entered inside the thick bordered area. This will automatically update all formulae to include the new column. This can be accomplished by right clicking on the appropriate column name and selecting `Insert`. A column will be inserted to the left of the selected column. See Fig. 8.12 for an example.

(3) **Formulae**: Most formulae can be dragged down or across to fill the new data. It is suggested that you drag from the first row (or column) down to the last to make sure that everything is properly aligned.

(4) **Macros**: A few of the templates contain macros that allow for easier and better implementation of the given procedures. When macros are present, make sure that the security is appropriately set so that they can be used (see Sect. 8.3.1: Security).

8.6.1 Normal Probability Plot Template

Requirements: Basic Excel® installation.

Goal: Create a normal probability plot in Excel®. Can be modified to deal with other distributions.

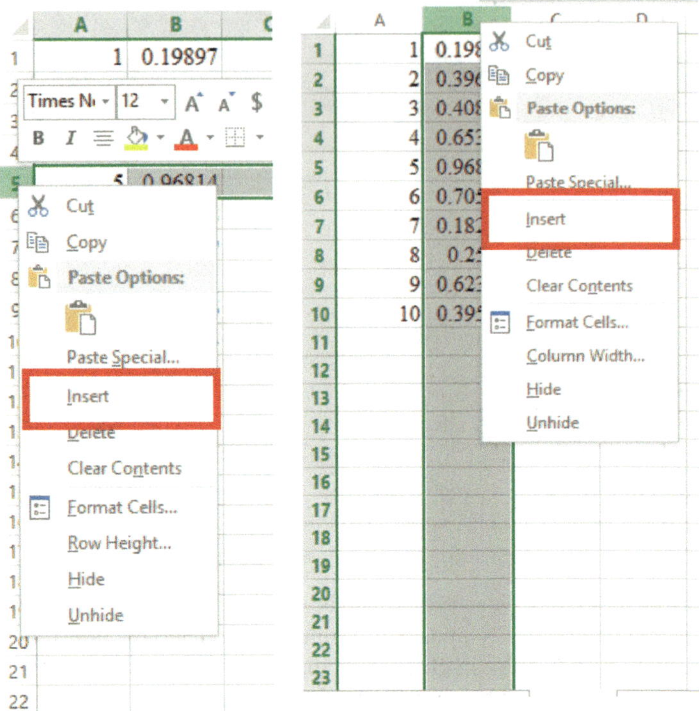

Fig. 8.12 Inserting a (left) row and (right) column (Excel® 2019)

File Name: `normplot.xltx`.

Description: A screenshot of the template with an explanation of the formulae used is shown in Fig. 8.13. The resulting normal probability plot is shown in Fig. 8.14. The steps for creating a normal probability plot can be summarised as follows:

(1)   Place the original data in Column A.
(2)   Obtain the order of the data in Column A in Column B. You can use the rank function.
(3)   In Column C, enter = **norm.s.inv((ColumnB1-0.5)/count (Column$A))**. The value of 0.5 is subtracted from the original ranked value in order to avoid asking the computer for the location for which the probability is 100% (it is $+\infty$!).
(4)   In Column D, compute the Z-score for each of the data points, that is, subtract the mean and divide the resulting value by the standard deviation of the values in Column A.
(5)   Plot a scatter plot of the data in Columns C and D.
(6)   The straight line can be added by plotting the data in Column C against itself.

Warnings: The axes of the plot are fixed to the range $[-3.0, 3.0]$. Should there be data outside this region, then it will be necessary to manually change the axis limits.

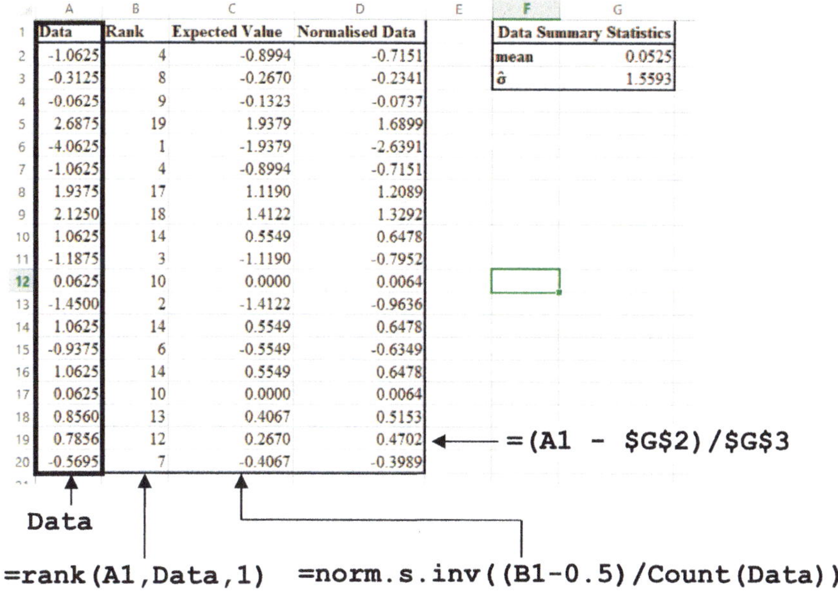

	A	B	C	D	E	F	G
1	Data	Rank	Expected Value	Normalised Data		Data Summary Statistics	
2	-1.0625	4	-0.8994	-0.7151		mean	0.0525
3	-0.3125	8	-0.2670	-0.2341		$\hat{\sigma}$	1.5593
4	-0.0625	9	-0.1323	-0.0737			
5	2.6875	19	1.9379	1.6899			
6	-4.0625	1	-1.9379	-2.6391			
7	-1.0625	4	-0.8994	-0.7151			
8	1.9375	17	1.1190	1.2089			
9	2.1250	18	1.4122	1.3292			
10	1.0625	14	0.5549	0.6478			
11	-1.1875	3	-1.1190	-0.7952			
12	0.0625	10	0.0000	0.0064			
13	-1.4500	2	-1.4122	-0.9636			
14	1.0625	14	0.5549	0.6478			
15	-0.9375	6	-0.5549	-0.6349			
16	1.0625	14	0.5549	0.6478			
17	0.0625	10	0.0000	0.0064			
18	0.8560	13	0.4067	0.5153			
19	0.7856	12	0.2670	0.4702			
20	-0.5695	7	-0.4067	-0.3989			

Fig. 8.13 Normal probability plot data (The formulae given are those placed in the first row, they would then be dragged down into each of the remaining rows.)

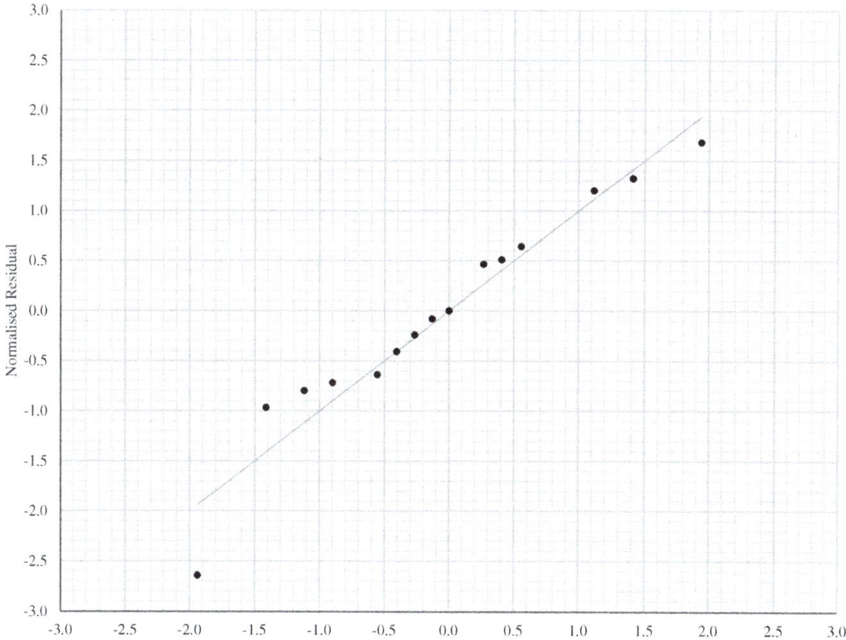

Fig. 8.14 Resulting normal probability plot

8.6.2   Box-and-Whisker Plot Template

Requirements: Basic Excel® installation.

Goal: Create a box-and-whisker plot in Excel®.

File Name: `boxplot.xltx`

Description: A screenshot of the box-and-whisker plot is shown in Fig. 8.15. The following steps can be used to create a box-and-whisker plot in Excel® from scratch:

(1)   Place the data in a column (or row) and call that range **data**.

(2)   Compute the minimum, first quartile, median, third quartile, and maximum values. This can be accomplished by using the following formulae: **min(data), quartile.inc (data,1), median(data), quartile.inc (data,3)**, and **max(data)**.

(3)   Create a column containing the following values in the specified order: Q1 − minimum, Q1, Median − Q1, Q3 − Median, and Maximum − Q3. This will allow the box-and-whisker plot to be properly created in Excel®.

(4)   Select the middle three items (Q1, Median − Q1, and Q3 − Median) and create a stacked column graph. The steps required are shown in Fig. 8.16. The arrows provide the sequence of steps that should be followed to create the graph. The initial graph that is obtained now needs to be formatted to look like a box-and-whisker plot.

(5)   Select the bottom blue box and add a negative error bar, which is equal to Q1 − minimum. Set the positive error bar equal to zero. The steps required are shown in Fig. 8.17. Select the top grey box and add a positive error bar, equal to Maximum − Q3. The negative error bar should be set to zero. The same procedure would be followed, *mutatis mutandis*.

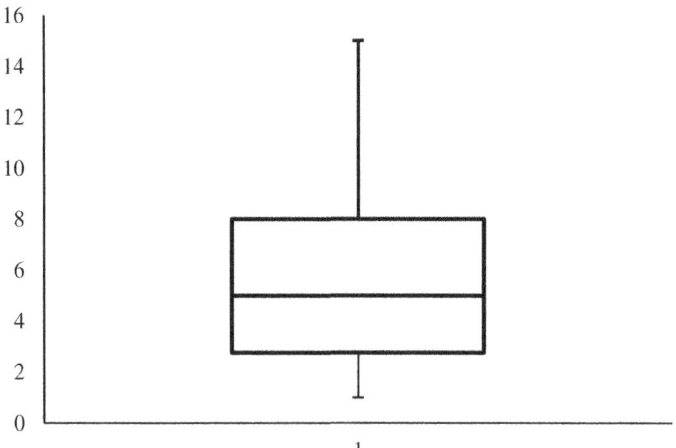

Fig. 8.15   Box-and-whisker plot in Excel®

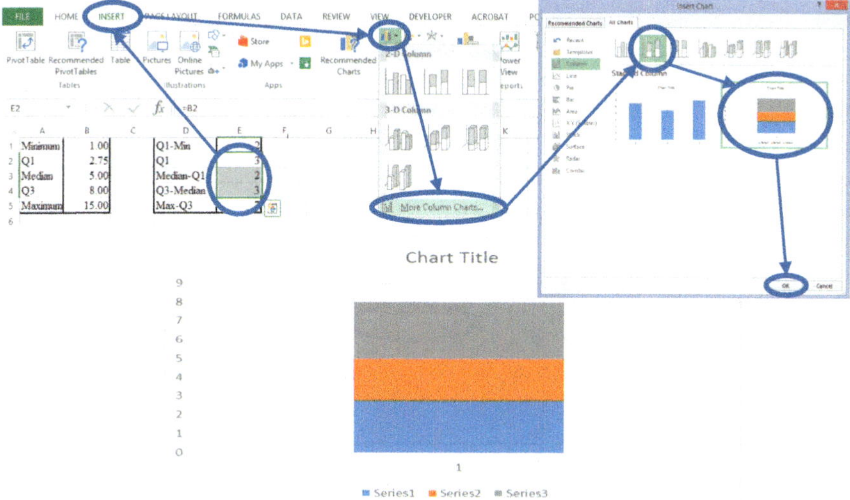

Fig. 8.16 Creating the initial graph for a box-and-whisker plot (Excel® 2019). The arrows provide the sequence of events to follow

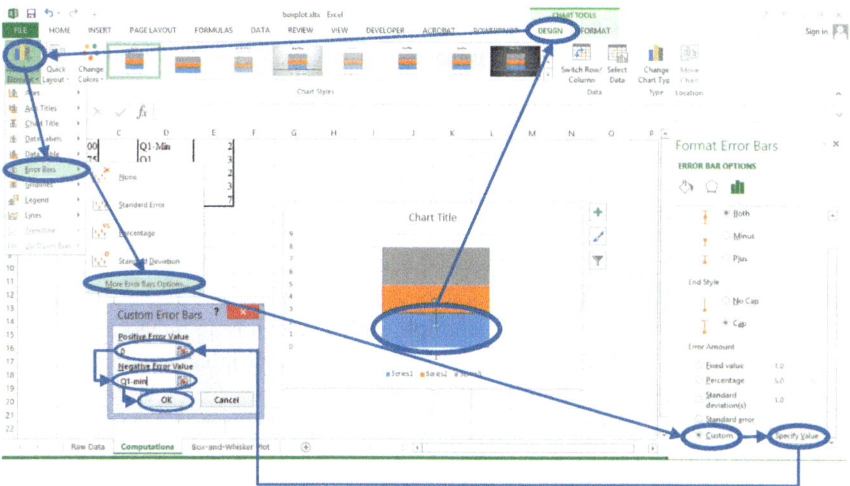

Fig. 8.17 Adding error bars (Excel® 2019). The arrows provide the sequence of events to follow

(6)   Once again, select the blue box and set the fill option to `no fill` and the border option to `no line`. The steps required are shown in Fig. 8.18. Select the orange box and set its fill to `no fill`. Repeat for the grey box. This should now look like a box-and-whisker plot. Additional formatting can be performed to obtain the final version.

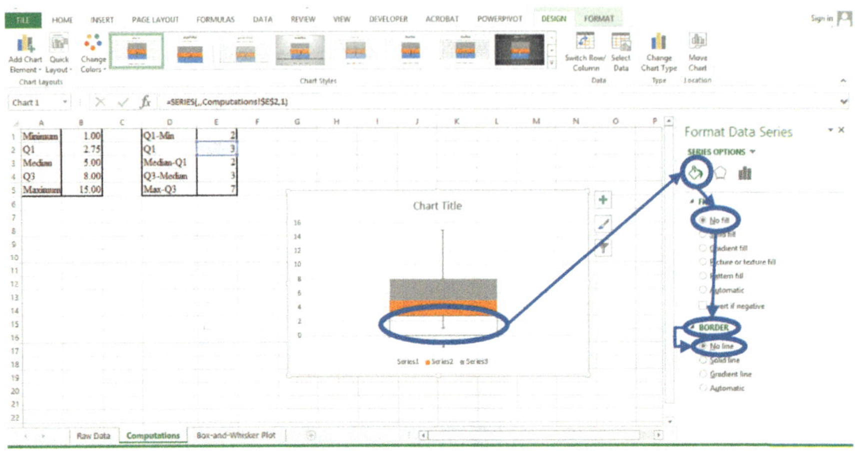

Fig. 8.18 Changing the fill and border options (Excel® 2019). The arrows provide the sequence of events to follow

8.6.3 Periodogram Template

Requirements: Basic Excel® installation plus installing the Data Analysis add-in (see Sect. 8.5: The Excel® Data Analysis for how to install it.)

Goal: Create both the full and half periodograms in Excel®.

File Name: `periodogram.xltx`.

Description: A screenshot of the template is shown in Fig. 8.19 with the resulting periodograms shown in Fig. 8.20. Note that every time new data are entered, it is necessary to rerun the Fourier analysis function in the Data Analysis add-in. The set-up of the Fourier transform window is shown as an inset in Fig. 8.19. As well, the number of data points must be a multiple of 2^n where n is an integer, that is 2, 4, 8, 16, 64, 128, 456,…. If the data set of interest is not a multiple, then it is necessary to add extra zeros to the end of the list to make it so.

An explanation of the columns is as follows:

(1) **Column A** contains a simple count of the sample number **starting from 1**.

(2) **Column B** contains the values corresponding to each sample number. This column is called **data** and must be a multiple of 2^n, where n is an integer.

(3) **Column C** contains the Fourier transform values as returned by the Fourier analysis function in Excel®. The values are complex numbers and should not be changed.

(4) **Column D** contains the magnitude of the values in Column C, that is, = `abs(C2)`. This column is used to construct the full periodogram.

(5) **Column E** contains the frequency corresponding to each sample, that is, = `(A2-1)/COUNT(data)`.

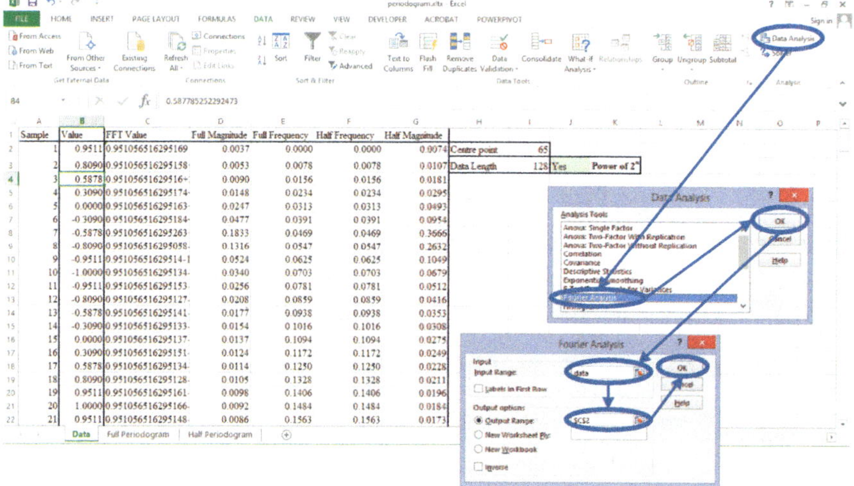

Fig. 8.19 Periodogram template layout (Excel® 2019). The inset shows how to initialise the Fourier analysis function

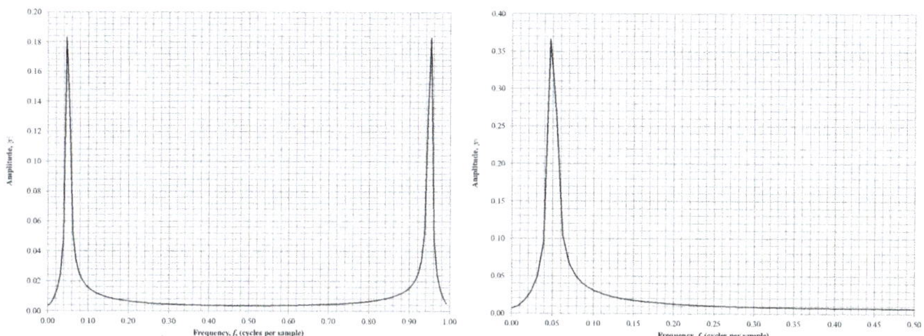

Fig. 8.20 Sample full and half periodograms

(6)   **Column F** contains the half periodogram frequencies, which is basically the first 2^{n-1} values from Column E with the remaining values set to #N/A, so that they will be ignored. The formula used is $=$ IF(A2-1 < I2,E2,#N/A). It should be noted that cell I2 contains the centre point value.

(7)   **Column G** contains the half periodogram magnitudes, which is basically twice the corresponding value in Column D, up to the centre point value, after which the values are arbitrarily set to #N/A. This allows the half periodogram to be plotted for an arbitrary number of values. The formula used is $=$ IF(A2-1 < I2,D2*2,"NaN").

(8)   **Full Periodogram**: The full periodogram is created by plotting Column D as the y-axis and Column E as the x-axis.

(9) **Half Periodogram**: The half periodogram is created by plotting Column G as the y-axis and Column F as the x-axis.

Warnings: The Fourier transform function must be rerun each time the data are changed. Furthermore, the data length must always be a multiple of 2^n, where n is an integer.

8.6.4 Linear Regression Template

Requirements: Basic Excel® installation plus ability to use macros.

Goal: Perform linear regression in Excel® in an easy and straightforward manner.

File Name: `linearregression.xltm`.

Description: A screenshot of the plain template is shown in Fig. 8.21. The yellow blocks are where the required data are entered. The green block represents the row in which an array formula needs to be entered. The complete green row should be selected and then the first cell highlighted. Finally, press `Ctrl+Shift+Enter` to copy the array formula to the entire green row. Adding additional parameters and data points will also require that the formulae be appropriately copied down. The spreadsheet automatically creates the normal probability plot for the residuals and plots of the residuals as a function of y and $\hat{y}$, as well as a time-series plot of the

y	a	b	c	d	$\hat{y}$	residuals	r^2		Ordered	Z-Rank	Z-Score
5	1.00	1	1.00	1	5.8333333	0.83	0.69		7	0	0.461266
6	1.00	1	1.00	-1	7.1666667	1.17	1.36		10	0.615141	0.645772
7	1.00	1	-1.00	1	4.0833333	-2.92	8.51		1	-1.76883	-1.61443
8	1.00	1	-1.00	-1	5.4166667	-2.58	6.67		2	-1.19838	-1.42992
2	1.00	-1	-1.00	1	2.8333333	0.83	0.69		7	0	0.461266
3	1.00	-1	-1.00	-1	4.1666667	1.17	1.36		10	0.615141	0.645772
4	1.00	-1	1.00	1	4.5833333	0.58	0.34		6	-0.19403	0.322886
5	1.00	-1	1.00	-1	5.9166667	0.92	0.84		9	0.395725	0.507392
2	1.00	1	-1.00	1	4.0833333	2.08	4.34		13	1.768825	1.153164
5	1.00	1	-1.00	-1	5.4166667	0.42	0.17		5	-0.39573	0.230633
7	1.00	-1	1.00	1	4.5833333	-2.42	5.84		3	-0.86942	-1.33767
8	1.00	-1	1.00	-1	5.9166667	-2.08	4.34		4	-0.61514	-1.15316
3	1.00	0	0.00	0	5	2.00	4.00		12	1.19838	1.107037

β		5	0.625	0.875	-0.66667	sum(r^2)	39.17	$\hat{\mu}_{residuals}$ 8.88E-16
$\pm\delta$		1.308845	1.444926	1.444926	1.362289	σ_{model}	2.09	$\sigma_{residuals}$ 1.806624

α	0.05
tinv(α)	2.262157
m	13
n	4
k	3
TSS	54
ESS	39.16667
SSR	14.83333
R^2	0.274691
F-test	1.13617
F-critical, model	3.862548
F-critical, factorial	5.117355

The normal probability plot is automatically created using the data provided in columns H, K, and L and displayed in the Normplot tab of this workbook.

Fig. 8.21 Linear regression template

residuals. Additional plots can be created by the user. An example of how to use the template is provided in Sect. 8.7.1: Linear Regression Example.

Warning: This template requires that the internal macros be enabled. As well, the array formulae need to be properly entered.

8.6.5 Nonlinear Regression Template

Requirements: Basic Excel® installation plus installation of Solver and the ability to use macros.

Goal: Perform nonlinear regression in Excel® in an easy and straightforward manner.

File Name: `nonlinearregression.xltm`.

Description: A screenshot of the plain template is shown in Fig. 8.22. The yellow blocks are where the required data are entered. Note that Solver needs to be used to obtain a solution to the problem. The configuration of Solver is shown as an inset in Fig. 8.22. The layout and formatting of the results are similar to the linear regression case. Two important differences are that the model and its Jacobian must be entered as a macro and that Solver must be used. The spreadsheet automatically creates the normal probability plot for the residuals and plots of the residuals as a function of y and $\hat{y}$, as well as a time-series plot of the residuals. Additional plots can be created by the user. An example of how to use the template is provided in Sect. 8.7.2: Nonlinear Regression Example.

The template comes with four predefined functions for creating the model and the corresponding Jacobian. Each function takes the same inputs: the range corresponding to the parameters and the range corresponding to the inputs. The four

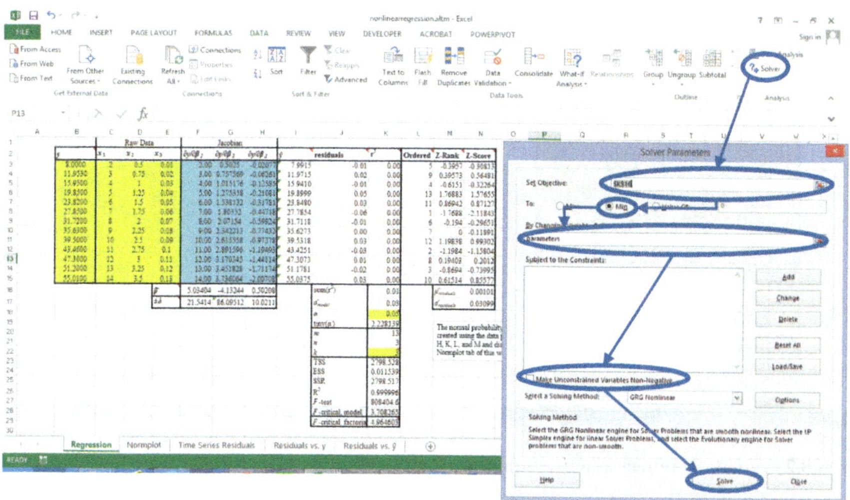

Fig. 8.22 Nonlinear regression template. The inset shows how to set up the Solver (Excel® 2019)

functions are: `model`, `dydb1`, `dydb2`, and `dydb3`. This approach is very similar to what MATLAB® requires and provides the most flexibility in defining the relevant functions.

Warning: This template requires that the internal macros be enabled and Solver installed.

8.6.6 Factorial Design Analysis Template

Requirements: Basic Excel® installation and appropriate macro security.

Goal: Perform the analysis of a factorial design experiment in Excel® in an easy and straightforward manner.

File Name: `factorialdesigntemplate.xltm`.

Description: A screenshot of the plain template is shown in Fig. 8.23. The yellow blocks are where the required data are entered. The green block represents the row in which an array formula needs to be entered. The complete green row should be selected and then the first cell highlighted. Finally, press `Ctrl+Shift+Enter` to copy the array formula to the entire green row. Adding additional parameters and data points will also require that the formulae be appropriately copied down.

The spreadsheet automatically creates the normal probability plot for the parameters and residuals as well as plots of the residuals as a function of y and $\hat{y}$ and a time-series plot of the residuals. Additional plots can be created by the user. An

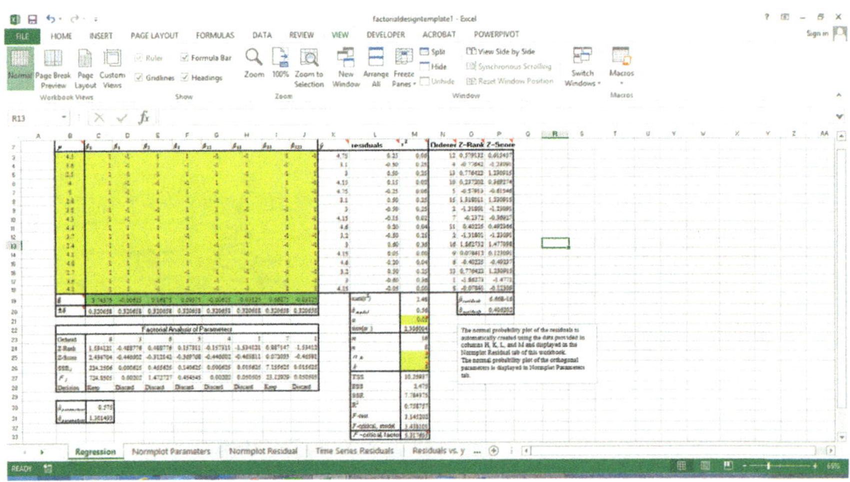

Fig. 8.23 Analysis of factorial experiments template

example of how to use the template is provided in Sect. 8.7.3: Factorial Design Examples.

Warning: This template requires that the internal macros be enabled. As well, the array formulae need to be properly entered.

8.7 Excel® Examples

This section presents three examples that show how to implement various forms of regression analysis in Excel®. The topics considered are linear regression, nonlinear regression, and analysis of factorial design. All examples are based on real data obtained from experiments. The procedures use the appropriate templates for solving the problem faster. The final form of the spreadsheet including all required information is provided as a reference for the user.

8.7.1 Linear Regression Example in Excel®

This example examines the problem of fitting a theoretical equation to experimental data in order to obtain the values of the different constants in the system. Detailed information about the problem can be found in Prickett et al. (2011), Elliott et al. (2007), Prickett et al. (2010), and Jochem and Körber (1987). Data used for this example come from Prickett, Elliott, and McGann (2011).

8.7.1.1 Problem Statement for Linear Regression Example

Consider the problem of obtaining the values of the parameters in a theoretical equation that describes the osmotic pressure of the sodium chloride (NaCl) salt and hydroxyethyl starch (HES, chemical formula $(C_6H_{10}O_5)_m(C_2H_5O)_n$). Based on the virial equation of state, it is assumed that the following equation can be used to describe the osmolality (Π) of such a mixture

$$\Pi = B_3 m_3^2 + B_3 k_{diss} m_2 m_3 + C_3 m_3^3 + k_c \tag{8.1}$$

where B_3 and C_3 are the virial parameters to be determined, m_2 is the molality of NaCl in millimol/kg of solvent, m_3 is the molality of HES in millimol/kg of solvent, k_{diss} is the disassociation constant that is equal to 1.678, and k_c is a known constant that depends on the system being analysed. An experiment was run where the ratio of the mass of HES to the mass of NaCl was fixed to 0.5. The results obtained are shown in Table 8.3.

Table 8.3 Fitting the virial equation (Excel® example)

m_2 (millimol/kg solv)	m_3 (millimol/kg solv)	k_c (milliosm/kg solv)	Π (milliosm/kg solv)
0	0.0000	0	0
600	0.0390	1,052	1,314
1,268	0.0823	2,326	2,267
2,013	0.1307	3,879	3,712
2,852	0.1852	5,792	5,496
3,803	0.2469	8,170	8,035
4,889	0.3175	11,161	11,513

8.7.1.2 Problem Solution for Linear Regression Example

Before linear regression can be applied, the above equation must be re-arranged so that all known constant information is on the left-hand side, and all the unknown variables are on the right-hand side. Thus, the equation would be rewritten as

$$\Pi - k_c = B_3\left(m_3^2 + k_{diss}m_2m_3\right) + C_3 m_3^3. \tag{8.2}$$

The required variables would be defined as

$$y = \Pi - k_c$$
$$\vec{x} = \left\langle m_3^2 + k_{diss}m_2m_3, m_3^3\right\rangle$$
$$\vec{\beta} = \langle B_3, C_3\rangle^T. \tag{8.3}$$

Entering all the data in the Excel® spreadsheet gives the results shown in Figs. 8.24 and 8.25.

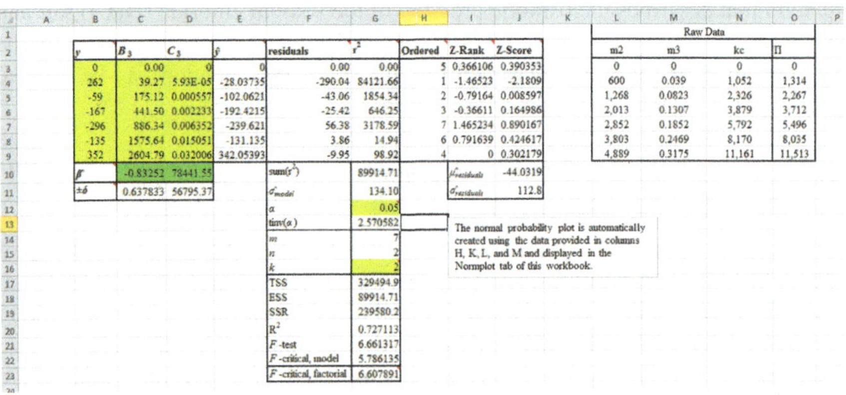

Fig. 8.24 Linear regression example: Data analysis results

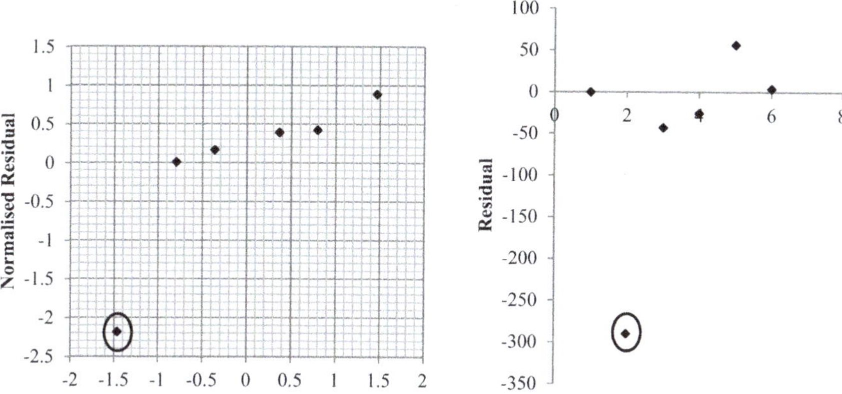

Fig. 8.25 Linear regression example: (left) Normal probability and (right) Time-series plots. The circled point is a potential outlier

	y	B_3	C_3	$\hat{y}$	residuals	r^2		Ordered	Z-Rank	Z-Score		m2	m3	kc	Π
											Raw Data			
	0	0.00	0	0	0.00	0.00		5	0.67449	0.231205		0	0	0	0
	-59	175.12	0.000557	-106.3545	-47.35	2242.45		1	-1.38299	-1.24741		1,268	0.0823	2,326	2,267
	-167	441.50	0.002233	-200.7887	-33.79	1141.68		2	-0.67449	-0.82383		2,013	0.1307	3,879	3,712
	-296	886.34	0.006352	-250.8708	45.13	2036.65		6	1.382994	1.640339		2,852	0.1852	5,792	5,496
	-135	1575.64	0.015051	-139.9814	-4.98	24.81		3	-0.21043	0.075665		3,803	0.2469	8,170	8,035
	352	2604.79	0.032006	348.56757	-3.43	11.78		4	0.210428	0.124029		4,889	0.3175	11,161	11,513
β		-0.86648	81408.59		sum(r^2)	5457.37		$\mu_{residuals}$	-7.40462						
$\pm\delta$		0.190134	16929.1		σ_{model}	36.94		$\sigma_{residuals}$	32.02622						
					a	0.05									
					tinv(a)	2.776445									
					m	6									
					n	2									
					k	2									
					TSS	245610.8									
					ESS	5457.365									
					SSR	240153.5									
					R^2	0.97778									
					F-test	88.01077									
					F-critical, model	6.944272									
					F-critical, factorial	7.708647									

The normal probability plot is automatically created using the data provided in columns H, K, L, and M and displayed in the Normplot tab of this workbook.

Fig. 8.26 Linear regression example: Data analysis results after removing the outlier

Using the original data shows that the second point ($\Pi = 1{,}314$) is potentially an outlier, since its residual is extremely large. Thus, the row corresponding to this point (row 4 in the original layout) was deleted and the regression analysis was redone. The results are shown in Figs. 8.26 and 8.27. The results are much better as there are now no clear outliers and the data confidence intervals, especially for C_3, are much smaller.

8.7.2 Nonlinear Regression Example in Excel®

This example examines the problem of fitting a theoretical equation to experimental data in order to obtain the values of the different constants in the system. Unlike the

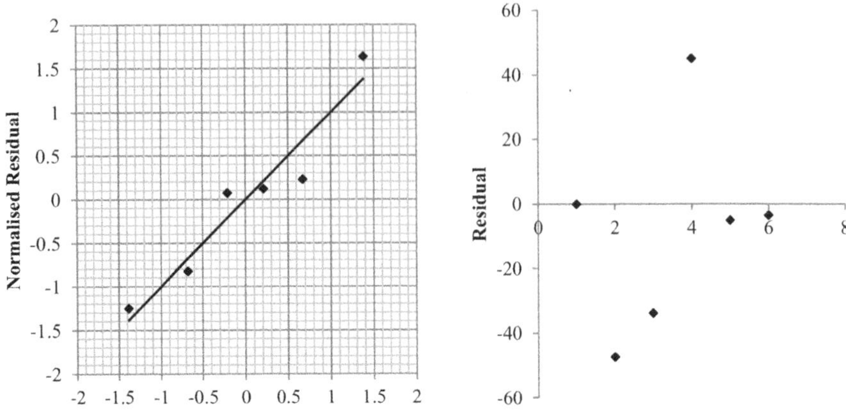

Fig. 8.27 Linear regression example: (left) Normal probability and (right) Time-series plots after removing outliers

previous case, nonlinear regression must be performed in order to obtain a result. Detailed information about the problem can be found in Ross-Rodriguez (2009). Data provided courtesy of Dr. Lisa Ross-Rodriguez.

8.7.2.1   Problem Statement for Nonlinear Regression Example

Consider the problem of obtaining a relationship for the ratio between the equilibrium and isotonic cell volumes given the osmotic pressure. The theoretical relationship can be written as

$$\frac{V}{V_0} = \left(1 - b^*\right)\frac{-1 + \sqrt{1 + 4B\Pi_0}}{-1 + \sqrt{1 + 4B\Pi}} + b^*, \tag{8.4}$$

where both B and b^* are the parameters to be determined and Π_0 is a known osmotic value. The experimental data are provided in Table 8.4. For this data set, Π_0 has a value of 0.293.

8.7.2.2   Problem Solution for Nonlinear Regression Example

Before we set up the problem in Excel®, it is first necessary to compute some preliminary information. First, we need to obtain the derivatives of Eq. (8.4) with respect to the parameters, that is,

$$\frac{d(V/V_0)}{db^*} = 1 - \frac{1 - \sqrt{1 + 4B\Pi_0}}{1 - \sqrt{1 + 4B\Pi}} \tag{8.5}$$

Table 8.4 Equilibrium cell volume data (Excel® example)

V/V_0	Π
1.000 34	0.292 78
0.804 65	0.571 72
0.753 58	0.855 14
0.715 48	1.135 95
0.685 88	1.433 49
0.666 00	1.729 08
0.659 13	2.028 15
0.640 04	2.326 60
0.626 61	2.667 04

$$\frac{d(V/V_0)}{dB} = 2(1-b^*)\left[\frac{\Pi_0}{\sqrt{1+4B\Pi_0}\left(-1+\sqrt{1+4B\Pi}\right)} - \frac{\Pi\left(-1+\sqrt{1+4B\Pi_0}\right)}{\sqrt{1+4B\Pi}\left(-1+\sqrt{1+4B\Pi}\right)^2}\right].$$

$$(8.6)$$

It can clearly be seen that this equation is nonlinear in the *parameters*. Thus, nonlinear regression using Solver will be performed. In order to obtain values for the parameter confidence intervals using Eq. (3.54), the grand Jacobian will be calculated using the "best" estimated values of the parameters and the above derivatives.

The nonlinear regression Excel® template used is set up identically to that of the linear regression template. The only difference is that now the estimated parameter values are not computed using a formula. Instead, they must be determined using Solver. Given the problem set-up, initial parameter estimates can be a bit of an issue, as the solution is sensitive to them. A recommended initial guess would be 0.5 for b and 2.5 for B. The macros are shown in Sect. 8.7.2.3: VB Macros.

The final results are shown in Fig. 8.28. Figure 8.29 shows the normal probability plot and a time-series plot of the residuals. It is easy to note that the B parameter is not significant and its value could be zero. Given the overall good fit and the relative well-behaved nature of the residuals, this would suggest that potentially not enough data have been collected to make an appropriate estimate. This situation partly explains why the Solver can have issues with obtaining a good value for B. The residual plots are shown in Fig. 8.29. Overall the results are decent, given the small sample. Since it has been assumed that the given equation holds, in order to obtain a better understanding of the data, additional experiments should be performed.

8.7.2.3   VB Macros

The macros required for performing the detailed regression analysis are shown here. There are three main macros: 1) to compute the model parameters (`model`); 2) to compute the derivative of the model with respect to b (`dydb1`); and 3) to compute the derivative of the model with respect to B (`dydb3`).

V/Vo	Π	Jacobian		y	residuals	r^2	Ordered	Z-Rank	Z-Score
		$\partial y/\partial\beta_1$	$\partial y/\partial\beta_2$						
1.0003	0.29278	0.00	-0.00001	1.0003	0.00	0.00	5	0	-0.05602
0.8047	0.57172	0.38	0.007262	0.8174	0.01	0.00	9	1.593219	1.924833
0.7536	0.85514	0.53	0.008435	0.7472	-0.01	0.00	1	-1.59322	-1.02829
0.7155	1.13595	0.61	0.008527	0.7095	-0.01	0.00	2	-0.96742	-0.9756
0.6859	1.43349	0.66	0.008329	0.6840	0.00	0.00	4	-0.28222	-0.33586
0.6660	1.72908	0.70	0.008053	0.6664	0.00	0.00	6	0.282216	0.015948
0.6591	2.02815	0.73	0.007763	0.6531	-0.01	0.00	3	-0.58946	-0.97267
0.6400	2.3266	0.75	0.007486	0.6429	0.00	0.00	7	0.589456	0.3941
0.6266	2.66704	0.77	0.007194	0.6336	0.01	0.00	8	0.967422	1.03356
β		0.524581	2.408129		sum(r^2)	0.00		$\mu_{residuals}$	0.000291
$\pm\delta$		0.043624	3.616267		σ_{model}	0.01		$\sigma_{residuals}$	0.006449
					α	0.05			
					tinv(α)	2.364624			
					m	9			
					n	2			
					k	2			
					TSS	0.109233			
					SSE	0.000333			
					SSR	0.1089			
					R^2	0.996947			
					F-test	1143.092			
					F-critical, model	4.737414			
					F-critical, factorial	5.591448			

The normal probability plot is automatically created using the data provided in columns H, K, and L and displayed in the Normplot tab of this workbook.

Fig. 8.28 Nonlinear regression example: Excel® spreadsheet results

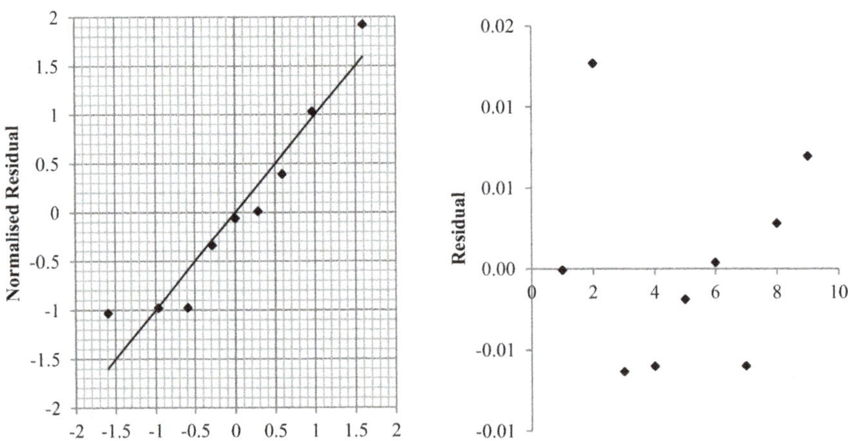

Fig. 8.29 Nonlinear regression example: (left) Normal probability plot and (right) Time-series plot of the residuals

```
Public Function model(parameter, x)
bs = parameter(1)
B = parameter(2)
```

```
model = (1 - bs) * (-1 + Sqr(1 + 4 * B * 0.293)) / (-1 + Sqr(1 +
4 * B * x(1))) + bs
End Function
```

```
Function dydb1(parameter As Range, x As Range)
bs = parameter(1)
B = parameter(2)
dydb1 = 1 - (-1 + Sqr(1 + 4 * B * 0.293)) / (-1 + Sqr(1 +
4 * B * x(1)))
End Function
```

```
Function dydb3(parameter As Range, x As Range)
bs = parameter(1)
B = parameter(2)
Pio = 0.293
so = Sqr(1 + 4 * B * Pio)
s = Sqr(1 + 4 * B * x(1))
dydb3 = 2 * (1 - bs) * (Pio / so / (s - 1) - x(1) * (so - 1) / s / (s -
1) ^ 2)
End Function
```

8.7.3   *Factorial Design Examples Using Excel®*

This section presents the Excel® spreadsheets for analysing some of the factorial design experiments presented in Chap. 4. The examples are all based on the factorial design template. The following examples have a corresponding Excel® spreadsheet:

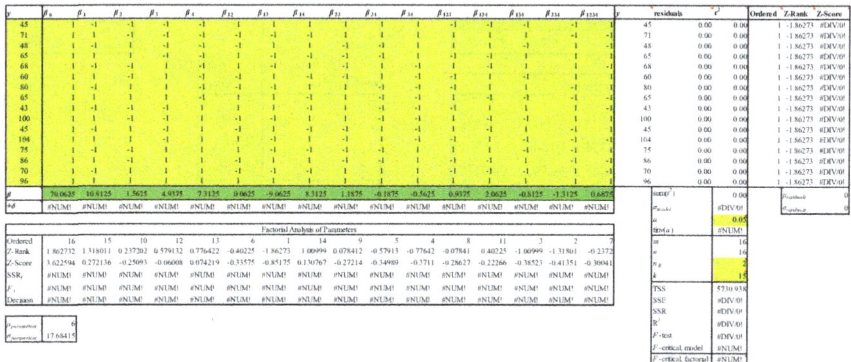

Fig. 8.30 Factorial design: Full factorial example

(1) Figure 8.30: Example 4.2: Analysis of a Full Factorial Experiment from Sect. 4.4.4: Projection;
(2) Figure 8.31: Sect. 4.7.4: Detailed Mixed-Level Example; and
(3) Figure 8.32: Sect. 4.8.2: Factorial Design with Centre Point Example.

Fig. 8.31 table — design matrix:

y	β_0	β_1	β_2	β_3	β_{12}	β_{13}	β_{23}	β_{123}	β_{11}	β_{112}	β_{113}	β_{1123}	y	residuals	r^2	Ordered	Z-Rank	Z-Score
-3	1	-1	-1	-1	1	1	1	-1	1	-1	-1	1	-2	1.00	1.00	23	1.534121	1.644957
-1	1	-1	-1	1	1	-1	-1	1	1	-1	1	-1	-0.5	0.50	0.25	15	0.264147	0.822478
-1	1	-1	1	-1	-1	1	-1	1	1	1	-1	-1	-0.5	0.50	0.25	21	1.054472	0.822478
1	1	-1	1	1	-1	-1	1	-1	1	1	1	1	1	0.00	0.00	12	-0.05225	-3.3E-16
0	1	0	-1	-1	0	0	1	0	-2	2	2	-2	0.5	0.50	0.25	22	1.258162	0.822478
2	1	0	-1	1	0	0	-1	0	-2	2	-2	2	1.5	-0.50	0.25	3	-1.25816	-0.82248
2	1	0	1	-1	0	0	-1	0	-2	-2	2	2	2.5	0.50	0.25	16	0.374095	0.822478
6	1	0	1	1	0	0	1	0	-2	-2	-2	-2	5.5	-0.50	0.25	5	-0.88715	-0.82248
5	1	1	-1	-1	-1	-1	1	1	1	-1	-1	1	4.5	-0.50	0.25	5	-0.88715	-0.82248
7	1	1	-1	1	-1	1	-1	-1	1	-1	1	-1	6.5	-0.50	0.25	5	-0.88715	-0.82248
7	1	1	1	-1	1	-1	-1	-1	1	1	-1	-1	8	1.00	1.00	24	2.036834	1.644957
10	1	1	1	1	1	1	1	1	1	1	1	1	10.5	0.50	0.25	16	0.374095	0.822478
-1	1	-1	-1	-1	1	1	1	-1	1	-1	-1	1	-2	-1.00	1.00	1	-2.03683	-1.64496
0	1	-1	-1	1	1	-1	-1	1	1	-1	1	-1	-0.5	-0.50	0.25	4	-1.05447	-0.82248
0	1	-1	1	-1	-1	1	-1	1	1	1	-1	-1	-0.5	-0.50	0.25	10	-0.26415	-0.82248
1	1	-1	1	1	-1	-1	1	-1	1	1	1	1	1	0.00	0.00	12	-0.05225	-3.3E-16
1	1	0	-1	-1	0	0	1	0	-2	2	2	-2	0.5	-0.50	0.25	11	-0.15731	-0.82248
1	1	0	-1	1	0	0	-1	0	-2	2	-2	2	1.5	0.50	0.25	14	0.157311	0.822478
3	1	0	1	-1	0	0	-1	0	-2	-2	2	2	2.5	-0.50	0.25	5	-0.88715	-0.82248
5	1	0	1	1	0	0	1	0	-2	-2	-2	-2	5.5	0.50	0.25	16	0.374095	0.822478
4	1	1	-1	-1	-1	-1	1	1	1	-1	-1	1	4.5	0.50	0.25	16	0.374095	0.822478
6	1	1	-1	1	-1	1	-1	-1	1	-1	1	-1	6.5	0.50	0.25	16	0.374095	0.822478
9	1	1	1	-1	1	-1	-1	-1	1	1	-1	-1	8	-1.00	1.00	2	-1.53412	-1.64496
11	1	1	1	1	1	1	1	1	1	1	1	1	10.5	-0.50	0.25	5	-0.88715	-0.82248
β	3.125	3.9375	1.375	0.958333	0.5625	0.1875	0.208333	0.0625	0.3125	-0.0625	-0.02083	-0.14583		sum(r^2)	8.50		$\mu_{residuals}$	2.04E-16
$\pm\delta$	0.374311	0.458436	0.374311	0.374311	0.458436	0.458436	0.374311	0.458436	0.264678	0.264678	0.264678	0.264678		σ_{model}	0.84		$\sigma_{residuals}$	0.607919

α	0.05
tinv(α)	2.178813
m	24
n	12
n_R	1
k	11
TSS	336.625
SSE	8.5
SSR	328.125
R^2	0.974749
F-test	42.1123
F-critical, model	2.717331
F-critical, factorial	4.747225

Factorial Analysis of Parameters:

Ordered	11	12	10	9	8	5	6	4	7	2	3	1
Z-Rank	1.150349	1.731664	0.812218	0.548522	0.318639	-0.31864	-0.10463	-0.54852	0.104633	-1.15035	-0.81222	-1.73166
Z-Score	1.693676	2.305281	0.376372	0.062729	-0.23523	-0.51751	-0.50183	-0.61161	-0.42342	-0.7057	-0.67433	-0.76843
SSR$_i$	234.375	248.0625	45.375	22.04167	5.0625	0.5625	1.041667	0.0625	4.6875	0.1875	0.020833	1.020833
F_i	330.8824	350.2059	64.05882	31.11765	7.147059	0.794118	1.470588	0.088235	6.617647	0.264706	0.029412	1.441176
Decision	Keep	Keep	Keep	Keep	Keep	Discard	Discard	Discard	Keep	Discard	Discard	Discard

$\mu_{parameters}$	0.875
$\sigma_{parameters}$	1.328471

Fig. 8.31 Factorial design: Mixed-level example

Fig. 8.32 table:

y	β_0	β_1	β_2	β_{12}	β_c	y	residuals	r^2	Ordered	Z-Rank	Z-Score
39.3	1	-1	-1	1	1	39.3	0.00	0.00	4	-0.28222	0
40	1	-1	1	-1	1	40	0.00	0.00	4	-0.28222	0
40.9	1	1	-1	-1	1	40.9	0.00	0.00	4	-0.28222	0
41.5	1	1	1	1	1	41.5	0.00	0.00	4	-0.28222	0
40.3	1	0	0	0	-0.8	40.46	0.16	0.03	8	0.967422	1.091191
40.5	1	0	0	0	-0.8	40.46	-0.04	0.00	3	-0.58946	-0.2728
40.7	1	0	0	0	-0.8	40.46	-0.24	0.06	1	-1.59322	-1.63679
40.2	1	0	0	0	-0.8	40.46	0.26	0.07	9	1.593219	1.773185
40.6	1	0	0	0	-0.8	40.46	-0.14	0.02	2	-0.96742	-0.95479
β	40.44444	0.775	0.325	-0.025	-0.01944	sum(r^2)	0.17		$\mu_{residuals}$	0	
$\pm\delta$	0.191912	0.287868	0.287868	0.287868	0.214564	σ_{model}	0.21		$\sigma_{residuals}$	0.146629	

α	0.05
tinv(α)	2.776445
m	9
n	5
n_R	2
k	4
TSS	3.002222
SSE	0.172
SSR	2.830222
R^2	0.942709
F-test	16.45478
F-critical, model	6.388233
F-critical, factorial	7.708647

Factorial Analysis of Parameters:

Ordered	5	4	3	1	2
Z-Rank	1.281552	0.524401	0	-1.28155	-0.5244
Z-Score	1.788558	-0.4187	-0.44374	-0.46321	-0.4629
SSR$_i$	14721.78	2.4025	0.4225	0.0025	0.002722
F_i	342366.9	55.87209	9.825581	0.05814	0.063307
Decision	Keep	Keep	Keep	Discard	Discard

$\mu_{parameters}$	8.3
$\sigma_{parameters}$	17.97227

Fig. 8.32 Factorial design: Combined factorial and centre point example

8.8   Further Reading

The following are references that provide additional information about the topic:

(1)   **General Excel**[®] *Help*:

   a.   Harvey, G. (2019). *Excel*[®] *2019 All-in-One For Dummies*. Hoboken, New Jersey, United States of America: John Wiley & Sons, Inc.

   b.   Brillo, J. (2007). *Excel*[®] *for Scientists and Engineers: Numerical Methods*. Hoboken, New Jersey, United States of America: John Wiley & Sons, Inc.

   c.   Schmuller, J. (2013). *Statistical Analysis with Excel*[®] *for Dummies* (3rd ed.). Hoboken, New Jersey, United States of America: Wiley & Sons, Inc.

(2)   **Linear Regression Data set**:

   a.   Elliott, J. A., Prickett, R. C., Elmoazzen, H. Y., Porter, K. R., & McGann, L. E. (2007). A Multisolute Osmotic Virial Equation for Solutions of Interest in Biology. *Journal of Physical Chemistry B, 111*, 1775–1785.

   b.   Prickett, R. C., Elliott, J. A., & McGann, L. E. (2010). Application of the osmotic virial equation in cyrobiology. *Cryobiology, 2010*, 30–42.

   c.   Prickett, R. C., Elliott, J. A., & McGann, L. E. (2011). Application of the Multisolute Osmotic Virial Equation to Solutions Containing Electrolytes. *The Journal of Physical Chemistry B, 115*, 14,531–14,543.

   d.   Jochem, M., & Körber, C. (1987). Extended Phase Diagrams for the Ternary Solutions H_2O − NaCl − Glycerol and H_2O − NaCl − Hydroxyethylstarch (HES) Determined by DSC. *Cryobiology, 24*, 513–536.

(3)   **Nonlinear Regression Data set**:

   a.   Ross-Rodriguez, L. U. (2009). *Cellular Osmotic Properties and Cellular Responses to Cooling*. Edmonton, Alberta, Canada: University of Alberta.

Correction to: Statistics for Chemical and Process Engineers

Correction to:
Y. A. W. Shardt, *Statistics for Chemical and Process Engineers,* **https://doi.org/10.1007/978-3-030-83190-5**

In the original version of the chapter, the following correction has been incorporated: There are some scientific errors in this publication from Chapters 2 to 7. The correction chapter and the book have been updated with the changes.

The updated version of the book can be found at
https://doi.org/10.1007/978-3-030-83190-5

Y. A. W. Shardt, *Statistics for Chemical and Process Engineers,*
https://doi.org/10.1007/978-3-030-83190-5_9

Appendix A
Solution Key

This appendix provides brief solutions to some of the problems in the book. Often only the final answer or value is provided without any explanation or justification. Given the sometimes subjective nature of regressionanalysis other equally valid answers can also exist. As well, although the author has strived to verify that the answers are correct, it is inevitable that one or two stray errorsmay appear. This should be borne in mind when comparing answers. If you believe that there is indeed an error in the solution key, please let the author know, so that appropriate corrections can be made.

Chapter 1

(1) F; (2) F; (3) T; (4) F; (5) T; (6) F; (7) T; (8) F; (9) T; (10) T; (11) F; (12) T; (13) F; (14) F; (15) T; (16) T; (17) F; (18) T; (19) T; (20) T.

(21) (a) $\mu = 4.3$; mode $= 5$; median $= 5$; (b) $\sigma^2 = 5.12$; MAD $= 1.84$; range $= 7$; (c) Q1 $= 2$; Q2 $= 5$; Q3 $= 5$.

(24) Maybe!

Chapter 2

(1) F; (2) F; (3) T; (4) T; (5) T; (6) F; (7) T; (8) T; (9) F; (10) F; (11) T; (12) T; (13) F; (14) T; (15) F; (16) T; (17) F.

(18) (a) $\mathbb{S} = \{Q\heartsuit, Q\spadesuit, Q\clubsuit\}$; $\mathbb{F} = \{\{\}, \{Q\heartsuit\}, \{Q\spadesuit\}, \{Q\clubsuit\}, \{Q\heartsuit, Q\spadesuit\}, \{Q\heartsuit, Q\clubsuit\}, \{Q\spadesuit, Q\clubsuit\}, \{Q\heartsuit, Q\spadesuit, Q\clubsuit\}\}$; $P(\{\}) = 0$; $P(\{Q\heartsuit\}) = \frac{1}{3}$, $P(\{Q\spadesuit\}) = \frac{1}{3}$; $P(\{Q\clubsuit\}) = \frac{1}{3}$; $P(\{Q\heartsuit, Q\spadesuit\}) = \frac{2}{3}$; $P(\{Q\heartsuit, Q\clubsuit\}) = \frac{1}{3}$; $P(\{Q\spadesuit, Q\clubsuit\}) = \frac{2}{3}$; $P(\{Q\heartsuit, Q\spadesuit, Q\clubsuit\}) = 1$; (b) $\frac{1}{3}$; (c) $\mu = 0$; $\sigma^2 = 2$.

(20) (a) $\mathbb{S} = \{HH, HT, TH, TT\}$; $\mathbb{F} = \{\{\}, \{HH\}, \{HT\}, \{TH\}, \{TT\}, \{HH, HT\}, \{HH, TH\}, \{HH, TT\}, \{HT, TH\}, \{HT, TT\}, \{TH, TT\}, \{HH, HT, TH\}, \{HH, HT, TT\}, \{HH, TH, TT\}, \{HT, TH, TT\}, \{HH, HT, TH, TT\}\}$; $P(\{\}) = 0$, $P(\{HH\}) = 0.3$, $P(\{HT\}) = 0.3$, $P(\{TH\}) = 0.2$, $P(\{TT\}) = 0.2$, $P(\{HH, HT\}) = 0.6$, $P(\{HH, TH\}) = P(\{HH, TT\}) = P(\{HT, TH\}) = P(\{HT, TT\}) = 0.5$, $P(\{TH, TT\}) = 0.4$,

Table A.1 Answers for Question 27 in Chap. 2

	$\hat{\mu}_1$	$\hat{\mu}_2$	$\hat{\mu}_3$	$\hat{\mu}_4$	$\hat{\mu}_5$
Bias	0	0	0.5μ	$-\frac{\mu}{3}$	0
σ^2	σ^2	$0.5\sigma^2$	$1.25\sigma^2$	$0.\bar{2}\sigma^2$	σ^2/N
MSE	σ^2	$0.5\sigma^2$	$1.25\sigma^2 + 0.25\mu$	$0.\bar{2}\sigma^2 + 0.\bar{1}\mu$	σ^2/N

$P(\{HH, HT, TH\}) = P(\{HH, HT, TT\}) = 0.8, P(\{HH, TH, TT\}) = P(\{HT, TH, TT\}) = 0.7, P(\{HH, HT, TH, TT\}) = 1.$

(21) (a) No; (b) No; (c) Yes, $\mu = 2.5$, $\sigma^2 = 25/12$, $E(|x|) = 2.5$.

(23) $E(2X - 4Y) = -6$; $E(3XY) = 6$; $E(X^2) = 3$.

(27) see Table A.1.

(29) (a) Yes; (b) No; (c) Yes; (d) No.

(34) (b) $\mu = 0.5\pi$; (c) $4.84 \le \psi \le 6.41$, yes; (d) sampled mean is not equal to the true value; (e) $P(0.25 < X < 0.5) = 0.046$.

Chapter 3

(1) F; (2) T; (3) F; (4) T; (5) F; (6) T; (7) T; (8) F; (9) F; (10) T; (11) F; (12) F; (13) T; (14) T; (15) F; (16) T; (17) F; (18) T; (19) T; (20) T.

(23) The solution is given as

$$\vec{y} = \langle y_3, y_4, \ldots, y_{100} \rangle^T$$

$$\mathcal{A} = \begin{bmatrix} -y_2 & -y_1 & u_2 & u_1 \\ -y_3 & -y_2 & u_3 & u_2 \\ \vdots & \vdots & \vdots & \vdots \\ -y_{99} & -y_{98} & u_{99} & u_{98} \end{bmatrix}$$

$$\vec{\beta} = \langle \alpha_1, \alpha_2, \beta_1, \beta_2 \rangle^T.$$

(24) (c) $\hat{R} = 2.15\ \Omega$, $\hat{\sigma} = 0.85$; (d) $2.1 \pm 0.2\ \Omega$; (e) Yes; (f) Fit is not good, as the errors are increasing with the current.

(25) (b) $\hat{R} = 2.044\ \Omega$; (c) An examination of the residuals (especially as a function of the current, I) and how close to normality they are; (d) Instrument error is often proportional to the magnitude of the measured value.

(27) (a + b) $\beta_0 = 276 \pm 6.8$; $\beta_1 = -1.98 \pm 1.0$; $\beta_2 = 0.012 \pm 0.041$; $\beta_3 = 0.00018 \pm 0.00045$; (c) 229.5 ± 3.3, 229.5 ± 8.5; (d) Yes; (e) $R^2 = 0.963$, F-score $= 87.23$, No.

(29) (b + c) $\ln \hat{K} = 5.3 \pm 0.45$, $\hat{b} = 0.54 \pm 0.28$; (d) $128 \le K \le 317$; $\hat{b}$ stays the same; (e) $\hat{\sigma} = 0.103$, $R^2 = 0.8789$; (g) Yes; (h) $R = 0.467\ \mathrm{kg \cdot m^{-0.5} \cdot s^{-1}}$; (i) $\hat{K} = 181 \pm 67$, $\hat{b} = 0.46 \pm 0.25$; $R^2 = 0.8905$, $\hat{\sigma} = 7.397$; (j) Yes, $R = 0.518\ \mathrm{kg \cdot m^{-0.5} \cdot s^{-1}}$.

(31) (a) No, plot the errors for the two runs using different symbols; (c) Yes.

Chapter 4

(1) T; (2) T; (3) F; (4) F; (5) F; (6) F; (7) F; (8) F; (9) T; (10) T; (11) T; (12) T; (13) T; (14) F; (15) T; (16) T; (17) F; (18) F; (19) F; (20) T;

(21) (a) No; (b) Yes. (23) (a) I = ACD = CDF = AF; (b) (two examples) A = F = CD = ACDF, B = ABF = ABCD = ABCDF; (c) III; (d) (one of many) I = ABCE = ACDF = BDEF.

(25) (a brief outline of the solution) A fractional factorial design with centre point replicates. Blockingand randomization should also be considered.

(26) (a) $I = x_1 x_2^2 x_3 = x_1^2 x_2 x_3^2$; (b) $x_1 = x_1^2 x_2^2 x_3 = x_2 x_3^2$, $x_2 = x_1 x_3 = x_1^2 x_2^2 x_3^2$, $x_3 = x_1 x_2^2 x_3^2 = x_1^2 x_2$, $x_1^2 = x_2^2 x_3 = x_1 x_2 x_3^2$, $x_2^2 = x_1 x_2 x_3 = x_1^2 x_3^2$, $x_3^2 = x_1 x_2^2 = x_1^2 x_2 x_3$, $x_1 x_2 = x_1^2 x_3 = x_2^2 x_3^2$, $x_2 x_3 = x_1 x_3^2 = x_1^2 x_2^2$, $x_1 x_2^2 x_3 = x_1^2 x_2 x_3^2 = I$; (c) $y = \beta_0 + \beta_1 x_1 + \beta_2 x_2 + \beta_3 x_3 + \beta_{11} x_1^2 + \beta_{22} x_2^2 + \beta_{33} x_3^2 + \beta_{12} x_1 x_2 + \beta_{23} x_2 x_3$.

(29) (b) E = −ABCD; (c) V; (d) BD, AE, DE, BE, and E.

(31) (a) $y = 100 - 4x_1^2 - 12x_2^2 - 9x_3^2$; (b) 100.

Chapter 5

(1) T; (2) F; (3) T; (4) T; (5) F; (6) F; (7) F; (8) T; (9) T; (10) F; (11) T; (12) T; (13) T; (14) F; (15) T; (16) F; (17) T; (18) F; (19) T; (20) F;

(24) AR(1). (25) seasonal MA(1) with $s = 3$ and a normal AR(4).

Chapter 6

(1) F; (2) T; (3) T; (4) F; (5) F; (6) t; (7) T; (8) T; (9) T; (10) T; (11) F; (12) F; (13) T; (14) F; (15) T; (16) F; (17) T; (18) F; (19) F; (20) T.

(24) left graph: 9, right graph: 5. (25) Model is not adequate. (26) Model is adequate.

References

Ashley R (1988) On the relative worth of recent macroeconomic forecasts. Int J Forecast 4:363–376

Barnett V, Lewis T (1994) Outliers in statistical data, 3rd edn. Wiley, Chichester, England, United Kingdom

Bloomfield P (2000) Fourier analysis of time series: an introduction, 2nd edn. Wiley, New York, New York, United States of America

Box GE, Jenkins GM (1970) Time series analysis, forecasting, and control. Holden-Day, Oakland, California, United States of America

Box GE, Hunter WG, Hunter JS (1978) Statistics for experimenters: an introduction to design, data analysis, and model building. Wiley, New York, New York, United States of America

Brillo J (2007) Excel for scientists and engineers: numerical methods. Wiley, Hoboken, New Jersey, United States of America

Chambers JM, Cleveland WS, Kleiner B, Tukey P (1983) Graphical methods for data analysis. Wadsworth International Group, Belmont, California, United States of America

Daniel C, Wood FS (1980) Fitting equations to data, 2nd edn. Wiley, New York, New York, United States of America

Davies L, Gather U (1993) The identification of multiple outliers. J Am Stat Assoc 88(423):782–792

Dean JA (1999) Lange's handbook of chemistry, 15th edn. McGraw-Hill Inc., New York, New York, United States of America

Elliott JA, Prickett RC, Elmoazzen HY, Porter KR, McGann LE (2007) A multisolute osmotic virial equation for solutions of interest in biology. J Phys Chem B 111:1775–1785

Franke J, Härdle WK, Hafner CM (2011) Statistics of financial markets: an introduction, 3rd edn. Springer, Heidelberg, Germany. https://doi.org/10.1007/978-3-642-16521-4

Gerhart PM, Gross RJ, Hochstein JI (1992) Fundamentals of fluid mechanics. Addison-Wesley Publication Co, Reading, Massachusetts, United States of America

Hald A (2003) A history of probability and statistics and their application before 1750. Wiley, Hoboken, New Jersey, United States of America

Hannan EJ, Deistler M (2012) The statistical theory of linear systems. Society of Industrial and Applied Mathematics

Hare LB (1988) In the soup: a case study to identify contributors to filling variability. J Qual Technol 20:36–43

Harvey AC (1991) Forecasting, structural time series models and the Kalman filter. Cambridge University Press, Cambridge

Harvey G (2019) Excel® 2019 all-in-one for dummies. Wiley, Hoboken, New Jersey, United States of America

Hassler, U (1994) The sample autocorrelation function of I(1) processes. Stat Papers Stat Hefte 35:1–16

Hawkins DM (1980) Identification of outliers. Chapman and Hall, London, England, United Kingdom

Hinkelmann K, Kempthorne O (2007) Design and analysis of experiments, vols I, II, III. Wiley, Hoboken, New Jersey, United States of America

Hodge VJ, Austin J (2004) A survey of outlier detection methodologies. Artif Intell Rev 22:85–126

Huang B, Kadali R (2008) Dynamic modeling, predictive control, and performance monitoring. Springer

Hunt BR, Lipsman RL, Rosenberg J (2014) A guide to MATLAB: for beginners and experienced users: updated for MATLAB 8 and Simulink 8, 3rd edn. Cambridge University Press, Cambridge, England, United Kingdom

Hyndman RJ, Fan Y (1996) Sample quantiles in statistical packages. Am Stat 50(4):361–365

Jochem M, Körber C (1987) Extended phase diagrams for the ternary solutions $H_2O-NaCl-glycerol$ and $H_2O-NaCl-hydroxyethylstarch$ (HES) determined by DSC. Cryobiology 24:513–536

Johnson RA, Wichern DW (2007) Applied multivariate statistical analysis, 6th edn. Prentice Hall, Upper Saddle River, New Jersey, United States of America

Kallenberg O (2002) Foundations of modern probability. Springer, New York, New York, United States of America

Kalman RE (1960) A new approach to linear filtering and prediction problems. Trans ASME J Basic Eng 82:35–45

Kalman RE, Bucy RS (1961) New results in filtering and prediction theory. Trans ASME J Basic Eng 83:95–108

Khintchine A (1934) Korrelationstheorie der stationären stochastischen Prozesse (Correlation Theory of Stocastic Processes). Math Ann 109(1):604–615. https://doi.org/10.1007/BF01449156

Lin B, Recke B, Knudsen JK, Jørgensen SB (2007) A systematic approach for soft sensor development. Comput Chem Eng 31:419–425

Ljung L (1999) System identification theory for the user. Prentice Hall Inc., Upper Saddle River, New Jersey, United States of America

Montgomery DC (1991) Design and analysis of experiments, 3rd edn. Wiley, New York, New York, United States of America

Montgomery DC, Peck EA (1982) Introduction to linear regression analysis, 1st edn. Wiley, New York, New York, United States of America

Montgomery DC, Runger GC (2007) Applied statistics and probability for engineers, 4th edn. Wiley, Hoboken, New Jersey, United States of America

Montgomery DC, Jennings CL, Kulahci M (2008) Introduction to time series analysis and forecasting. Wiley, Hoboken, New Jersey, United States of America

Myers RH (1971) Response surface methodology. Allyn and Bacon Inc., Boston, Massachusetts, United States of America

Nelson CR (1972) The prediction performance of the FRB-MIT-PENN model of the U.S. Economy. Am Econ Rev 62(5):902–917

Ogunnaike BA (2010) Random phenomena: fundamentals of probability and statistics for engineers. CRC Press, Boca Raton, Florida, United States of America

Prickett RC, Elliott JA, McGann LE (2010) Application of the osmotic virial equation in cyrobiology. Cryobiology 2010:30–42

Prickett RC, Elliott JA, McGann LE (2011) Application of the multisolute osmotic virial equation to solutions containing electrolytes. J Phys Chem B 115:14531–14543

Priestley MB (1981) Spectral analysis and time series: vol. 1: univariate series and vol 2: multivariate series, prediction, and control. Academic Press, New York, New York, United States of America

Ross-Rodriguez LU (2009) Cellular osmotic properties and cellular responses to cooling. University of Alberta, Edmonton, Alberta, Canada

Schlittgen R (2012) Einführung in die Statistik Analyse und Modellierung von Daten, 12th edn. Oldenbourg Wissenschaftsverlag GmbH, Munich, Bavaria, Germant

Schmuller J (2013) Statistical analysis with excel® for dummies, 3rd edn. Wiley, Hoboken, New Jersey, United States of America

Seber GA, Wild CJ (1989) Nonlinear regression. Wiley, New York, New York, United States of America

Shardt YAW (2012) Data quality assessment for closed-loop system identification and forecasting with application to soft sensors. Doctoral Thesis, University of Alberta, Department of Chemical and Materials Engineering, Edmonton, Alberta, Canada. doi: http://hdl.handle.net/10402/era.29018

Shardt YAW (2012). Data quality assessment for closed-loop system identification and forecasting with application to soft sensors. Edmonton, Alberta, Canada: University of Alberta. doi: http://hdl.handle.net/10402/era.29018

Shardt YAW, Huang B (2011) Closed-loop identification with routine operating data: effect of time delay and sampling time. J Process Control 21:997–1010

Shardt YAW, Huang B (2014) Minimal required excitation for closed-loop identification: implications for PID control loops. In: ADCONIP conference proceedings. Hiroshima, Japan, pp 296–301. http://www.nt.ntnu.no/users/skoge/prost/proceedings/adconip-2014/pdf/SUBS61 TO80/0069/0069_FI.pdf

Sheynin O (2004) History of the theory of probability to the beginning of the 20th century. NG Verlag, Berlin, Germany

Shumway RH, Stoffer DS (2011) Time series analysis and its applications with R examples, 3rd edn. Springer, New York, New York, United States of America. https://doi.org/10.1007/978-1-4419-7865-3

Sizemore J, Mueller JP (2015) MATLAB for dummies. Wiley, Hoboken, New Jersey, United States of America

Söderström T, Gustavsson I, Ljung L (1975) Identifiability conditions for linear systems operating in closed loop. Int J Control 21(2):243–255

Sokal RR, Rohlf FJ (1995) Biometry: the principles and practice of statisics in biological research, 3rd edn. W. H. Freeman and Company, New York, New York, United States of America

Steel RG, Torrie JH (1980) Principles and procedures of statistics: a biometrical approach, 2nd edn. McGraw-Hill Kogakusha Ltd., Tokyo, Japan

Stoica P, Moses R (2005) Spectral analysis of signals. Prentice Hall, Upper Saddle River, New Jersey, United States of America

Tufte ER (1997) Visual and statistical thinking: displays of evidence for making decisions. Graphics Press LLC, Cheshire, Connecticut, United States of America

Tufte ER (2001) The visual display of quantitative information. Graphics Press LLC, Cheshire, Connecticut, United States of America

Varberg DE (1963) The development of modern statistics. Math Teach 56(4):252–257

Welch PD (1967) The use of fast fourier transform for the estimation of power spectra: a method based on time averaging over short, modified periodograms. IEEE Trans Audio Electroacoust AU-15(2):70–73

Wichern DW (1973) The behaviour of the sample autocorrelation function for an integrated moving average process. Biometrika 60(2):235–239. doi: http://www.jstor.org/stable/2334535?origin=JSTOR-pdf

Zhu Y (2001) Multivariable system identification for process control. Elsevier Science Ltd., Oxford, United Kingdom

Zorn ME, Gibbons RD, Sonzogni WC (1997) Weighted least-squares approach to calculating limits of detection and quantification by modeling variability as a function of concentration. Anal Chem 69(15):3069–3075

Index

MATLAB and Excel Functions

Made in the USA
Monee, IL
07 July 2026

56544672R00256